Birkhäuser

Progress in Mathematical Physics

Volume 77

More information about this series at http://www.springer.com/series/4813

Markus Kunze

A Birman-Schwinger Principle in Galactic Dynamics

Birkhäuser

Markus Kunze
Mathematisches Institut
Universität zu Köln
Cologne, Germany

ISSN 1544-9998 ISSN 2197-1846 (electronic)
Progress in Mathematical Physics
ISBN 978-3-030-75188-3 ISBN 978-3-030-75186-9 (eBook)
https://doi.org/10.1007/978-3-030-75186-9

Mathematics Subject Classification: 85A05, 35Q83, 35B35, 81V55, 47A10

This book is published under the imprint Birkhäuser, www.birkhauser-science.com by the registered company Springer Nature Switzerland AG
The registered company address is: Gewerbestrasse 11, 6330 Cham, Switzerland

Of all the ways of acquiring books,
writing them oneself is regarded
as the most praiseworthy method.[1]

[1] Benjamin W.: *Illuminations*, translated by Harry Zohn, Schocken Books, New York 1968, p. 61.

Preface

There are many stability problems in mathematical physics that ultimately can be reduced to estimates on eigenvalues. In this respect, one of the best-known examples concerns the stability of matter, which first asks the question of why point-like electrons do not simply fall into point-like nuclei (called 'stability of the first kind'); see [53] and the references cited therein for more information. A related problem (called 'stability of the second kind') is if matter with a large number N of atoms has the property that energy and volume of a system of $2N$ atoms are twice the energy and volume of a system of N atoms. In 1967–8, Dyson and Lenard [16] could answer this question in the positive, using a very intricate argument. An essential improvement came in 1975 with the work of Lieb and Thirring [54], who invented a fermionic kinetic energy inequality (now known as the Lieb-Thirring inequality) to derive the desired lower energy bound. The main tool in their work (and in many others) is provided by the Birman-Schwinger principle, which relates eigenvalue estimates for Schrödinger operators $H = -\Delta + V$ on $L^2(\mathbb{R}^n)$ to the study of a certain family of non-negative compact operators. These operators are expressible in terms of an explicitly known integral kernel, at least for dimensions three and one.

Turning to galactic dynamics, a main line of research is to determine the stability properties of steady state solutions that correspond to static distributions (of stars or galaxies). Upon linearizing an appropriate energy-Casimir functional for the underlying gravitational Vlasov-Poisson system about such a steady state, the so-called Antonov functional is obtained as the 'Hessian' at the steady state; it was a fundamental observation of Antonov [4, 5] that this functional is strictly coercive when restricted to an appropriate subclass of functions, thus providing a general way for deriving stability.

It is the main purpose of this book to link both subjects and point out that there is a Birman-Schwinger principle that relates spectral questions for the self-adjoint operator L, which corresponds to the Antonov functional, to a certain family of non-negative compact operators. As we will see, this perfectly parallels the Schrödinger case in every respect. As an application, the Birman-Schwinger principle is used to characterize in which cases the 'best constant' in the Antonov stability estimate is attained; the best constant will then be the principal eigenvalue of L. This amounts to solving a quite non-standard variational problem, and it will be important for understanding the nonlinear dynamics of the system close to a steady state.

Cologne, Germany Markus Kunze

Acknowledgments The author is grateful to H. Andréasson, Y. Guo, M. Kiessling and G. Rein for ongoing discussions.

Contents

Chapter 1
Introduction

1.1 The Birman-Schwinger Principle

The Birman-Schwinger principle is a widely used and well-established tool in mathematical quantum mechanics. It was introduced through the independent works of Birman [10] and Schwinger [81], with the idea of counting or at least estimating the number of eigenvalues of Schrödinger operators on $L^2(\mathbb{R}^n)$. To be more specific, consider (only formal at this point)

$$H = -\Delta + V;$$

to avoid introducing negative parts, we will assume that $V \le 0$. Then it is not difficult to calculate that

(a) $-e$ is a (negative) eigenvalue of H if and only if 1 is an eigenvalue of the Birman-Schwinger operator

$$B_e = \sqrt{-V}(-\Delta + e)^{-1}\sqrt{-V}; \tag{1.1}$$

see [53, Section 4.3.1].

Furthermore,

(b) if ϕ is an eigenfunction of H for the eigenvalue $-e$, then $\psi = \sqrt{-V}\phi$ is an eigenfunction of B_e for the eigenvalue 1;
(c) if ψ is an eigenfunction of B_e for the eigenvalue 1, then $\phi=(-\Delta + e)^{-1}(\sqrt{-V}\psi)$ is an eigenfunction of H for the eigenvalue $-e$.

The operators B_e are non-negative Hilbert-Schmidt operators (if V decays sufficiently fast and $n \le 3$), and in particular, they are compact. Their eigenvalues can be ordered: $\lambda_1(e) \ge \lambda_2(e) \ge \cdots \to 0$, and the eigenvalue curves are decreasing in e, in that $\tilde{e} \ge e$ implies that $\lambda_k(e) \le \lambda_k(\tilde{e})$ for all k. This implies that the number of eigenvalues of

M. Kunze, *A Birman-Schwinger Principle in Galactic Dynamics*,
Progress in Mathematical Physics 77,
https://doi.org/10.1007/978-3-030-75186-9_1

H less than or equal to $-e$ agrees with the number of eigenvalues of B_e greater than or equal to 1, counting multiplicities in both cases; cf. [53, Figure 4.1, p. 78] for an illustration. In this way, not only the number of eigenvalues of H can be bounded, but for instance also eigenvalue moments like $\sum_j |-e_j|^\gamma$, where the sum extends over all negative eigenvalues $-e_j$ of H. This fact lies at the heart of many important results in the field. Let us only mention here the Lieb-Thirring bound

$$\sum_j |-e_j| \leq L_{1,3} \int_{\mathbb{R}^3} |V(x)|^{5/2}\, dx$$

in three dimensions for an absolute constant $L_{1,3} > 0$ and $V \in L^{5/2}(\mathbb{R}^3)$. It is used in those authors' proof of the stability of matter [54], which has found many generalizations [53] and which is much easier to follow than the original argument by Dyson and Lenard [16]. Good general textbooks that cover the Birman-Schwinger principle are [53, Section 4.3], [71, 82, 83] or [86, Section 7.9]. A classical reference is [46], and the papers [8, 34, 68] provide an extensive list of related literature. There is also a large number of further applications of the Birman-Schwinger principle in a variety of different contexts. For instance, complex-valued potentials are treated in [1, 20, 22, 23], Dirac operators in [12], the Bardeen-Cooper-Schrieffer model of superconductivity in [33] and the linearized 2D Euler equations in [47].

1.2 Non-Relativistic Galactic Dynamics and the Vlasov-Poisson System

In order to explain how the Birman-Schwinger principle will turn out to be useful in galactic dynamics, we are going to introduce the gravitational Vlasov-Poisson system. It is a standard PDE system to describe the time evolution of a self-gravitating system that consists of a large number of objects (like stars or galaxies), which interact via gravitational forces.

Galactic dynamics in general refers to the modeling of the time evolution of self-gravitating matter such as galaxies or, on an even larger scale, clusters of galaxies. One attempt to do so is to write down an N-body problem, with N quite large: $N \sim 10^6 - 10^{11}$ for galaxies and $N \sim 10^2 - 10^3$ for clusters of galaxies. This N-body problem consists of coupled Newtonian equations, one for each individual object (the 'objects' in a galaxy are stars, those in a cluster of galaxies are galaxies), to study the collective behavior of the system. While results may be obtainable numerically in this way, the mathematical complexity of even the three-body problem prevents one from rigorously addressing deeper questions (concerning for instance galaxy formation or stability) for such stellar systems.

Therefore, from the early days of the field, a statistical description of the evolution has been proposed by Vlasov [91] in 1938 for plasmas (in this case a related equation is satisfied) and by Jeans [41] in 1915 for gravitational systems; see [36] for an interesting historical discussion of the origins of the equation. It is also known as the 'collisionless Boltzmann equation', which refers to the fact that collisions among the stars or galaxies are sufficiently rare to be neglected. A standard source of information on galactic dynamics is [9].

The time evolution of such a system is then governed by a distribution function $f = f(t, x, v)$ that depends on time $t \in \mathbb{R}$, position $x \in \mathbb{R}^3$ and velocity $v \in \mathbb{R}^3$. The quantity $\int_{\mathcal{X}} dx \int_{\mathcal{V}} dv\, f(t, x, v)$ should be thought of as the number of objects (henceforth called 'particles') at time t, which are located at some point $x \in \mathcal{X} \subset \mathbb{R}^3$ and which have velocities $v \in \mathcal{V} \subset \mathbb{R}^3$. Each individual particle follows a trajectory $(X(s), V(s))$ in phase space $\mathbb{R}^3 \times \mathbb{R}^3$ such that $(X(t), V(t)) = (x, v)$ at time t and

$$\dot{X}(s) = V(s), \quad \dot{V}(s) = F(s, X(s)), \tag{1.2}$$

where F denotes the force field that is collectively generated by all particles. The requirement that f be constant along the curves defined by (1.2) then leads to the relation

$$\begin{aligned} 0 &= \frac{d}{ds}\left[f(s, X(s), V(s))\right] \\ &= \partial_t f(s, X(s), V(s)) + V(s) \cdot \nabla_x f(s, X(s), V(s)) \\ &\quad + F(s, X(s)) \cdot \nabla_v f(s, X(s), V(s)) \end{aligned}$$

for all s. Evaluated at time t, this yields

$$\partial_t f(t, t, v) + v \cdot \nabla_x f(t, x, v) + F(t, x) \cdot \nabla_v f(t, x, v) = 0$$

for all (t, x, v), which is usually called the Vlasov equation (despite the historic inadequacy of this terminology). The next step is to express the force field F in terms of the distribution function f. Since we are aiming at describing gravitational binding, we need to have $F \sim -\nabla_x V_C$ for the Coulomb potential $V_C(x) = -\frac{1}{|x|}$ at large distances. This suggests to use the field $F = -\nabla_x U$ induced by the Poisson equation

$$\begin{aligned} &\Delta_x U_f(t, x) = 4\pi \rho_f(t, x), \quad \lim_{|x| \to \infty} U_f(t, x) = 0, \\ &\text{where} \quad \rho_f(t, x) = \int_{\mathbb{R}^3} f(t, x, v)\, dv \end{aligned} \tag{1.3}$$

denotes the charge density induced by f. Observe that $\int_{\mathcal{X}} dx\, \rho_f(t, x)$ represents the number of particles at time t, of any velocity, which are located at some point $x \in \mathcal{X}$. Then

$$U_f(t, x) = -\int_{\mathbb{R}^3} \frac{\rho_f(t, y)}{|y - x|}\, dy \tag{1.4}$$

is Coulomb-like as $|x| \to \infty$.

To summarize, the Vlasov-Poisson system in the gravitational case is

$$\partial_t f(t,x,v) + v \cdot \nabla_x f(t,x,v) - \nabla_x U_f(t,x) \cdot \nabla_v f(t,x,v) = 0 \tag{1.5}$$

together with (1.3), and the equations are supposed to hold for $(t,x,v) \in \mathbb{R} \times \mathbb{R}^3 \times \mathbb{R}^3$. Initial data $f(0,x,v) = f_0(x,v)$ at time $t = 0$ have to be specified for f only, since then (1.4) determines the initial data $U_f(0,x)$. We will exclusively be interested in classical solutions of (1.5) and (1.3), whose global-in-time existence is ensured, under reasonable assumptions on f_0, by [55, 67, 80]. For a mathematical overview of the system and more background material, the reader may wish to consult [27, 63, 73].

The gravitational Vlasov-Poisson system is widely used to describe non-relativistic galactic dynamics. When it comes to relativistic galactic dynamics, the appropriate model is the Einstein-Vlasov system [2]. In the present book, we will not be dealing with this more general system, but of course it will be tempting to determine which results could be transferred to the Einstein-Vlasov system; see [18, 19, 31, 32, 38–40] for work in this context that is related to the so-called Antonov bound.

1.3 Steady State Solutions

The Vlasov-Poisson system possesses an abundance of solutions $Q = Q(x,v)$ that are independent of time. It is therefore of interest to study the stability of those steady states and, more ambitiously, the dynamics close to a steady state. Let $e_Q(x,v) = \frac{1}{2}|v|^2 + U_Q(x)$ denote the particle energy and let $\ell^2 = |L|^2 = |x|^2|v|^2 - (x \cdot v)^2$ be the square of the angular momentum $L = x \wedge v$. Then both e_Q and ℓ^2 are conserved along solutions of the characteristic equations $\ddot{X}(s) = -\nabla U_Q(X(s))$, which result from (1.2) for $F = -\nabla U_Q$; note that also U_Q is independent of time. Next, recall that a function $g = g(x,v)$ is said to be spherically symmetric if $g(Ax, Av) = g(x,v)$ for all $A \in \mathrm{SO}(3)$ and $x, v \in \mathbb{R}^3$. Expressed in more sophisticated terms, g needs to be equivariant w.r. to the group action $\mathrm{SO}(3) \times (\mathbb{R}^3 \times \mathbb{R}^3) \to \mathbb{R}^3 \times \mathbb{R}^3$, $(A,x,v) \mapsto (Ax, Av)$. Now, it is the content of Jeans's theorem that the distribution function Q of every spherically symmetric steady state solution has to be of the form $Q = Q(e_Q, \ell^2)$; see [7, Section 2] for a precise formulation. Such steady state solutions are called non-isotropic, in contrast to the isotropic ones, which can be written as $Q = Q(e_Q)$; a solution of the latter form will necessarily be spherically symmetric [25, 72]. Observe that we are going to systematically abuse notation in that we consider $Q = Q(x,v)$ to be a function of (x,v) and at the same time write $Q = Q(e_Q, \ell^2)$ or $Q = Q(e_Q)$, which indicates that Q is a function of two or of one scalar variable(s); in general, no confusion will result from this simplification.

To precisely state our results later, we will focus on the isotropic case, and we need to introduce the following assumptions (Q1)–(Q4) that we are going to impose throughout the book on the profile function $Q : \mathbb{R} \to [0, \infty[$ and the (radial) density

$\rho_Q : [0, \infty[\to [0, \infty[$. The diligent reader is invited to check which parts of this work remain valid under less restrictive hypotheses (there are several) or for non-isotropic steady states.

(Q1) The support $K = \operatorname{supp} Q$ of the steady state solution Q is compact and its mass $\|Q\|_{L^1(\mathbb{R}^6)}$ is finite.

(Q2) $Q \in L^\infty_{\text{loc}}(\mathbb{R})$ satisfies $Q \geq 0$, and there exists a cut-off energy $e_0 < 0$ such that $Q(e) = 0$ for $e \geq e_0$, $Q \in C^1(]-\infty, e_0[)$ and $Q > 0$ in some interval $[e_1, e_0[$, where $e_1 < e_0$. For $\hat{e} \in]U_Q(0), e_0[$, there exists $\varepsilon > 0$ such that

$$\inf\{|Q'(e)| : e \in [\hat{e} - \varepsilon, \hat{e} + \varepsilon]\} > 0.$$

(Q3) $Q' \in L^\infty_{\text{loc}}(\mathbb{R})$ and $Q'(e) \leq 0$ a.e.

(Q4) ρ_Q is continuous and has compact support $\operatorname{supp} \rho_Q = [0, r_Q]$. In addition, $\rho_Q \in C^1([0, r_Q])$.

For one result (Corollary 4.17), we will need more precise information on the behavior of Q' close to $e = e_0$.

(Q5) There are constants $C > 0$ and $\alpha > 0$ such that

$$|Q'(e)| \leq C(e_0 - e)^\alpha, \quad e \in [U_Q(0), e_0[.$$

1.4 Examples

To illustrate that the general assumptions on Q as stated in Sect. 1.3 are verified in many cases, we consider the steady state solution class of the polytropes and the King models in some more detail. It should be remarked that many further examples could be given, for instance by using [69] or [74, Theorem 3.1(a)], which basically says that under mild technical assumptions on Q and if $Q(e) = C(e_0 - e)_+^k + \mathcal{O}((e_0 - e)_+^{k+\delta})$ as $e \to e_0-$ for some $e_0 < 0$, $k \in]-\frac{1}{2}, \frac{3}{2}[$, $C > 0$ and $\delta > 0$, then the resulting steady state solution will have a finite radius and finite mass.

1.4.1 Polytropes

We consider the polytropes

$$Q(e_Q) = (e_0 - e_Q)_+^k \tag{1.6}$$

for a cut-off energy $e_0 < 0$ and $k \in]-\frac{1}{2}, \frac{7}{2}[$. Then

$$\rho_Q(r) = c_n(e_0 - U_Q(r))_+^n, \quad n = k + \frac{3}{2} \in]1, 5[, \quad c_n = (2\pi)^{3/2} \frac{\Gamma(k+1)}{\Gamma(k+\frac{5}{2})};$$

see [7, Example 4.1]. All these steady state solutions do have finite radius r_Q (i.e., compact support) and finite mass $M_Q = \int_{\mathbb{R}^3} \rho_Q(x)\,dx = 4\pi \int_0^{r_Q} r^2 \rho_Q(r)\,dr = \int_{\mathbb{R}^3}\int_{\mathbb{R}^3} Q(x,v)\,dx\,dv$. The limiting case $k = 7/2$ is called the Plummer sphere, where M_Q is still finite, but $r_Q = \infty$. We have $Q'(e) = -k(e_0 - e)_+^{k-1} \le 0$ (outside of $e = e_0$ for $k \le 1$) and $\rho_Q \in C^1([0, r_Q])$. Thus, if we take $k > 1$ for simplicity, then assumptions (Q1)–(Q5) are satisfied.

1.4.2 King models

The ansatz function for the King model [9, pp. 307–311] is given by

$$Q(e_Q) = (\exp(e_0 - e_Q) - 1)_+$$

for some cut-off energy $e_0 < 0$. Then $Q \in C^1(]-\infty, e_0[)$ and $Q'(e) = -\exp(e_0 - e) \le 0$ for $e < e_0$. The associated steady state solution does exist and has finite radius and finite mass; see [74, Theorem 3.1(a) and Sect. 4]. The density is found to be

$$\begin{aligned}\rho_Q(r) &= \int_{\mathbb{R}^3} Q(x,v)\,dv \\ &= \int_{\mathbb{R}^3} \Big(\exp\Big(e_0 - \frac{1}{2}|v|^2 - U_Q(r)\Big) - 1\Big)_+ dv \\ &= (\sqrt{2\pi})^3\Big(e^s \operatorname{erf}(\sqrt{s}) - \sqrt{\frac{4s}{\pi}}\Big(1 + \frac{2s}{3}\Big)\Big), \quad s = e_0 - U_Q(r),\end{aligned}$$

where $\operatorname{erf}(x) = \frac{2}{\sqrt{\pi}}\int_0^x e^{-t^2}\,dt$ denotes the error function, which has the asymptotic expansion $\operatorname{erf}(x) = \frac{2x}{\sqrt{\pi}} - \frac{2x^3}{3\sqrt{\pi}} + \mathcal{O}(x^5)$ as $x \to 0$. For $\varphi(s) = e^s \operatorname{erf}(\sqrt{s}) - \sqrt{\frac{4s}{\pi}}(1 + \frac{2s}{3})$, this yields the asymptotic expansion $\varphi(s) = \frac{8s^{5/2}}{15\sqrt{\pi}} + \mathcal{O}(s^{7/2})$ as $s \to 0^+$. Since $U_Q \in C^2([0,\infty[)$, we infer that in particular $\rho_Q'(r_Q) = 0$ holds and it follows that assumptions (Q1)–(Q4) are satisfied. However, since $Q(e) = (e_0 - e) + \mathcal{O}((e_0 - e)^2)$ as $e \to e_0-$, assumption (Q5) does not hold for the King model.

1.5 Linearization and the Antonov Stability Estimate

Without being too precise about its properties, we consider an isotopic steady state solution $Q = Q(e_Q)$. To study the stability of Q, we will closely follow [30] and write $f(t) = Q + g(t)$ with g 'small'. The total energy

$$\mathcal{H}(f(t)) = \frac{1}{2}\int_{\mathbb{R}^3} |v|^2 f(t,x,v)\,dx\,dv - \frac{1}{8\pi}\int_{\mathbb{R}^3} |\nabla U_{f(t)}(t,x)|^2\,dx$$

is conserved along solutions, so it could be suspected to be a Lyapunov function. The expansion about Q then yields

$$\mathcal{H}(f(t)) = \mathcal{H}(Q) + \int_{\mathbb{R}^3}\int_{\mathbb{R}^3} \Big(\frac{1}{2}|v|^2 + U_Q\Big)\, g(t)\,dx\,dv \\ -\frac{1}{8\pi}\int_{\mathbb{R}^3} |\nabla U_{g(t)}|^2\,dx + \mathcal{O}(g^3); \tag{1.7}$$

note that $f \mapsto U_f$ is linear. The linear term on the right-hand side of (1.7) does not vanish, i.e., Q is not a critical point of $\mathcal{H}$. However, this defect can be remedied by making use of the fact that every 'Casimir functional'

$$\mathcal{C}_\Phi(f(t)) = \int_{\mathbb{R}^3}\int_{\mathbb{R}^3} \Phi(f(t,x,v))\,dx\,dv$$

is also conserved along solutions, provided that Φ is sufficiently well-behaved. Passing from $\mathcal{H}$ to

$$\mathcal{H}_\Phi = \mathcal{H} + \mathcal{C}_\Phi$$

and repeating the expansion, one arrives at

$$\mathcal{H}_\Phi(f(t)) = \mathcal{H}_\Phi(Q) + \int_{\mathbb{R}^3}\int_{\mathbb{R}^3} (e_Q + \Phi'(Q))\, g(t)\,dx\,dv \\ +\frac{1}{2}\int_{\mathbb{R}^3}\int_{\mathbb{R}^3} \Phi''(Q)\, g(t)^2\,dx\,dv - \frac{1}{8\pi}\int_{\mathbb{R}^3} |\nabla U_{g(t)}|^2\,dx + \mathcal{O}(g^3). \tag{1.8}$$

Writing $e = e_Q$, since $Q = Q(e)$, the equation $e + \Phi'(Q(e)) = 0$ can be (formally) solved by taking $\Phi'(\xi) = -Q^{-1}(\xi)$, at least if for instance $Q'(e) < 0$ is verified for the relevant e in the support of Q. Then Q becomes a critical point of this $\mathcal{H}_\Phi$, and due to $1 + \Phi''(Q(e))Q'(e) = 0$ and $Q'(e) < 0$, the expansion (1.8) simplifies to

$$\mathcal{H}_\Phi(f(t)) = \mathcal{H}_\Phi(Q) + \frac{1}{2}\mathcal{A}(g(t), g(t)) + \mathcal{O}(g^3), \\ \mathcal{A}(g,g) = \int_{\mathbb{R}^3}\int_{\mathbb{R}^3} \frac{dx\,dv}{|Q'(e_Q)|}|g|^2 - \frac{1}{4\pi}\int_{\mathbb{R}^3} |\nabla_x U_g|^2\,dx. \tag{1.9}$$

Thus, one can expect that the stability of Q will be determined by the properties of the quadratic (second variation) part $\mathcal{A} = 2\,D^2\mathcal{H}_\Phi(Q)$, which we will call the Antonov functional. It should also be noted that $\mathcal{A}(g(t), g(t))$ is conserved along solutions $g(t)$ of the system that is linearized about Q; see [63, Prop. 3.2] and (1.21) below.

If we now consider functions $u = u(x, v)$ that are spherically symmetric and odd in v, i.e., they satisfy $u(x, -v) = -u(x, v)$, then the celebrated Antonov stability estimate [4, 5] is

$$\mathcal{A}(\mathcal{T}u, \mathcal{T}u) \geq c\|u\|_Q^2 \tag{1.10}$$

for some constant $c > 0$ that only depends on Q, where

$$\mathcal{T}g = \{g, e_Q\} = v \cdot \nabla_x g - \nabla_v g \cdot \nabla_x U_Q \tag{1.11}$$

for the standard Poisson bracket $\{g, h\} = \nabla_x g \cdot \nabla_v h - \nabla_v g \cdot \nabla_x h$. The weighted inner product

$$(g, h)_Q = \iint_K \frac{1}{|Q'(e_Q)|} \overline{g(x, v)}\, h(x, v)\, dx\, dv \tag{1.12}$$

induces the norm $\|\cdot\|_Q$, and $K = \operatorname{supp} Q \subset \mathbb{R}^6$ denotes the support of the steady state solution Q, which is compact, if (Q1) holds. Perturbations of the form $g = \mathcal{T}u$ are called 'dynamically accessible', for reasons explained in [62]; also see [66]. Antonov [4, 5] could prove that the positive definiteness (1.10) is equivalent to the linear stability of Q. Many works followed these pioneering observations, and until to date, almost all stability proofs, linear or nonlinear, use the Antonov stability estimate in one way or another. The bound (1.10), or variations thereof, is applied in a number of papers, both in the physics and in the mathematics community, to address a variety of stability issues; see [15, 26, 28, 30, 42, 43, 50, 51, 60, 89] and many further.

1.6 The Best Constant in the Antonov Stability Estimate

In this section, we will explain the connection of the functional $u \mapsto \mathcal{A}(\mathcal{T}u, \mathcal{T}u)$ from (1.10) to a certain self-adjoint operator L. Before doing so, we need to introduce some relevant notation, function spaces, etc. Since we restrict ourselves to isotropic steady states, the solutions will be spherically symmetric. Thus, we will consider (1.5) and (1.3) in the spherical symmetric framework only, and it is well-known [7] that then the system can be written as

$$\partial_t f(t, r, p_r, \ell^2) + p_r\, \partial_r f(t, r, p_r, \ell^2) + \Big(\frac{\ell^2}{r^3} - \partial_r U_f(t, r)\Big)\, \partial_{p_r} f(t, r, p_r, \ell^2) = 0$$

and

$$U_f''(t, r) + \frac{2}{r}\, U_f'(t, r) = 4\pi \rho_f(t, r), \quad \lim_{r \to \infty} U_f(t, r) = 0,$$
$$\rho_f(t, r) = \frac{2\pi}{r^2} \int_0^\infty d\ell\, \ell \int_{\mathbb{R}} dp_r\, f(t, r, p_r, \ell^2), \tag{1.13}$$

the $'$ indicating $\frac{d}{dr}$ or ∂_r, and $p_r = \frac{x \cdot v}{r}$. If $g = g(x, v)$ is spherically symmetric, then $\rho_g(x) = \rho_g(r)$ and $U_g(x) = U_g(r)$ are radially symmetric, and we will in general denote

$$\rho_g(x) = \int_{\mathbb{R}^3} g(x, v)\, dv, \quad U_g(x) = -\int_{\mathbb{R}^3} \frac{\rho_g(y)}{|y - x|}\, dy. \tag{1.14}$$

Also $g = g(x, v)$ can be identified with a function $g = g(r, p_r, \ell)$ or $g = g(r, p_r, \ell^2)$; see Appendix I, Section A.1.

Next, define the linear operator $\mathcal{K}$ by

$$\mathcal{K}g = \{Q, U_g\};$$

it should be mentioned that both $\mathcal{T}$ from (1.11) and $\mathcal{K}$ do arise naturally upon linearizing the Vlasov-Poisson system about Q; see (1.21) below. Since $U_g(x) = U_g(|x|) = U_g(r)$, we obtain

$$\mathcal{K}g = \{Q, U_g\} = -\nabla_v Q \cdot \nabla_x U_g = -Q'(e_Q)\, v \cdot \frac{x}{r}\, U_g'(r) = -Q'(e_Q)\, p_r\, U_g'(r). \tag{1.15}$$

The operator L is introduced as

$$Lu = -\mathcal{T}^2 u - \mathcal{K}\mathcal{T}u. \tag{1.16}$$

For what concerns the appropriate function spaces, we will pass to action-angle variables as follows. On $K = \operatorname{supp} Q$, we consider the equation

$$\ddot{r} = -U_{\text{eff}}'(r, \ell), \tag{1.17}$$

where $U_{\text{eff}}(r, \ell) = U_Q(r) + \frac{\ell^2}{2r^2}$ is the effective potential that occurs in the energy function

$$e_Q = e_Q(r, p_r, \ell) = \frac{1}{2}\,|v|^2 + U_Q(r) = \frac{1}{2}\, p_r^2 + U_{\text{eff}}(r, \ell),$$

where $p_r = \dot{r}$ is the radial velocity and ℓ should be thought of as fixed. By standard Hamiltonian system theory (see Section A.1 for details), it is then possible to write spherically symmetric functions $g = g(x, v) = g(r, p_r, \ell)$ in the form $g = g(\theta, I, \ell)$ if we apply a canonical transformation $(\theta, I) \mapsto (r, p_r)$ at fixed ℓ. Working in action-angle variables has many advantages. First of all, it turns out that e_Q becomes a function of (I, ℓ) alone, $e_Q = E(I, \ell)$. Secondly, the functions g are 2π-periodic in θ, so they can be conveniently represented as a Fourier series

$$g(\theta, I, \ell) = \sum_{k \in \mathbb{Z}} g_k(I, \ell)\, e^{ik\theta},$$

where

$$g_k(I, \ell) = \frac{1}{2\pi} \int_0^{2\pi} g(\theta, I, \ell)\, e^{-ik\theta}\, d\theta$$

are the Fourier coefficients. The spaces $X^{\alpha}_{\rm odd}$ (cf. Appendix II, Sect. B.1) are defined in terms of this series representation by means of the norms

$$\|g\|^2_{X^\alpha} \sim \sum_{k\in\mathbb{Z}} (1+k^2)^\alpha \, \|g_k\|^2_{L^2_{\frac{1}{|Q'|}}(D)},$$

where $L^2_{\frac{1}{|Q'|}}(D)$ is a weighted L^2-space on the domain D of the variables (I, ℓ). The subscript 'odd' in $X^{\alpha}_{\rm odd}$ indicates that the functions are odd in v, which translates into the condition $g_{-k} = -g_k$ for $k \in \mathbb{Z}$ on the coefficients (so that in particular $g_0 = 0$).

Now we can give a precise meaning to the fact that $\mathcal{A}(\mathcal{T}u, \mathcal{T}u) = (Lu, u)_Q$ is the quadratic form associated with the operator L from (1.16). We have the following result.

Lemma 1.1 *L is self-adjoint on the domain $\mathcal{D}(L) = X^2_{\rm odd}$ in $X^0_{\rm odd}$. In addition, $(Lu, u)_Q = \mathcal{A}(\mathcal{T}u, \mathcal{T}u)$ holds for $u \in X^2_{\rm odd}$.*

Proof Most of this will be shown later; see Corollary B.19 for the properties of L. At this point, let us just mention that by (B.44) in Corollary B.19 the term $(\mathcal{K}\mathcal{T}u, u)_Q$ can be written as $\frac{1}{4\pi}\int_{\mathbb{R}^3} |\nabla_x U_{\mathcal{T}u}|^2\, dx$. Hence, we deduce that

$$\begin{aligned}(Lu, u)_Q &= (-\mathcal{T}^2 u, u)_Q - (\mathcal{K}\mathcal{T}u, u)_Q \\ &= \int_{\mathbb{R}^3}\int_{\mathbb{R}^3} \frac{dx\, dv}{|Q'(e_Q)|} |\mathcal{T}u|^2 - \frac{1}{4\pi}\int_{\mathbb{R}^3} |\nabla_x U_{\mathcal{T}u}|^2\, dx \\ &= \mathcal{A}(\mathcal{T}u, \mathcal{T}u); \end{aligned} \tag{1.18}$$

recall (1.9). □

As a consequence, we can re-express (1.10) as follows.

Theorem 1.2 (Antonov stability estimate) *If $u \in X^2_{\rm odd}$, then*

$$(Lu, u)_Q = \mathcal{A}(\mathcal{T}u, \mathcal{T}u) \geq c\, \|u\|^2_Q \tag{1.19}$$

for $c = \frac{1}{r_Q^3}\|Q\|_{L^1(\mathbb{R}^6)} > 0$, where $\operatorname{supp}\rho_Q = [0, r_Q]$.

We will indicate a proof of Theorem 1.2 in Chapter 2. Therefore,

$$\lambda_* = \inf\{(Lu, u)_Q : u \in X^2_{\rm odd},\ \|u\|_Q = 1\} > 0 \tag{1.20}$$

is well-defined; it is the 'best constant' in the Antonov stability estimate and a main object of study in the present work. We will derive many results related to λ_*, as will be described in Section 1.8. In particular, we will be able to characterize the cases where λ_* is attained, in the sense that $\lambda_* = (Lu_*, u_*)_Q$ for some minimizing function $u_* \in X^2_{\rm odd}$ such that $\|u_*\|_Q = 1$. It turns out that then u_* will be an eigenfunction of L corresponding to the eigenvalue λ_*, so that $Lu_* = \lambda_* u_*$. The quantity λ_* will be of fundamental importance for the dynamics of the gravitational Vlasov-Poisson system.

Lemma 1.3 *Let $u_* \in X^2_{\mathrm{odd}}$ be a minimizer and define*

$$g_*(t,x,v) = \cos(\sqrt{\lambda_*}t)\, u_*(x,v) - \frac{1}{\sqrt{\lambda_*}} \sin(\sqrt{\lambda_*}t)\, (\mathcal{T} u_*)(x,v).$$

Then g_* is a $\frac{2\pi}{\sqrt{\lambda_*}}$-periodic solution of the equation

$$\partial_t g + \mathcal{T} g + \mathcal{K} g = 0 \tag{1.21}$$

that is obtained by linearizing (1.5) and (1.3) about Q.

Proof To linearize the system about Q, let $f = Q + g$ as before. As a consequence of the fact that $v \cdot \nabla_x f - \nabla_x U_f \cdot \nabla_v f = \{f, e_f\}$ for $e_f(x,v) = \frac{1}{2}|v|^2 + U_f(x)$, we may write

$$\begin{aligned} 0 = \partial_t f + \{f, e_f\} &= \partial_t g + \left\{Q + g, \frac{1}{2}|v|^2 + U_Q + U_g\right\} \\ &= \partial_t g - \nabla_v Q \cdot \nabla_x U_g + v \cdot \nabla_x g - \nabla_v g \cdot \nabla_x U_Q - \nabla_v g \cdot \nabla_x U_g, \end{aligned}$$

which is equivalent to

$$\partial_t g + \mathcal{T} g + \mathcal{K} g = \nabla_v g \cdot \nabla_x U_g. \tag{1.22}$$

Thus, (1.21) is indeed the linearization. Next, note that u_* is odd in v. Hence, $\rho_{u_*}(x) = \int_{\mathbb{R}^3} u_*(x,v)\,dv = 0$ implies that $U_{u_*} = 4\pi\Delta^{-1}\rho_{u_*} = 0$ and therefore $\mathcal{K}u_* = 0$ by (1.15). Consequently,

$$\begin{aligned} \partial_t g_* + \mathcal{T} g_* + \mathcal{K} g_* &= -\sqrt{\lambda_*}\sin(\sqrt{\lambda_*}t)\, u_* - \cos(\sqrt{\lambda_*}t)\, \mathcal{T} u_* \\ &\quad + \cos(\sqrt{\lambda_*}t)\, \mathcal{T} u_* - \frac{1}{\sqrt{\lambda_*}} \sin(\sqrt{\lambda_*}t)\, \mathcal{T}^2 u_* \\ &\quad + \cos(\sqrt{\lambda_*}t)\, \mathcal{K} u_* - \frac{1}{\sqrt{\lambda_*}} \sin(\sqrt{\lambda_*}t)\, \mathcal{K}\mathcal{T} u_* \\ &= -\sqrt{\lambda_*}\sin(\sqrt{\lambda_*}t)\, u_* + \frac{1}{\sqrt{\lambda_*}} \sin(\sqrt{\lambda_*}t)\, L u_* = 0, \end{aligned}$$

as claimed. □

At present, it is not known if periodic solutions to (1.5) and (1.3) close to steady state solutions do exist; see [17, 56, 70]. However, in this case, $\frac{2\pi}{\sqrt{\lambda_*}}$ will conceivably be the limiting period of the oscillations, a fact for which there is some numerical evidence [70]. To give a heuristic argument, suppose that g_ε is an ε-small and T_ε-periodic solution to (1.22) such that $T_\varepsilon \to T_0$ as $\varepsilon \to 0$. Then, $\tilde{g}_\varepsilon = \varepsilon^{-1} g_\varepsilon$ will be of order one, T_ε-periodic and satisfies

$$\partial_t \tilde{g}_\varepsilon + \mathcal{T}\tilde{g}_\varepsilon + \mathcal{K}\tilde{g}_\varepsilon = \varepsilon \nabla_v \tilde{g}_\varepsilon \cdot \nabla_x U_{\tilde{g}_\varepsilon}.$$

Assuming now that $\tilde{g}_\varepsilon \to \tilde{g}_*$ in a suitable norm, $\tilde{g}_* \neq 0$ will be T_0-periodic and such that

$$\partial_t \tilde{g}_* + \mathcal{T}\tilde{g}_* + \mathcal{K}\tilde{g}_* = 0.$$

If $\partial_t + \mathcal{T} + \mathcal{K}$ does have a one-dimensional kernel, then $\tilde{g}_*$ is proportional to g_* from Lemma 1.3, and hence $T_0 = \frac{2\pi}{\sqrt{\lambda_*}}$.

1.7 Domains in Action-Angle Variables

Before we will be able to describe our main results and their connection to the Birman-Schwinger principle in Sect. 1.8, we have to take a closer look at the domains that occur as the supports of steady state solutions, expressed in action-angle variables (θ, I, ℓ); recall that the particle energy $e_Q = E(I, \ell)$ is a function of (I, ℓ) alone.

The frequency functions associated with the energy E are

$$\omega_1(I, \ell) = \frac{\partial E(I, \ell)}{\partial I}, \quad \omega_2(I, \ell) = \frac{\partial E(I, \ell)}{\partial L_3} = 0, \quad \omega_3(I, \ell) = \frac{\partial E(I, \ell)}{\partial \ell},$$

where (I, L_3, ℓ) are the action variables. We would like to emphasize that ω_1, together with the corresponding period function $T_1(I, \ell) = \frac{2\pi}{\omega_1(I, \ell)}$, will be a main player in the game, and understanding its properties will be of central importance. This is due to the fact that in action-angle variables the operator $\mathcal{T}$ from (1.11) is found to be very simple:

$$(\mathcal{T}g)(\theta, I, \ell) = \omega_1(I, \ell)\, \partial_\theta g(\theta, I, \ell),$$

or $g_k \mapsto ik\omega_1\, g_k$ in terms of the Fourier coefficients. Since ω_1 is independent of θ, this also yields $-\mathcal{T}^2 g = -\omega_1^2\, \partial_\theta^2 g$ or $g_k \mapsto k^2\omega_1^2\, g_k$. It will turn out (see Section 3.1) that ω_1 is strictly positive, so that, at fixed ℓ, the map $I \mapsto E(I, \ell)$ is strictly increasing. Therefore, it can be inverted as a map $E \mapsto I(E, \ell)$, and accordingly functions $g = g(\theta, I, \ell)$ can be viewed as functions $\tilde{g}(\theta, E, \ell) = g(\theta, I(E, \ell), \ell)$ and vice versa.

From (Q1)–(Q4) in Sect. 1.3, the following can be shown (cf. the argument in Sect. 1.7.1 below):
In action-angle variables, one has

$$K = \{(\theta, E, \beta) : \theta \in [0, 2\pi], \beta \in [0, \beta_*], E \in [e_{\min}(\beta), e_0]\},$$

where $E = E(I, \ell)$ and $I = I(E, \ell)$. Furthermore, $\beta = \ell^2$, $\beta_* > 0$, and

$$e_{\min}(\beta) = U_{\text{eff}}(r_0(\beta), \beta)$$

is the minimal energy of the effective potential $U_{\text{eff}}(\cdot, \beta)$, which is attained at the unique point $r_0(\beta)$; see Appendix I, Sect. A.1. Also, $e_{\min}(\cdot)$ is non-decreasing and

$$\min\{e_{\min}(\beta) : \beta \in [0, \beta_*]\} = U_Q(0) < e_0,$$
$$\max\{e_{\min}(\beta) : \beta \in [0, \beta_*]\} = e_0.$$

We will always denote

$$D = \{(E, \beta) : \beta \in [0, \beta_*], E \in [e_{\min}(\beta), e_0]\},$$

which at times will be expressed in terms of ℓ as

$$D = \{(E, \ell) : \ell \in [0, \ell_*], E \in [e_{\min}(\ell), e_0]\}, \tag{1.23}$$

and similarly, we will write

$$K = \{(\theta, E, \ell) : \theta \in [0, 2\pi], \ell \in [0, l_*], E \in [e_{\min}(\ell), e_0]\}. \tag{1.24}$$

It is also understood that K and D can be written in terms of the variables (θ, I, β), (θ, I, ℓ) and (I, β), (I, ℓ), respectively, without this being reflected by renaming the sets. In any case, we will always have $K = [0, 2\pi] \times D$.

For illustration, we are going to determine K and D for the polytropes and the King models, respectively. A general domain D is shown in Fig. 1.1.

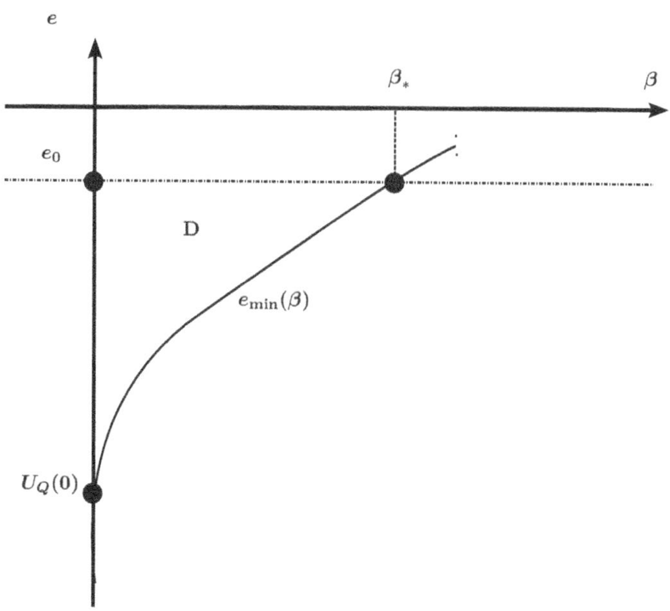

Fig. 1.1 The domain D in coordinates $(e, \beta) = (E, \beta)$

1.7.1 Polytropes Revisited

We wish to determine the support

$$K = \text{supp}\, Q = \{e_0 - e_Q \geq 0\}$$

of the polytropes in terms of $\beta = \ell^2$ and $e = e_Q$. More precisely, since always $\theta \in [0, 2\pi]$ on K for the angular variable θ, we have to exhibit a set D of (e, β) such that $K = [0, 2\pi] \times D$. On this domain D, we need to have

$$e_0 \geq e \geq U_{\text{eff}}(r, \beta) \geq U_{\text{eff}}(r_0(\beta), \beta) = U_Q(r_0(\beta)) + \frac{\beta}{2r_0(\beta)^2}, \tag{1.25}$$

with $r_0(\beta)$ denoting the unique point where the effective potential $U_{\text{eff}}(r, \beta) = U_Q(r) + \frac{\beta}{2r^2}$ attains its minimum value $e_{\min}(\beta) = U_{\text{eff}}(r_0(\beta), \beta)$. From (1.25), we get

$$2r_0(\beta)^2\,(e_0 - U_Q(r_0(\beta)) \geq \beta.$$

Let

$$J = \{\beta \geq 0 : 2r_0(\beta)^2\,(e_0 - U_Q(r_0(\beta)) \geq \beta\}.$$

First, we claim that J is an interval. To see this, note that

$$2r^2(e_0 - U_{\text{eff}}(r, \beta)) + \beta = 2r^2\Big(e_0 - U_Q(r) - \frac{\beta}{2r^2}\Big) + \beta = 2r^2(e_0 - U_Q(r)).$$

Therefore,

$$2r^2(e_0 - U_Q(r)) \geq \beta \iff U_{\text{eff}}(r, \beta) \leq e_0,$$

which implies that

$$J = \{\beta \geq 0 : e_{\min}(\beta) \leq e_0\}. \tag{1.26}$$

Now $\beta \mapsto e_{\min}(\beta)$ is increasing by Lemma A.7(c) below (which is a general result), and thus J has to be an interval.

The next aim is to show that $[0, \varepsilon] \subset J$ for some $\varepsilon > 0$ small enough. For, by Lemma A.7(f), we have

$$r_0(\beta)^4 = \frac{1}{A(0)}\,\beta + \mathcal{O}(\beta^2) \quad \text{and} \quad e_{\min}(\beta) = U_Q(0) + \mathcal{O}(\beta^{1/2})$$

as $\beta \to 0^+$. Since $U_Q(0) < e_0$ (the cut-off energy), the condition $e_{\min}(\beta) \leq e_0$ from the characterization of J in (1.26) is satisfied with strict inequality at $\beta = 0$. It follows that $[0, \varepsilon] \subset J$ if $\varepsilon > 0$ is sufficiently small.

Now, we are going to show that J is bounded. First, if $\beta \in J$, then $r_0(\beta) \leq r_Q$, where $\text{supp}\, \rho_Q = [0, r_Q]$. Otherwise, we would have $r_0(\beta) > r_Q$ for some

$\beta \in J \setminus \{0\}$. Since r_Q is characterized by $U_Q(r_Q) = e_0$, this gives $U_Q(r_0(\beta)) > e_0$, and consequently $\beta \le 2r_0(\beta)^2\,(e_0 - U_Q(r_0(\beta)) \le 0$, which is a contradiction. Then, $r_0(\beta) \le r_Q$ for $\beta \in J$ in turn leads to the boundedness of J, owing to

$$\beta \le 2r_0(\beta)^2\,(e_0 - U_Q(r_0(\beta)) \le 2r_Q^2\,(e_0 - U_Q(r_0(\beta)) \le 2r_Q^2\,(e_0 - U_Q(0))$$

uniformly for $\beta \in J$.

Lastly, we will check that $\beta_* = \max J$ satisfies $e_{\min}(\beta_*) = e_0$. In fact, at β_*, we must have $2r_0(\beta_*)^2\,(e_0 - U_Q(r_0(\beta_*)) = \beta_*$. Thus,

$$e_{\min}(\beta_*) = U_{\text{eff}}(r_0(\beta_*), \beta_*) = U_Q(r_0(\beta_*)) + \frac{\beta_*}{2r_0(\beta_*)^2} = e_0. \tag{1.27}$$

To summarize, since the condition on e is $e_0 \ge e \ge e_{\min}(\beta)$, we have shown that

$$D = \{(\beta, e) : \beta \in [0, \beta_*], e \in [e_{\min}(\beta), e_0]\}$$

and $K = [0, 2\pi] \times D$ for the support K of Q in terms of e and β, and the lower boundary curve $[0, \beta_*] \ni \beta \mapsto e_{\min}(\beta)$ strictly increases from $U_Q(0)$ to e_0.

We would also like to point out that $r_0(\beta_*) \in]0, r_Q[$. By construction, one has $r_0(\beta_*) \le r_Q$, so suppose that we had $r_0(\beta_*) = r_Q$. Since $r_0(\beta)^3 U_Q'(r_0(\beta)) = \beta$, (1.27) yields

$$e_0 = U_Q(r_0(\beta_*)) + \frac{\beta_*}{2r_0(\beta_*)^2} = U_Q(r_0(\beta_*)) + \frac{1}{2}\,r_0(\beta_*)\,U_Q'(r_0(\beta_*)). \tag{1.28}$$

But $U_Q(r_Q) = e_0$, whence $0 = U_Q'(r_Q) = \frac{4\pi}{r_Q^2}\int_0^{r_Q} s^2 \rho_Q(s)\,ds$, which is a contradiction. The relation (1.28) characterizes $r_0(\beta_*)$, since $\varphi(r) = U_Q(r) + \frac{1}{2}\,rU_Q'(r)$ satisfies $\varphi'(r) = \frac{1}{2}\,rB(r)$ for $B(r) = \frac{U_Q'(r)}{r} + 4\pi\rho_Q(r) > 0$ from Lemma A.6(b). In addition, $\varphi(0) = U_Q(0) < e_0$ and $\varphi(r_Q) = e_0 + \frac{1}{2}\,r_Q U_Q'(r_Q) > e_0$.

Finally, observe that the reasoning in this section did not depend on the specific form of the polytropic ansatz function (1.6), but only on the general properties of the functions $r_0(\beta)$ and $e_{\min}(\beta)$.

1.7.2 King Models Revisited

Exactly as in Sect. 1.7.1, here we also get

$$K = \operatorname{supp} Q = [0, 2\pi] \times D, \quad D = \{(\beta, e) : \beta \in [0, \beta_*], e \in [e_{\min}(\beta), e_0]\},$$

for the corresponding functions $r_0(\beta)$ and $e_{\min}(\beta) = U_{\text{eff}}(r_0(\beta), \beta)$. In addition, we have $r_0(\beta_*) \in]0, r_Q[$.

1.8 Summary of the Main Results

Now, we are in a position to outline the main results of this book. In Chap. 3, we will study the properties of ω_1 or equivalently of T_1 in some detail. First, it is shown (Theorem 3.2) that

$$\delta_1 = \inf\{\omega_1(e, \ell) : (e, \ell) \in \mathring{D}\} > 0. \tag{1.29}$$

This fact has been mentioned above and it will be used many times. The number δ_1, or more precisely δ_1^2, is intimately related to the spectrum of L, since $\delta_1^2 = \min \sigma_{\rm ess}(L)$ is the minimum of the essential spectrum of L. In this connection, let us also mention that the essential spectrum of L can be determined explicitly, and it is large in the sense that $[\lambda_c, \infty[\subset \sigma_{\rm ess}(L)$ for some $\lambda_c > \delta_1^2$. Furthermore, $\lambda_* \leq \delta_1^2$ is satisfied (Section 3.4). Along with (1.29), we will also prove that

$$\Delta_1 = \sup\{\omega_1(e, \ell) : (e, \ell) \in \mathring{D}\} < \infty;$$

see Theorem 3.5. Concerning the regularity of ω_1 or T_1, it is not very difficult to see that $T_1 \in C^1(\mathring{D})$, as will be derived in Theorem 3.6. It is considerably harder to verify that $T_1 \in C(D)$, i.e., that T_1 can be continuously extended to the boundary ∂D of D. This will be done in a series of lemmas, and the results are summarized in Theorem 3.13; the most challenging part is to make sure that T_1 is continuous at $(e, \beta) = (U_Q(0), 0)$, which is the lower left corner of D. It will also turn out that T_1 is increasing on the lower boundary curve of D (Lemma 3.14) and on the left boundary part of D (Lemma 3.15).

In Chapter 4, we are going to make the connection of the spectral problem for L to the Birman-Schwinger principle. We will be using an approach to reformulate the problem that is inspired by the physics reference [61], although this paper does neither use the operator L nor realize the underlying Birman-Schwinger principle. Let L_r^2 denote the L^2-Lebesgue space of radially symmetric functions $\Psi(x) = \Psi(r)$ on $\mathbb{R}^3$ with inner product

$$\langle \Psi, \Phi \rangle = \int_{\mathbb{R}^3} \overline{\Psi(x)}\, \Phi(x)\, dx = 4\pi \int_0^\infty r^2\, \overline{\Psi(r)}\, \Phi(r)\, dr.$$

It will be shown that one can define a family $\mathcal{Q}_\lambda$ of non-negative Hilbert-Schmidt operators on L_r^2 with the following properties for $\lambda < \delta_1^2$:

(a) λ is an eigenvalue of L if and only if 1 is an eigenvalue of $\mathcal{Q}_\lambda$.

This observation provides a natural way for showing that λ_* is an eigenvalue of L, provided that one has $\lambda_* < \delta_1^2$ (i.e., there is a spectral gap). The first eigenvalue function $\mu_1(\lambda)$ of $\mathcal{Q}_\lambda$ turns out to be increasing in λ, and one has to locate the value of λ, where μ_1 becomes 1; in this way, we will be able to show that λ_* is attained. Furthermore, we will also prove:

(b) if $u \in X^2_{\mathrm{odd}}$ is an eigenfunction of L for the eigenvalue λ, then $\Psi = 4\pi \int_{\mathbb{R}^3} p_r\, u\, dv \in L^2_r$ is an eigenfunction of $\mathcal{Q}_\lambda$ for the eigenvalue 1;
(c) if $\Psi \in L^2_r$ is an eigenfunction of $\mathcal{Q}_\lambda$ for the eigenvalue 1, then $u = (-\mathcal{T}^2 - \lambda)^{-1}(|Q'(e_Q)|\, p_r \Psi) \in X^2_{\mathrm{odd}}$ is an eigenfunction of L for the eigenvalue λ.

Thus, if we compare (a)–(c) for our galactic dynamics setup to (a)–(c) from the Schrödinger case in Sect. 1.1, then we see that both are formally identical if we associate $p_r \sim \sqrt{-V}$ and $-\Delta \sim -\mathcal{T}^2$ and furthermore disregard the velocity average $\int_{\mathbb{R}^3} dv$; the appearance of $|Q'(e_Q)|$ in $|Q'(e_Q)|\, p_r \Psi$ is due to the $(\cdot, \cdot)_Q$ that is used. There is yet another fact that supports the analogy of both approaches. One of the ways to represent $\mathcal{Q}_\lambda$ is

$$\mathcal{Q}_\lambda \Psi = 4\pi \int_{\mathbb{R}^3} p_r\, (-\mathcal{T}^2 - \lambda)^{-1}\, (|Q'(e_Q)|\, p_r \Psi)\, dv. \tag{1.30}$$

Comparing this relation to (1.1), it turns out that both relations do agree if we apply the same identifications as before.

Throughout the book, we are going to exploit this Birman-Schwinger principle in galactic dynamics to deal with the question in which cases λ_* from (1.20) is attained. However, there seems to be a wide range of further possible applications that could for instance be related to a limiting absorption principle or L^p–L^q-estimates on the 'free resolvent' $(-\mathcal{T}^2 - \lambda)^{-1}$, in the spirit of [45] for the Laplacian. One advantage when dealing with (1.1) is that, in three dimensions, the operator has the explicit integral kernel

$$B_e(x, y) = \sqrt{-V(x)}\, \frac{1}{4\pi|x-y|}\, \exp(-\sqrt{e}\,|x-y|)\, \sqrt{-V(y)},$$

which allows for hands-on estimates. It would also be desirable to obtain something similar for (1.30).

The explicit form of the operator Q_λ is

$$\mathcal{Q}_\lambda : L^2_r \to L^2_r,$$

$$(\mathcal{Q}_\lambda \Psi)(r) = \frac{16\pi}{r^2} \sum_{k \neq 0} \int_0^\infty d\tilde{r}\, \Psi(\tilde{r}) \iint_D d\ell\, \ell\, de\, \mathbf{1}_{\{r_-(e,\ell) \le r, \tilde{r} \le r_+(e,\ell)\}}\, \frac{\omega_1(e,\ell)\, |Q'(e)|}{k^2 \omega_1^2(e,\ell) - \lambda}$$
$$\times \sin(k\theta(r, e, \ell))\, \sin(k\theta(\tilde{r}, e, \ell)),$$

where $r_\pm(e, \ell)$ are the maximal resp. minimal value of r along the orbit of (1.17) that has energy e, and $\theta(r, e, \ell)$ is the associated angle. Note that $\lambda < \delta_1^2$ implies $k^2\omega_1^2(e, \ell) - \lambda \geq \delta_1^2 - \lambda > 0$ for $k \neq 0$, so the denominators do not vanish. It turns out that the family $\mathcal{Q}_\lambda$ can be analytically continued to $\mathcal{Q}_z$ for $z \in \Omega = \mathbb{C} \setminus [\delta_1^2, \infty[$, by simply replacing λ with z. In addition, we can write $(\mathcal{Q}_z \Psi)(r) = \langle K_{\bar{z}}(r, \cdot), \Psi\rangle$ for some $L^2 \times L^2$-integral kernel K, which allows us to show that each $\mathcal{Q}_z$ is a Hilbert-Schmidt operator on L^2_r. Furthermore, $\langle \mathcal{Q}_\lambda \Psi, \Psi\rangle \geq 0$ and $\lambda \to \langle \mathcal{Q}_\lambda \Psi, \Psi\rangle$

are increasing for real λ. Then, the spectrum of $\mathcal{Q}_\lambda$ consists of $\mu_1(\lambda) \geq \mu_2(\lambda) \geq \ldots \to 0$ (the eigenvalues are listed according to their multiplicities). In addition,

$$\mu_1(\lambda) = \|\mathcal{Q}_\lambda\| = \sup\{\langle \mathcal{Q}_\lambda \Psi, \Psi\rangle : \|\Psi\|_{L^2_r} \leq 1\},$$

where $\|\cdot\| = \|\cdot\|_{\mathcal{B}(L^2_r)}$, and every function

$$\mu_k(\cdot) :]-\infty, \delta_1^2[\to]0, \infty[$$

for $k \in \mathbb{N}$ is monotone increasing and locally Lipschitz continuous. According to the Birman-Schwinger characterization of an eigenvalue λ for L, we have to determine those k and λ, where $\mu_k(\lambda) = 1$. Since we expect $\lambda_* \leq \delta_1^2$ to be the principal eigenvalue of L, more specifically we need to find λ such that $\mu_1(\lambda) = 1$. In this respect, the quantity

$$\mu_* = \lim_{\lambda \to \delta_1^2 -} \mu_1(\lambda) = \sup\{\mu_1(\lambda) : \lambda \in [0, \delta_1^2[\} \in [\mu_1(0), \infty]$$

will be important, and in what follows, we are going to outline our results, depending on μ_*.

Let us first recall that $\delta_1^2 = \min \sigma_{\mathrm{ess}}(L)$, and if $\lambda_* < \delta_1^2$ and λ_* were an eigenvalue of L, then there would exist a spectral gap. We are going to prove in Theorem 4.13 that the conditions $\lambda_* < \delta_1^2$ and $\mu_* > 1$ are equivalent, and in this case, $\mu_1(\lambda_*) = 1$ and λ_* is an eigenvalue of L. The difficult part of the argument is to show that a spectral gap $\lambda_* < \delta_1^2$ forces λ_* to be an eigenvalue. This is accomplished by studying (at great length in Appendix C) a certain evolution equation, for which $\lambda_* < \delta_1^2$ translates into a compactness condition; the argument is summarized in Section C.1.

Next, we turn to the case where $\mu_* < 1$. Then necessarily $\lambda_* = \delta_1^2$, so there is no spectral gap and we cannot use the Birman-Schwinger principle. Nevertheless, it is possible to prove (Theorem 4.14) that now $\lambda_* = \delta_1^2$ is not an eigenvalue, provided that the following condition is satisfied:

(ω_1-1) $\{(I, \ell) \in D : \omega_1(I, \ell) = \delta_1\}$ has the Lebesgue measure zero.

This excludes (Lemma B.12) that δ_1^2 is an eigenvalue of $-\mathcal{T}^2$. The proof works by deriving suitable estimates for the operators $\mathcal{Q}_{\delta_1^2 - \varepsilon + i\varepsilon^3}$ in the limit $\varepsilon \to 0^+$. We would not be surprised if the case $\mu_* < 1$ could not occur at all, but we were not able to verify this.

The most pathological case seems to be $\mu_* = 1$. Then once again $\lambda_* = \delta_1^2$, there is no spectral gap and the Birman-Schwinger principle does not apply. To see that here one needs to add another condition on ω_1, let us change the perspective and ask where, in D, $\delta_1 = \inf_{\mathring{D}} \omega_1 = \min_D \omega_1$ is attained. If this happens at an interior point $(\hat{e}, \hat{\beta}) \in \mathring{D}$, then $\nabla \omega_1(\hat{e}, \hat{\beta}) = (0, 0)$ and the following condition will be verified:

(ω_1-2) There are a point $(\hat{e}, \hat{\beta}) \in \mathring{D}$, a neighborhood U of $(\hat{e}, \hat{\beta})$ and a constant $C_1 > 0$ such that $\omega_1(\hat{e}, \hat{\beta}) = \delta_1$ and

$$|\omega_1(e,\beta)-\delta_1| \le C_1\,|(e,\beta)-(\hat{e},\hat{\beta})|^2, \quad (e,\beta)\in U. \tag{1.31}$$

But then Corollary 4.16 implies that $\mu_* = \infty$, which is not compatible with $\mu_* = 1$. Hence, we can assume that the minimum is attained at some point $(\hat{e},\hat{\beta}) \in \partial D$, the boundary of D. According to Corollary 3.16, then $(\hat{e},\hat{\beta})$ lies on the 'upper line' $\{(e,\beta) : e = e_0, \beta \in [0,\beta_*]\}$ of the boundary and one needs to have more precise information on the behavior of ω_1 close to $(\hat{e},\hat{\beta}) = (e_0,\hat{\beta})$. If $\nabla\omega_1(e_0,\hat{\beta}) \sim (0,0)$ (the following motivation is not rigorous since we don't know that ω_1 is differentiable on ∂D), then we would be in a similar situation as what has been described before. Therefore, we can assume that $\nabla\omega_1(e_0,\hat{\beta}) \nsim (0,0)$ in the sense that at least one of the derivatives $\frac{\partial\omega_1}{\partial e}$ and $\frac{\partial\omega_1}{\partial\beta}$ does not vanish at $(e_0,\hat{\beta})$. If it is exactly one of the two derivatives that does not vanish, one could also derive a bunch of results, with techniques that are similar to the ones outlined below. Hence, we are going to assume that both derivatives do not vanish, in a weak sense that does not need the differentiability, as formulated in the following condition:

(ω_1-3) There are a point $(e_0,\hat{\beta}) \in D$ and a constant $c_1 > 0$ such that $\omega_1(e_0,\hat{\beta}) = \delta_1$ and

$$|\omega_1(e,\beta)-\delta_1| \ge c_1|(e,\beta)-(e_0,\hat{\beta})|, \quad (e,\beta)\in D;$$

it would be sufficient to require (ω_1-3) only locally in a neighborhood of $(e_0,\hat{\beta})$. Supposing that (ω_1-3) holds, we can show in Theorem 4.15 for $\mu_* = 1$ that $\lambda_* = \delta_1^2$ is an eigenvalue of L if and only if

$$\|\mu_1'\|_{L^\infty(]-\infty,\delta_1^2[)} < \infty \tag{1.32}$$

is verified; since $\mu_1(\cdot)$ is differentiable a.e., this condition is meaningful. The proof works by first observing that, as a consequence of (ω_1-3), the operator $\mathcal{Q}_{\delta_1^2} = \lim_{\lambda\to\delta_1^2-}\mathcal{Q}_\lambda$ does exist in the Hilbert-Schmidt norm (Lemma 4.9) and hence is a Hilbert-Schmidt operator itself. In addition, $\mu_* = 1$ is its first eigenvalue $\mu_1(\delta_1^2)$. Due to the compactness of $\mathcal{Q}_{\delta_1^2}$, if $\Psi_j \in L_r^2$ is a normalized eigenfunction of $\mathcal{Q}_{\lambda_j}$ for $\mu_1(\lambda_j)$ and $\lambda_j \to \delta_1^2-$, then a subsequence will converge to a normalized eigenfunction Ψ_* of $\mathcal{Q}_{\delta_1^2}$ for the eigenvalue $\mu_* = 1$ (Corollary 4.11, no need to assume (1.32)). Once again, the situation is very much analogous to what is known for Schrödinger operators, cf. [82, pp. 83–85] and [84, Section 2] for instance: a threshold eigenvalue and eigenfunction of the Birman-Schwinger operator do not immediately give rise to a threshold eigenvalue and eigenfunction of the Schrödinger operator, but in fact the existence of the latter is characterized by an additional condition, which is (1.32) in our case. To understand its meaning, suppose for simplicity of the presentation that there is $\varepsilon > 0$ such that $]\delta_1^2-\varepsilon,\delta_1^2[\ni\lambda\mapsto\mu_1(\lambda)$ is real analytic, and in addition that there are $\Psi_\lambda \in L_r^2$ satisfying $\|\Psi_\lambda\|_{L_r^2} = 1$, $\mathcal{Q}_\lambda\Psi_\lambda = \mu_1(\lambda)\Psi_\lambda$, so that also $]\delta_1^2-\varepsilon,\delta_1^2[\ni\lambda\mapsto\Psi_\lambda$ is real analytic. This will follow from the Kato-Rellich perturbation theory if μ_* is known to be a simple eigenvalue of $\mathcal{Q}_{\delta_1^2}$. In the general case, which is much more technical, one needs to work with appropriate sequences

$\lambda_j \to \delta_1^2-$ that are constructed using an appropriate generalization of the standard Kato-Rellich perturbation theory (Appendix IV). In the real analytic case, define $\psi_\lambda(r, p_r, \ell) = |Q'(e_Q)|\, p_r \Psi_\lambda(r)$ and $g_\lambda = (-\mathcal{T}^2 - \lambda)^{-1}\psi_\lambda$. Then it is found that

$$\|g_\lambda\|_{X^0}^2 = \frac{1}{4\pi}\langle \mathcal{Q}'_\lambda \Psi_\lambda, \Psi_\lambda\rangle = \frac{1}{4\pi}\mu_1'(\lambda)$$

and μ_1' is increasing. Thus, (1.32) is equivalent to the condition $\sup \|g_\lambda\|_{X^0} < \infty$, i.e., to the boundedness of $(g_\lambda) \subset X^0$. In addition, one can prove that

$$Lg_\lambda = (1 - \mu_1(\lambda))\psi_\lambda + \lambda g_\lambda,$$

cf. Lemma 4.7(c). Since $\mu_1(\lambda) \to \mu_1(\delta_1^2) = \mu_* = 1$, the weak convergence $g_\lambda \rightharpoonup g_*$ is seen to be sufficient to ensure that $g_* \neq 0$ and $Lg_* = \delta_1^2 g_*$, i.e., g_* is the wanted eigenfunction of L. To establish the converse assertion, i.e., that the existence of an eigenfunction of L for $\lambda_* = \delta_1^2$ leads to (1.32), a different argument has to be used; see Theorem 4.15. Corollary 4.17 contains an example of a situation where (1.32) can be shown to hold. For this, we add (Q5) from Sect. 1.3 as an additional condition on Q. It should not be surprising that the regularity of Q' close to $e = e_0$ will become important in this respect since we are dealing with integrals of the form

$$\sum_{k\neq 0} \iint_D d\beta\, de\, \frac{\omega_1(e,\beta)\,|Q'(e)|}{k^2\omega_1^2(e,\beta) - \lambda}\,(\ldots)$$

many times. If $\lambda \sim \delta_1^2$ and $k = \pm 1$, then the behavior of ω_1 close to $(e, \beta) = (e_0, \hat{\beta})$ gets important; this is addressed by condition (ω_1-3). On the other hand, there is an interplay with the term $|Q'(e)|$ for e close to e_0, which could compensate for possible losses (or it could be bad itself). Generally speaking, many different results could be derived for $\mu_* = 1$ by combining assumptions of ω_1 with assumptions on Q' close to e_0.

Let us remark that we don't see an immediate path to calculate μ_* for a given steady state solution Q. However, there might be a smart way to settle this question, and in any case μ_*, together with additional important quantities like λ_* and δ_1, for sure could be determined numerically. Another notable fact is as follows. The Vlasov-Poisson system (1.5) and (1.3) has many invariances; see Chap. 6; quantities that remain invariant under the scaling could be expected to be of 'fundamental' importance. It turns out that μ_* is such a quantity, but λ_* and δ_1 are not. On the contrary, the conditions $\lambda_* < \delta_1^2$ and $\lambda_* = \delta_1^2$ are both invariant. We will deduce several other invariants in Chap. 6, among them the "Eddington-Ritter relation", which says that

$$\frac{2\pi}{\sqrt{\lambda_*}}\sqrt{\rho_Q(0)}$$

is invariant; note that $\frac{2\pi}{\sqrt{\lambda_*}}$ is the "linear period" from Lemma 1.3.

There are several other operators around that are used to assist (by means of their coercivity) stability proofs for stellar systems, among them the "Hartree-Fock exchange operator" by Lynden-Bell [57, 58] and the "Guo-Lin operator" [29, 88]. Concerning the latter, we are able to make a connection to the operators $\mathcal{Q}_\lambda$ that we are using, more precisely to $\mathcal{Q}_0$. Let $\lambda_{\mathrm{GL}} > 0$ denote the best constant for the Guo-Lin operator; see (5.2). Then we have

$$\lambda_{\mathrm{GL}} + \mu_1(0) = 1$$

by Lemma 5.1, and $0 < \mu_1(0) < 1$ implies that $\lambda_{\mathrm{GL}} > 0$ will always be attained (Corollary 5.2). Of course, the clear advantage of the operators $\mathcal{Q}_\lambda$ is the underlying Birman-Schwinger principle, as they can be used to detect the λ_* that will be the eigenvalue.

Finally, there are four appendices. Appendix I and Appendix II contain the necessary background material for what concerns the change of coordinates to action-angle variables, function spaces and operators. Appendix III is independent and provides a proof (using a new evolution equation) of the fact that $\lambda_* < \delta_1^2$ implies that λ_* is an eigenvalue of L; this will enter into the theorems obtained in Sect. 4.2. Lastly, Appendix IV concerns some specifics of the Kato-Rellich perturbation theory that are also used to study the properties of $\mathcal{Q}_\lambda$ as $\lambda \to \delta_1^2-$.

Chapter 2
The Antonov Stability Estimate

The purpose of this short chapter is to provide some more details on the Antonov stability estimate, Theorem 1.2.

Proof of Theorem 1.2 We (formally) repeat the argument from [50, Prop. 4.1(ii)]; also see [30, Appendix]. Define $q(x, v) = \frac{u(x,v)}{x \cdot v}$. Then a direct calculation shows that

$$|\mathcal{T}u|^2 = (x \cdot v)^2\, |\mathcal{T}q|^2 + \mathcal{T}\Big((x \cdot v)|q|^2\mathcal{T}(x \cdot v)\Big) - (x \cdot v)\, |q|^2\mathcal{T}^2(x \cdot v).$$

Moreover $\mathcal{T}(x \cdot v) = |v|^2 - rU_Q'(r)$ and, using $\Delta U_Q = 4\pi\rho_Q$,

$$\mathcal{T}^2(x \cdot v) = -(x \cdot v)\Big(4\pi\rho_Q(r) + \frac{U_Q'(r)}{r}\Big). \tag{2.1}$$

Therefore,

$$|\mathcal{T}u|^2 - 4\pi\rho_Q(r)|u|^2 = (x \cdot v)^2\, |\mathcal{T}q|^2 + \mathcal{T}\Big((x \cdot v)\, |q|^2\mathcal{T}(x \cdot v)\Big) + \frac{U_Q'(r)}{r}\, |u|^2.$$

Now integration by parts yields $\int_{\mathbb{R}^3}\int_{\mathbb{R}^3} \frac{dx\,dv}{|Q'(e_Q)|}\, \mathcal{T}(\ldots) = 0$, cf. Lemma B.9(a). So what remains after integration is

$$\begin{aligned}(Lu, u)_Q &= \int_{\mathbb{R}^3}\int_{\mathbb{R}^3} \frac{dx\,dv}{|Q'(e_Q)|}\, |\mathcal{T}u|^2 - \frac{1}{4\pi}\int_{\mathbb{R}^3} |\nabla_x U_{\mathcal{T}u}|^2\, dx \\ &= \int_{\mathbb{R}^3}\int_{\mathbb{R}^3} \frac{dx\,dv}{|Q'(e_Q)|}\, \Big[4\pi\rho_Q(r)|u|^2 + (x \cdot v)^2\, |\mathcal{T}q|^2 + \frac{U_Q'(r)}{r}\, |u|^2\Big] \\ &\quad - \frac{1}{4\pi}\int_{\mathbb{R}^3} |\nabla_x U_{\mathcal{T}u}|^2\, dx.\end{aligned}$$

M. Kunze, *A Birman-Schwinger Principle in Galactic Dynamics*,
Progress in Mathematical Physics 77,
https://doi.org/10.1007/978-3-030-75186-9_2

From Lemma 2.4 below, we have $U'_{\mathcal{T}u}(r) = 4\pi \int_{\mathbb{R}^3} p_r\, u\, dv$. Therefore,

$$\int_{\mathbb{R}^3} |\nabla_x U_{\mathcal{T}u}|^2\, dx = \int_{\mathbb{R}^3} \left|\frac{x}{r}\, U'_{\mathcal{T}u}\right|^2 dx = 16\pi^2 \int_{\mathbb{R}^3} \left|\int_{\mathbb{R}^3} p_r\, u\, dv\right|^2 dx,$$

and this leads to

$$\begin{aligned}(Lu, u)_Q = &\int_{\mathbb{R}^3}\int_{\mathbb{R}^3} \frac{dx\, dv}{|Q'(e_Q)|}\left[(x\cdot v)^2 \left|\mathcal{T}\frac{u}{x\cdot v}\right|^2 + \frac{U'_Q(r)}{r}\,|u|^2\right] \\ &+4\pi \int_{\mathbb{R}^3} dx \left[\rho_Q(r)\int_{\mathbb{R}^3} \frac{dv}{|Q'(e_Q)|}\,|u|^2 - \left|\int_{\mathbb{R}^3} p_r\, u\, dv\right|^2\right].\end{aligned} \tag{2.2}$$

To obtain the lower bound, note that by Lemma 2.5 below

$$\begin{aligned}\left|\int_{\mathbb{R}^3} p_r\, u\, dv\right| &= \left|\int_{\mathbb{R}^3} p_r |Q'(e_Q)|^{1/2}\, |Q'(e_Q)|^{-1/2}\, u\, dv\right| \\ &\le \left(\int_{\mathbb{R}^3} p_r^2\, |Q'(e_Q)|\, dv\right)^{1/2}\left(\int_{\mathbb{R}^3} \frac{dv}{|Q'(e_Q)|}\,|u|^2\right)^{1/2} \\ &= \rho_Q(r)^{1/2}\left(\int_{\mathbb{R}^3} \frac{dv}{|Q'(e_Q)|}\,|u|^2\right)^{1/2}.\end{aligned}$$

Therefore, (2.2) yields

$$(Lu, u)_Q \ge \int_{\mathbb{R}^3}\int_{\mathbb{R}^3} \frac{dx\, dv}{|Q'(e_Q)|}\,\frac{U'_Q(r)}{r}\,|u|^2. \tag{2.3}$$

The function $A(r) = \frac{U'_Q(r)}{r}$, together with $B(r) = 4\pi\rho_Q(r) + A(r)$, will be considered in Lemma A.6. It turns out that this function is strictly decreasing, positive and such that $A(r_Q) = \frac{1}{r_Q^3}\|Q\|_{L^1(\mathbb{R}^6)}$. Hence (1.19) follows. □

Example 2.1 We are going to show that

$$L\left(|Q'(e_Q)|(x\cdot v)\right) = A(r)\,|Q'(e_Q)|\,(x\cdot v), \tag{2.4}$$

and in particular $u(x, v) = |Q'(e_Q)|\,(x\cdot v)$ is not an eigenfunction for λ_* from (1.20). Regarding (2.4), (2.1) says that $\mathcal{T}^2(x\cdot v) = -(x\cdot v)(4\pi\rho_Q(r) + A(r)) = -B(r)(x\cdot v)$, cf. Lemma A.6. Hence, using $\mathcal{T}|Q'(e_Q)| = 0$ from Lemma B.9(a), also $\mathcal{T}^2 u = -B(r)u$ holds. Therefore, due to (B.37), Lemma 2.4 and Lemma 2.5,

$$\begin{aligned}\mathcal{K}\mathcal{T}u &= 4\pi\,|Q'(e_Q)|\, p_r \int_{\mathbb{R}^3} \tilde{p}_r\, u\, d\tilde{v} = 4\pi\,|Q'(e_Q)|\, p_r\, r \int_{\mathbb{R}^3} \tilde{p}_r^2\, |Q'(e_Q)|\, d\tilde{v} \\ &= 4\pi\,|Q'(e_Q)|\,(x\cdot v)\rho_Q(r) = 4\pi\rho_Q(r)u,\end{aligned}$$

where $p_r = x \cdot v/r$ and $\tilde{p}_r = x \cdot \tilde{v}/r$. As a consequence,

$$Lu = -\mathcal{T}^2 u - \mathcal{K}\mathcal{T} u = B(r)u - 4\pi\rho_Q(r)u = A(r)u,$$

so that (2.4) is established. ◇

Corollary 2.2 *We have*

$$0 < A(r_Q) \le \lambda_* < A(0) \quad \text{and} \quad 0 < B(r_Q) \le \lambda_* < B(0).$$

Proof The lower bound $\lambda_* \ge A(r_Q)$ follows from (2.3) and $A(r) \ge A(r_Q)$. Regarding the upper bound $\lambda_* < A(0)$, consider as in Example 2.1 the function $u(x, v) = |Q'(e_Q)|(x \cdot v)$, which is odd in v. Then $Lu = A(r)u$, and thus

$$(Lu, u)_Q = \int_{\mathbb{R}^3}\int_{\mathbb{R}^3} \frac{dx\,dv}{|Q'(e_Q)|}\, A(r)\,|u|^2 \le A(0) \int_{\mathbb{R}^3}\int_{\mathbb{R}^3} \frac{dx\,dv}{|Q'(e_Q)|}\,|u|^2 = A(0)\|u\|_Q^2. \tag{2.5}$$

Hence, it follows that $\lambda_* \le A(0)$, and it remains to be shown that $\lambda_* = A(0)$ is impossible. Suppose that in fact $\lambda_* = A(0)$. Then, for the same u,

$$\begin{aligned}\int_{\mathbb{R}^3}\int_{\mathbb{R}^3} \frac{dx\,dv}{|Q'(e_Q)|}\, A(0)\,|u|^2 &= A(0)\|u\|_Q^2 = \lambda_*\|u\|_Q^2 \\ &\le \left(L\frac{u}{\|u\|_Q}, \frac{u}{\|u\|_Q}\right)_Q \|u\|_Q^2 = (Lu, u)_Q.\end{aligned}$$

Using

$$(Lu, u)_Q = \int\int_K \frac{dx\,dv}{|Q'(e_Q)|}\, A(r)\,|u|^2$$

from (2.5), this leads to

$$\int_{\mathbb{R}^3}\int_{\mathbb{R}^3} \frac{dx\,dv}{|Q'(e_Q)|}\,(A(r) - A(0))\,|u|^2 \ge 0.$$

But this is not possible since A is strictly decreasing by Lemma A.6(a). For the estimates in terms of B, it is sufficient to note that $B(r) \ge A(r)$ and $B(r_Q) = A(r_Q)$, cf. Lemma A.6(b). □

Remark 2.3 Observe that $0 < A(r_Q) \le \lambda_* < A(0)$ and $0 < B(r_Q) \le \lambda_* < B(0)$ from Corollary 2.2 together with Lemma A.6 imply that $\lambda_* = A(r_A)$ and $\lambda_* = B(r_B)$ for certain $r_A, r_B \in]0, r_Q[$. It would be interesting to understand whether these radii r_A and r_B do have a special meaning. ◇

The following observation has been made before; see [30, p. 507] and [50, (B.1)] for instance.

Lemma 2.4 *For appropriate spherically symmetric functions* g,

$$U'_{Tg}(r) = 4\pi \int_{\mathbb{R}^3} p_r\, g\, dv.$$

Proof Since $\nabla_x U_Q$ is independent of v, by (1.11) the density is found to be

$$\rho_{Tg}(x) = \int_{\mathbb{R}^3} (v \cdot \nabla_x g - \nabla_v g \cdot \nabla_x U_Q)\, dv = \operatorname{div}_x \int_{\mathbb{R}^3} v\, g\, dv.$$

Therefore, using (A.2) below and Gauss's theorem, it follows that

$$\begin{aligned} U'_{Tg}(r) &= \frac{1}{r^2} \int_{|x|\le r} \rho_{Tg}(x)\, dx = \frac{1}{r^2} \int_{|x|\le r} dx\, \operatorname{div}_x \left(\int_{\mathbb{R}^3} v\, g\, dv \right) \\ &= \frac{1}{r^2} \int_{|x|=r} dS(x)\, \frac{x}{r} \cdot \left(\int_{\mathbb{R}^3} v\, g\, dv \right) = \frac{1}{r^2} \int_{|x|=r} dS(x) \left(\int_{\mathbb{R}^3} p_r\, g\, dv \right), \end{aligned}$$

where $g = g(x, v) = g(r, p_r, \ell)$. As g is spherically symmetric, the function $G(x) = \int_{\mathbb{R}^3} p_r\, g\, dv$ is invariant under rotation, i.e., $G(x) = G(|x|)$. Thus, $U'_{Tg}(r) = 4\pi \int_{\mathbb{R}^3} p_r\, g\, dv$ as desired. □

For the next lemma, cf. [30, p. 507] and [50, (B.3)].

Lemma 2.5 *We have*

$$\int_{\mathbb{R}^3} |Q'(e_Q)|\, p_r^2\, dv = \rho_Q(r).$$

Proof Since $\frac{d}{dp_r}[Q(e_Q)] = Q'(e_Q) p_r$ due to $e_Q = \frac{1}{2} p_r^2 + U_Q(r) + \frac{\ell^2}{2r^2}$, we have from (A.40):

$$\begin{aligned} \int_{\mathbb{R}^3} |Q'(e_Q)|\, p_r^2\, dv &= - \int_{\mathbb{R}^3} Q'(e_Q)\, p_r^2\, dv \\ &= -\frac{2\pi}{r^2} \int_{\mathbb{R}} dp_r \int_0^\infty d\ell\, \ell\, Q'(e_Q)\, p_r^2 \\ &= -\frac{2\pi}{r^2} \int_0^\infty d\ell\, \ell \int_{\mathbb{R}} dp_r\, \frac{d}{dp_r} [Q(e_Q)]\, p_r \\ &= \frac{2\pi}{r^2} \int_0^\infty d\ell\, \ell \int_{\mathbb{R}} dp_r\, Q(e_Q) \\ &= \int_{\mathbb{R}^3} Q(e_Q)\, dv = \rho_Q(r), \end{aligned}$$

where once again (A.40) has been used. Note that the boundary term vanishes in the integration by parts above. Indeed, if we use the notation from Lemma B.7, cf. (B.10), (B.11) and (B.12), then for fixed r and $\ell \in [0, \hat{l}(r)]$, we see that

$$(Q(e_Q)\, p_r)\Big|_{-\hat{p}}^{\hat{p}} = Q(e_0)\hat{p} - Q(e_0)(-\hat{p}) = 2\hat{p}\, Q(e_0) = 0,$$

since $Q(e_0) = 0$ by (Q2). □

Chapter 3
On the Period Function T_1

Associated with every effective potential $U_{\rm eff}(r,\ell) = U_Q(r) + \frac{\ell^2}{2r^2}$ is a period function $T_1(\cdot, \ell)$ that is defined for certain energies $e \in [e_{\min}(\ell), e_0]$, for which periodic solutions of $\ddot{r} = -U'_{\rm eff}(r,\ell)$ do exist; see Appendix I, Sect. A.1, for more information. According to (A.20), this period function is given by

$$T_1(e,\ell) = 2\int_{r_-(e,\ell)}^{r_+(e,\ell)} \frac{dr}{\sqrt{2(e - U_{\rm eff}(r,\ell))}},$$

where $r_\pm = r_\pm(e,\ell)$ are the zeros of $0 = 2(e - U_{\rm eff}(r,\ell))$ and satisfy $0 < r_-(e,\ell) < r_+(e,\ell)$. In addition, for every $\ell > 0$, the potential minimum $\inf\{U_{\rm eff}(r,\ell) : r \geq 0\}$ is attained at some unique $r_0(\ell) \in]r_-(e,\ell), r_+(e,\ell)[$. The corresponding frequency function is $\omega_1(e,\ell) = \frac{2\pi}{T_1(e,\ell)}$.

3.1 Upper Boundedness of T_1

Recall that

$$D = \{(e,\beta) : \beta \in [0,\beta_*], e \in [e_{\min}(\beta), e_0]\}, \tag{3.1}$$

and

$$\mathring{D} = \{(e,\beta) : \beta \in]0,\beta_*[, e \in]e_{\min}(\beta), e_0[\}$$

is its interior. We are going to show that T_1 is bounded from above (or equivalently, ω_1 is bounded from below), uniformly in $\mathring{D}$, which is the set of relevant (e,ℓ), where T_1 is defined. As T_1 will be shown to be continuous in D (see Theorem 3.13 below), this is of course for free, but since the direct argument in Theorem 3.2 could be of

M. Kunze, *A Birman-Schwinger Principle in Galactic Dynamics*,
Progress in Mathematical Physics 77,
https://doi.org/10.1007/978-3-030-75186-9_3

general interest, we include it anyhow; the same remark applies to Theorem 3.5 on the lower boundedness of T_1.

We start with an auxiliary lemma that will be useful for the proof of Theorem 3.2 and beyond.

Lemma 3.1 *The following assertions are verified.*

(a) If $r > s > 0$, *then*

$$\frac{2\pi}{3}\rho_Q(r)(r^2 - s^2) \le U_Q(r) - U_Q(s) \le \frac{2\pi}{3}\rho_Q(0)(r^2 - s^2). \tag{3.2}$$

Moreover, for $r_Q \ge r > s > 0$,

$$U_Q(r) - U_Q(s) \ge \frac{\pi}{12}\rho_Q\left(\frac{r_Q}{2}\right)(r^2 - s^2). \tag{3.3}$$

(b) One has

$$\frac{\pi}{6}\rho_Q\left(\frac{r_Q}{2}\right) r_-^2 r_+^2 \le \ell^2 \le \frac{4\pi}{3}\rho_Q(0)\, r_-^2 r_+^2.$$

(c) One has

$$r_0 \le \left(\frac{6}{\pi\rho_Q(\frac{r_Q}{2})}\right)^{1/4}\sqrt{\ell}.$$

Proof (a) According to (A.2), we have by changing variables $s = r\tau$, $ds = r d\tau$,

$$U_Q'(r) = \frac{4\pi}{r^2}\int_0^r s^2\rho_Q(s)\, ds = 4\pi r\int_0^1 \tau^2\rho_Q(r\tau)\, d\tau. \tag{3.4}$$

In particular, $U_Q'(r) \ge 0$. Furthermore, for $r > s > 0$ and putting $t = \sigma/r$, $dt = d\sigma/r$,

$$\begin{aligned} U_Q(r) - U_Q(s) &= \int_s^r U_Q'(\sigma)\, d\sigma = 4\pi\int_s^r d\sigma\sigma\int_0^1 d\tau\tau^2\rho_Q(\sigma\tau) \\ &= 4\pi\int_0^1 d\tau\tau^2\int_s^r d\sigma\sigma\rho_Q(\sigma\tau) \qquad (3.5) \\ &= 4\pi r^2\int_0^1 d\tau\tau^2\int_{\frac{s}{r}}^1 dt\, t\, \rho_Q(r\tau t). \qquad (3.6) \end{aligned}$$

Due to (A.32), we have that $\rho_Q'(r) \le 0$, i.e., ρ_Q is radially decreasing. Thus, if $\tau \in [0, 1]$ and $\sigma \in [s, r]$, then $\rho_Q(r) \le \rho_Q(\sigma\tau) \le \rho_Q(0)$ and (3.2) follows from (3.5). To establish (3.3), we use (3.6). To begin with, since $\rho_Q \ge 0$,

$$U_Q(r) - U_Q(s) \ge 4\pi r^2\int_0^{\frac{1}{2}} d\tau\tau^2\int_{\frac{s}{r}}^1 dt\, t\, \rho_Q(r\tau t).$$

Owing to $r \le r_Q$, $\tau \in [0, \frac{1}{2}]$, and $t \le 1$, we have $r\tau t \le \frac{r_Q}{2}$, so that $\rho_Q(r\tau t) \ge \rho_Q(\frac{r_Q}{2})$. It follows that

$$U_Q(r) - U_Q(s) \ge 4\pi r^2 \rho_Q\Big(\frac{r_Q}{2}\Big) \int_0^{\frac{1}{2}} d\tau \tau^2 \int_{\frac{s}{r}}^1 dt\, t = 4\pi r^2 \rho_Q\Big(\frac{r_Q}{2}\Big) \frac{1}{48}\Big(1 - \Big(\frac{s}{r}\Big)^2\Big),$$

which is (3.3). (b) The condition $U_{\text{eff}}(r_\pm, \ell) = e$ means that $U_Q(r_\pm) + \frac{\ell^2}{2r_\pm^2} = e$, and hence

$$2r_\pm^2 U_Q(r_\pm) + \ell^2 = 2r_\pm^2 e. \tag{3.7}$$

Therefore,

$$\begin{aligned} 2(r_+^2 - r_-^2)e &= 2(r_+^2 U_Q(r_+) - r_-^2 U_Q(r_-)) \\ &= 2(r_+^2 - r_-^2) U_Q(r_+) + 2r_-^2 (U_Q(r_+) - U_Q(r_-)), \end{aligned}$$

so that

$$(r_+^2 - r_-^2)\frac{\ell^2}{r_+^2} = 2(r_+^2 - r_-^2)(e - U_Q(r_+)) = 2r_-^2 (U_Q(r_+) - U_Q(r_-)).$$

It remains to use (3.3) and the upper bound from (3.2). (c) First note that $\rho_Q(\frac{r_Q}{2}) > 0$, as otherwise $\operatorname{supp} \rho_Q \subset [0, \frac{r_Q}{2}]$. By Lemma A.7(a), (3.4) and since ρ_Q is non-negative and radially decreasing,

$$\begin{aligned} \ell^2 = r_0^3\, U_Q'(r_0) &= 4\pi r_0^4 \int_0^1 \tau^2 \rho_Q(r_0\tau)\, d\tau \ge 4\pi r_0^4 \int_0^{\frac{1}{2}} \tau^2 \rho_Q(r_0\tau)\, d\tau \\ &\ge 4\pi r_0^4 \int_0^{\frac{1}{2}} \tau^2 \rho_Q\Big(\frac{r_Q}{2}\Big)\, d\tau = \frac{\pi}{6}\, \rho_Q(\frac{r_Q}{2})\, r_0^4. \end{aligned}$$

We will derive a more precise asymptotics of r_0 as $\ell \to 0^+$ below in (A.34). □

Now, we are in a position to derive a uniform lower bound on ω_1 or equivalently a uniform upper bound on T_1.

Theorem 3.2 *We have*

$$\delta_1 = \inf\{\omega_1(e, \ell) : (e, \ell) \in \mathring{D}\} > 0.$$

Proof Put $a_Q = \rho_Q(\frac{r_Q}{2}) > 0$. Then in particular $a_Q \le \rho_Q(0)$, so that

$$\delta_Q = 1 - \sqrt{\frac{a_Q}{16\rho_Q(0)}} \in \Big[\frac{1}{2}, 1\Big[.$$

Let $r_\pm = r_\pm(e, \ell)$ and $r_0 = r_0(\ell)$ be as before. From Lemma 3.1(c), we recall that

$$r_0 \leq \left(\frac{6}{\pi a_Q}\right)^{1/4} \sqrt{\ell}. \tag{3.8}$$

Case 1: $r_0 \geq (1-\delta_Q)r_+$. Then Lemma A.10(b) in conjunction with (3.8) implies that

$$T_1(e,\ell) \leq \pi \frac{\sqrt{r_- r_+}}{\ell}(r_- + r_+) \leq 2\pi \frac{r_+^2}{\ell} \leq \frac{2\pi}{(1-\delta_Q)^2}\frac{r_0^2}{\ell} \leq \frac{2\pi}{(1-\delta_Q)^2}\left(\frac{6}{\pi a_Q}\right)^{1/2}. \tag{3.9}$$

Case 2: $r_0 \leq (1-\delta_Q)r_+$. This is the nontrivial part of the argument. Here, we split up the integral as

$$\begin{aligned} T_1(e,\ell) &= 2\int_{r_-}^{r_+} \frac{dr}{\sqrt{2(e-U_{\mathrm{eff}}(r,\ell))}} \\ &= 2\int_{r_-}^{r_0} \frac{dr}{\sqrt{2(e-U_{\mathrm{eff}}(r,\ell))}} + 2\int_{r_0}^{r_+} \frac{dr}{\sqrt{2(e-U_{\mathrm{eff}}(r,\ell))}} \\ &=: T_1^-(e,\ell) + T_1^+(e,\ell). \end{aligned}$$

Using Lemma A.10(a), we can bound T_1^- as

$$\begin{aligned} T_1^-(e,\ell) &\leq 2\frac{\sqrt{r_- r_+}}{\ell}\int_{r_-}^{r_0} \frac{r\,dr}{\sqrt{(r-r_-)(r_+-r)}} \\ &\leq 2\frac{\sqrt{r_- r_+}}{\ell}\frac{r_0}{\sqrt{r_+-r_0}}\int_{r_-}^{r_0}\frac{dr}{\sqrt{r-r_-}} \\ &= 4\frac{\sqrt{r_- r_+}}{\ell}\frac{r_0}{\sqrt{r_+-r_0}}\sqrt{r_0-r_-}. \end{aligned} \tag{3.10}$$

It follows from $r_0 \leq (1-\delta_Q)r_+$ that $\sqrt{r_+} \leq \delta_Q^{-1/2}\sqrt{r_+-r_0}$. Thus, by (3.8) and since $\delta_Q \geq 1/2$,

$$T_1^-(e,\ell) \leq 4\delta_Q^{-1/2}\frac{\sqrt{r_-}}{\ell} r_0\sqrt{r_0-r_-} \leq 4\delta_Q^{-1/2}\frac{r_0^2}{\ell} \leq 4\sqrt{2}\left(\frac{6}{\pi a_Q}\right)^{1/2}. \tag{3.11}$$

Regarding T_1^+, we can invoke Lemma A.7(a) to get for $r \in [r_0, r_+]$ by also using Lemma A.6(a),

$$\begin{aligned} \frac{\ell^2}{2r_+^2 r^2} &\leq \frac{\ell^2}{2r_+^2 r_0^2} = \frac{r_0\,U_Q'(r_0)}{2r_+^2} \leq \frac{r_0\,U_Q'(r_0)}{2r_0^2}(1-\delta_Q)^2 = \frac{1}{2}(1-\delta_Q)^2 A(r_0) \\ &\leq \frac{1}{2}(1-\delta_Q)^2 A(0) = \frac{2\pi}{3}(1-\delta_Q)^2 \rho_Q(0). \end{aligned} \tag{3.12}$$

We then deduce from (3.3) in Lemma 3.1(a) and (3.12) that for $r \in [r_0, r_+]$,

$$
\begin{aligned}
e - U_{\text{eff}}(r,\ell) &= U_{\text{eff}}(r_+,\ell) - U_{\text{eff}}(r,\ell) = U_Q(r_+) + \frac{\ell^2}{2r_+^2} - U_Q(r) - \frac{\ell^2}{2r^2} \\
&= U_Q(r_+) - U_Q(r) - \frac{\ell^2}{2r_+^2 r^2}(r_+^2 - r^2) \\
&\geq \left[\frac{\pi}{12} a_Q - \frac{2\pi}{3}(1-\delta_Q)^2 \rho_Q(0)\right](r_+^2 - r^2) \\
&= \frac{\pi}{24} a_Q (r_+^2 - r^2),
\end{aligned}
$$

the latter owing to the choice of δ_Q. This in turn yields

$$
\begin{aligned}
T_1^+(e,\ell) &= 2\int_{r_0}^{r_+} \frac{dr}{\sqrt{2(e - U_{\text{eff}}(r,\ell))}} \leq \sqrt{2}\sqrt{\frac{24}{\pi a_Q}} \int_{r_0}^{r_+} \frac{dr}{\sqrt{r_+^2 - r^2}} \\
&\leq \frac{4\sqrt{3}}{\sqrt{\pi a_Q}} \frac{1}{\sqrt{r_+}} \int_{r_0}^{r_+} \frac{dr}{\sqrt{r_+ - r}} = \frac{8\sqrt{3}}{\sqrt{\pi a_Q}} \frac{1}{\sqrt{r_+}} \sqrt{r_+ - r_0} \leq \frac{8\sqrt{3}}{\sqrt{\pi a_Q}}.
\end{aligned}
$$

Adding this to (3.11), we have shown that

$$
T_1(e,\ell) \leq 4\sqrt{2}\left(\frac{6}{\pi a_Q}\right)^{1/2} + \frac{8\sqrt{3}}{\sqrt{\pi a_Q}} = \frac{16\sqrt{3}}{\sqrt{\pi a_Q}}. \tag{3.13}
$$

Hence, the boundedness of T_1 from above is a consequence of (3.9) and (3.13). □

Observe that in the proof of Theorem 3.2 actually no properties of the sets $\mathring{D}$ or D from (3.1) have been used, apart from the fact that $T_1(e,\ell)$ is defined for $(e,\ell) \in \mathring{D}$.

3.2 Lower Boundedness of T_1

It is the purpose of this section to verify that T_1 is bounded from below (or equivalently, ω_1 is bounded from above), uniformly in $\mathring{D}$.

In some cases, it will be convenient to be able to re-express the period function

$$
T_1(e,\beta) = 2\int_{r_-(e,\beta)}^{r_+(e,\beta)} \frac{dr}{\sqrt{2(e - U_{\text{eff}}(r,\beta))}} \tag{3.14}
$$

from (A.20), written in terms of $\beta = \ell^2$, by means of an integral with fixed limits of integration; this is more or less taken from [11, Section 2].

Lemma 3.3 *We have*

$$
T_1(e,\beta) = \sqrt{2}\int_{-\pi/2}^{\pi/2} \frac{d\theta}{\frac{\partial h}{\partial s}(s(\sqrt{\hat{e}(\beta)}\sin\theta,\beta),\beta)},
$$

where

$$h(s,\beta) = s\left(\frac{V(s,\beta)}{s^2}\right)^{1/2}, \quad h(0,\beta) = 0,$$

for

$$V(s,\beta) = U_{\text{eff}}(r_0(\beta)+s,\beta) - e_{\min}(\beta).$$

Also, $\hat{e}(\beta) = e - e_{\min}(\beta)$*, and* $R \mapsto s(R,\beta) = s$ *denotes the inverse mapping to* $s \mapsto h(s,\beta) = R$*. Explicitly,*

$$\frac{\partial h}{\partial s}(s,\beta) = \frac{\operatorname{sgn}(s)}{2} \frac{U'_{\text{eff}}(r_0(\beta)+s,\beta)}{\sqrt{U_{\text{eff}}(r_0(\beta)+s,\beta) - e_{\min}(\beta)}} \geq 0, \tag{3.15}$$

so that also

$$T_1(e,\beta) = 2\sqrt{2}\int_{-\pi/2}^{\pi/2} d\theta \, \frac{[\int_0^1 (1-\rho)\, U''_{\text{eff}}(r_0(\beta)+\rho s(\sqrt{\hat{e}(\beta)}\sin\theta,\beta),\beta)\, d\rho]^{1/2}}{\int_0^1 U''_{\text{eff}}(r_0(\beta)+\rho s(\sqrt{\hat{e}(\beta)}\sin\theta,\beta),\beta)\, d\rho}. \tag{3.16}$$

Proof Let $s_\pm(e,\beta) = r_\pm(e,\beta) - r_0(\beta)$. Setting $s = r - r_0(\beta)$, $ds = dr$, we obtain

$$\begin{aligned} T_1(e,\beta) &= 2\int_{s_-(e,\beta)}^{s_+(e,\beta)} \frac{ds}{\sqrt{2(e - e_{\min}(\beta) - [U_{\text{eff}}(r_0(\beta)+s,\beta) - e_{\min}(\beta)])}} \\ &= 2\int_{s_-(e,\beta)}^{s_+(e,\beta)} \frac{ds}{\sqrt{2(\hat{e}(\beta) - V(s,\beta))}}. \end{aligned} \tag{3.17}$$

Note that $V(\cdot,\beta)$ is increasing in $[0, s_+(e,\beta)]$, decreasing in $[s_-(e,\beta), 0]$ and such that

$$V(s_\pm(e,\beta),\beta) = e - e_{\min}(\beta) = \hat{e}(\beta).$$

Furthermore, $V(0,\beta) = U_{\text{eff}}(r_0(\beta),\beta) - e_{\min}(\beta) = 0$ by definition and $\frac{\partial V}{\partial s}(0,\beta) = U'_{\text{eff}}(r_0(\beta),\beta)$ by (A.35), i.e., $V(\cdot,\beta)$ is at least quadratic about $s = 0$. The next change of variables to be applied is

$$s \mapsto R = h(s,\beta), \quad dR = \frac{\partial h}{\partial s}\, ds, \quad R^2 = V(s,\beta).$$

Then (3.17) transforms into

$$T_1(e,\beta) = 2\int_{-\sqrt{\hat{e}(\beta)}}^{\sqrt{\hat{e}(\beta)}} \frac{dR}{\frac{\partial h}{\partial s}(s(R,\beta),\beta)\sqrt{2(\hat{e}(\beta) - R^2)}}.$$

Finally, put $R = \sqrt{\hat{e}(\beta)}\sin\theta$, $dR = \sqrt{\hat{e}(\beta)}\cos\theta\, d\theta$. This yields

$$T_1(e,\beta)=\sqrt{2}\int_{-\pi/2}^{\pi/2}\frac{d\theta}{\frac{\partial h}{\partial s}(s(\sqrt{\hat{e}(\beta)}\sin\theta,\beta),\beta)},$$

and thus the claimed formula for T_1. The relation (3.15) is straightforward, whereas (3.16) follows from Lemma A.9. □

Corollary 3.4 *If* $s\in[r_-(e,\beta)-r_0(\beta),0]$, *then*

$$0\le\frac{\partial h}{\partial s}(s,\beta)\le\frac{1}{\sqrt{2B(r_Q)}}\left(3\beta\int_0^1\frac{d\rho}{(r_0(\beta)+\rho s)^4}+\frac{28\pi}{3}\rho_Q(0)\right).$$

Proof Let $s_-=s_-(e,\beta)-r_0(\beta)$. If $s\in[s_-,0]$, then

$$0\le\frac{\partial h}{\partial s}(s,\beta)=\left|\frac{\partial h}{\partial s}(s,\beta)\right|=\frac{1}{2}\frac{|U'_{\rm eff}(r_0(\beta)+s,\beta)|}{\sqrt{U_{\rm eff}(r_0(\beta)+s,\beta)-e_{\min}(\beta)}}$$

by (3.15) in Lemma 3.3. Thus, it remains to use (A.37) and (A.38) from Lemma A.9. □

Theorem 3.5 *We have*

$$\Delta_1=\sup\{\omega_1(e,\ell):(e,\ell)\in\mathring{D}\}<\infty.$$

Proof As above, we write $r_\pm=r_\pm(e,\beta)$ and $r_0=r_0(\beta)$. If $r\in[r_-,r_+]$, then by Lemma 3.1(a),

$$\begin{aligned}e-U_{\rm eff}(r,\beta)&=U_Q(r_+)-U_Q(r)-\frac{\beta}{2r_+^2r^2}(r_+^2-r^2)\\&\le U_Q(r_+)-U_Q(r)\le\frac{2\pi}{3}\rho_Q(0)(r_+^2-r^2)\le\frac{4\pi}{3}\rho_Q(0)r_+(r_+-r).\end{aligned}\tag{3.18}$$

<u>Case 1:</u> $r_+/2\ge r_0$. Here (3.18) implies that

$$\begin{aligned}T_1(e,\beta)&=2\int_{r_-}^{r_+}\frac{dr}{\sqrt{2(e-U_{\rm eff}(r,\beta))}}\ge\sqrt{2}\int_{r_0}^{r_+}\frac{dr}{\sqrt{e-U_{\rm eff}(r,\beta)}}\\&\ge\sqrt{\frac{3}{2\pi\rho_Q(0)}}\frac{1}{\sqrt{r_+}}\int_{r_0}^{r_+}\frac{dr}{\sqrt{r_+-r}}=2\sqrt{\frac{3}{2\pi\rho_Q(0)}}\sqrt{\frac{r_+-r_0}{r_+}}\ge\sqrt{\frac{3}{\pi\rho_Q(0)}}.\end{aligned}$$

<u>Case 2:</u> $r_-\le r_+/2\le r_0$. Similarl to the first case, we obtain

$$\begin{aligned}T_1(e,\beta)&=2\int_{r_-}^{r_+}\frac{dr}{\sqrt{2(e-U_{\rm eff}(r,\beta))}}\ge\sqrt{2}\int_{r_+/2}^{r_+}\frac{dr}{\sqrt{e-U_{\rm eff}(r,\beta)}}\\&\ge\sqrt{\frac{3}{2\pi\rho_Q(0)}}\frac{1}{\sqrt{r_+}}\int_{r_+/2}^{r_+}\frac{dr}{\sqrt{r_+-r}}=\sqrt{\frac{6}{\pi\rho_Q(0)}}.\end{aligned}$$

Case 3: $0 \le r_+/2 \le r_-$. Then $r_+/2 \le r_- \le r_+$ and also $r_- \le r_0 \le r_+ \le 2r_-$ as well as $r_0 \le r_+ \le 2r_- \le 2r_0$, so all of r_-, r_0 and r_+ are of comparable size. In particular, if $r \in [r_-, r_+]$, then $r_0/2 \le r \le 2r_0$. In the following, we are going to use the notation from the proof of Lemma 3.3. Let $R = \sqrt{\hat{e}(\beta)} \sin\theta$. If $\theta \in [-\pi/2, 0]$, then $R \in [-\sqrt{\hat{e}(\beta)}, 0]$ and hence $s(R, \beta) \in [s_-, 0]$. Thus, if furthermore $\rho \in [0, 1]$, then $r_0 + \rho s(R, \beta) \in r_0 + [s_-, 0] = [r_-, r_0]$, so that

$$\frac{1}{2} r_0 \le r_0 + \rho s(R, \beta) \le 2r_0. \tag{3.19}$$

Since $s(R, \beta) \in [s_-, 0]$, Corollary 3.4 and (3.19) imply that

$$\begin{aligned} 0 \le \frac{\partial h}{\partial s}(s(R, \beta), \beta) &\le \frac{1}{\sqrt{2B(r_Q)}} \left(3\beta \int_0^1 \frac{d\rho}{(r_0 + \rho s(R, \beta))^4} + \frac{28\pi}{3} \rho_Q(0) \right) \\ &\le \frac{1}{\sqrt{2B(r_Q)}} \left(\frac{48\beta}{r_0^4} + \frac{28\pi}{3} \rho_Q(0) \right) \end{aligned} \tag{3.20}$$

for $\theta \in [-\pi/2, 0]$. By (A.34) from Lemma A.7, we have

$$r_0^4 = \frac{1}{A(0)} \beta + \mathcal{O}(\beta^{5/4}) = \beta \Big(\frac{1}{A(0)} + \mathcal{O}(\beta^{1/4}) \Big)$$

as $\beta \to 0^+$. Hence, there is $\beta_0 \in]0, \beta_*[$ such that

$$\frac{\beta}{2A(0)} \le r_0^4 \le \frac{2\beta}{A(0)}, \quad \beta \in]0, \beta_0].$$

Accordingly, owing to Lemma A.7(a), we can find a constant $c_0 > 0$ so that $r_0 \ge c_0$ for $\beta \in [\beta_0, \beta_*]$. If we now distinguish the cases $\beta \in]0, \beta_0]$ and $\beta \in [\beta_0, \beta_*]$, by using the foregoing estimates, we deduce that in any case

$$\frac{\beta}{r_0^4} \le \max \left\{ 2A(0), \frac{\beta_*}{c_0^4} \right\}.$$

Upon going back to (3.20), it follows that

$$0 \le \frac{\partial h}{\partial s}(s(R, \beta), \beta) \le \frac{1}{\sqrt{2B(r_Q)}} \left(48 \max \left\{ 2A(0), \frac{\beta_*}{c_0^4} \right\} + \frac{28\pi}{3} \rho_Q(0) \right) =: C_1$$

for $\theta \in [-\pi/2, 0]$. Since generally $\frac{\partial h}{\partial s} \ge 0$, we finally get from Lemma 3.3

$$T_1(e, \beta) = \sqrt{2} \int_{-\pi/2}^{\pi/2} \frac{d\theta}{\frac{\partial h}{\partial s}(s(R, \beta), \beta)} \ge \sqrt{2} \int_{-\pi/2}^{0} \frac{d\theta}{\frac{\partial h}{\partial s}(s(R, \beta), \beta)} \ge \frac{\pi}{\sqrt{2}\, C_1},$$

which completes the proof, as we have found a positive lower bound on T_1 in all three cases. □

3.3 Further Properties of T_1

First, we discuss some regularity properties of T_1.

Theorem 3.6 *We have $T_1 \in C^1(\mathring{D})$.*

Proof The continuity of T_1 may be shown directly from (3.14), as we already know that $r_\pm \in C^2(\mathring{D})$ by Remark A.3; we omit the details. To prove the differentiability, we use a method that is known and that we learned from R. Ortega. It is considerably less painful than differentiating an explicit relation for T_1 like (3.14). For $(e, \beta) \in \mathring{D}$, we consider

$$\ddot{r} = -U'_{\text{eff}}(r, \beta), \quad r(0) = r_-(e, \beta), \quad \dot{r}(0) = 0,$$

where $r(t) = r(t, e, \beta)$. Defining

$$F : \mathbb{R} \times \mathring{D} \to \mathbb{R}, \quad F(t, e, \beta) = \dot{r}(t, e, \beta),$$

we have $F \in C^1(\mathbb{R} \times \mathring{D})$ by Lemma A.11(a). Next observe that $F(t, e, \beta) = 0$ exactly for

$$t = 0, \; t = \pm\frac{1}{2}\, T_1(e, \beta), \; t = \pm T_1(e, \beta), \; t = \pm\frac{3}{2}\, T_1(e, \beta), \; \ldots.$$

Fix $(\tilde{e}, \tilde{\beta}) \in \mathring{D}$ and define $\tilde{t} = T_1(\tilde{e}, \tilde{\beta})$. Then $F(\tilde{t}, \tilde{e}, \tilde{\beta}) = 0$ by the above. Furthermore,

$$\frac{\partial F}{\partial t}(t, e, \beta) = \ddot{r}(t, e, \beta) = -U'_{\text{eff}}(r(t, e, \beta), \beta)$$

and $r(\tilde{t}, \tilde{e}, \tilde{\beta}) = r(T_1(\tilde{e}, \tilde{\beta}), \tilde{e}, \tilde{\beta}) = r(0, \tilde{e}, \tilde{\beta}) = r_-(\tilde{e}, \tilde{\beta})$ in conjunction with Lemma A.4 imply that

$$\frac{\partial F}{\partial t}(\tilde{t}, \tilde{e}, \tilde{\beta}) = -U'_{\text{eff}}(r_-(\tilde{e}, \tilde{\beta}), \tilde{\beta}) > 0.$$

Hence, the implicit function theorem yields the existence of a C^1-function $t = t(e, \beta)$ that is defined for (e, β) in a neighborhood $U \subset \mathring{D}$ of $(\tilde{e}, \tilde{\beta})$, such that

$$F(t(e, \beta), e, \beta) = 0 \text{ for } (e, \beta) \in U \quad \text{and} \quad t(\tilde{e}, \tilde{\beta}) = \tilde{t} = T_1(\tilde{e}, \tilde{\beta}).$$

According to our previous remarks, for every $(e, \beta) \in U$, we must have

$$t(e, \beta) = k(e, \beta) \frac{1}{2} T_1(e, \beta)$$

for some $k(e, \beta) \in \mathbb{Z}$. Then k is continuous in U and such that $k(\tilde{e}, \tilde{\beta}) = 2$, which means that $k = 2$ throughout U. Thus, $T_1 = t$ in U shows that $T_1 \in C^1(U)$. □

Remark 3.7 If $\rho_Q \in C^k$, then $U_Q \in C^{k+2}$. As a consequence, $r_- \in C^{k+2}(\mathring{D})$ by the argument from Remark A.3. Comparing to Lemma A.11(a), this entails $F = \dot{r} \in C^{k+1}(\mathbb{R} \times \mathring{D})$, so that $t = t(e, \beta) \in C^{k+1}(U)$ for the solution function in the proof of Theorem 3.6. Hence, we get $T_1 \in C^{k+1}(\mathring{D})$ in this case. ◇

Now, we are going to show that T_1 can be extended continuously from $\mathring{D}$ to D. We start with the continuous extension to $\{(e, \beta) : \beta \in]0, \beta_*], e = e_{\min}(\beta)\}$.

Lemma 3.8 *Let $\hat{\beta} \in]0, \beta_*]$. Then*

$$T_1(e, \beta) \to \frac{2\pi}{\sqrt{B(r_0(\hat{\beta}))}} \quad \text{as} \quad \mathring{D} \ni (e, \beta) \to (e_{\min}(\hat{\beta}), \hat{\beta}). \tag{3.21}$$

Proof This relies on the representation (3.16) of $T_1(e, \beta)$, which we recall as

$$T_1(e, \beta) = 2\sqrt{2} \int_{-\pi/2}^{\pi/2} d\theta \, \frac{[\int_0^1 (1-\rho)\, U''_{\text{eff}}(r_0(\beta) + \rho s(\sqrt{\hat{e}(\beta)} \sin\theta, \beta), \beta)\, d\rho]^{1/2}}{\int_0^1 U''_{\text{eff}}(r_0(\beta) + \rho s(\sqrt{\hat{e}(\beta)} \sin\theta, \beta), \beta)\, d\rho}. \tag{3.22}$$

Here, $h(s, \beta) = s(\frac{V(s,\beta)}{s^2})^{1/2}$ and $h(0, \beta) = 0$ for $V(s, \beta) = U_{\text{eff}}(r_0(\beta) + s, \beta) - e_{\min}(\beta)$. Furthermore, $\hat{e}(\beta) = e - e_{\min}(\beta)$ and $R \mapsto s(R, \beta) = s$ denotes the inverse mapping to $s \mapsto h(s, \beta) = R$. Due to $\beta \to \hat{\beta} > 0$, we can assume that $\beta \geq \hat{\beta}/2$ throughout the argument. If $r \in [r_-, r_+]$ and $\beta \in]0, \beta_*]$, then Lemma A.6(c) and (A.28) yields

$$\begin{aligned} U'''_{\text{eff}}(r, \beta) &= -\frac{12\beta}{r^5} + B'(r) - 3A'(r) = -\frac{12\beta}{r^5} + 4\pi\rho'_Q(r) - 2A'(r) \\ &= -\frac{12\beta}{r^5} + 4\pi\rho'_Q(r) - \frac{8\pi}{r^4} \int_0^r s^3 \rho'_Q(s)\, ds. \end{aligned}$$

Therefore, (Q4) gives the bound

$$|U'''_{\text{eff}}(r, \beta)| \leq C\Big(1 + \frac{1}{r_-^5}\Big), \quad r \in [r_-, r_+], \quad \beta \in]0, \beta_*], \quad e \in [e_{\min}(\beta), e_0]. \tag{3.23}$$

By definition, we have $U_Q(r_-) + \frac{\beta}{2r_-} = e$. Hence, $U'_Q(r) \geq 0$ leads to

$$\frac{\beta}{2r_-^2} \leq U_Q(r_-) - U_Q(0) + \frac{\beta}{2r_-^2} = e - U_Q(0),$$

and thus

$$r_- \geq \sqrt{\frac{\beta}{2(e - U_Q(0))}} \geq \sqrt{\frac{\hat{\beta}}{4(e - U_Q(0))}}$$

for $\beta \in [\hat{\beta}/2, \beta_*]$ and $e \in [e_{\min}(\beta), e_0]$; note that we will have $e \to e_{\min}(\hat{\beta}) > U_Q(0)$. Going back to (3.23), we obtain

$$|U'''_{\rm eff}(r, \beta)| \leq C, \quad r \in [r_-, r_+], \quad \beta \in [\hat{\beta}/2, \beta_*], \quad e \in [e_{\min}(\beta), e_0]. \tag{3.24}$$

Next, we assert that

$$\lim_{\beta \to \hat{\beta},\, e \to e_{\min}(\hat{\beta})} \; \sup_{\theta \in [-\pi/2, \pi/2]} |s(\sqrt{\hat{e}(\beta)} \sin\theta, \beta)| = 0. \tag{3.25}$$

Otherwise, there would be $\varepsilon_0 > 0$ and sequences (β_j), (e_j) and (θ_j) such that $\beta_j \to \hat{\beta}$, $\theta_j \to \hat{\theta} \in [-\pi/2, \pi/2]$, $\hat{e}(\beta_j) = e_j - e_{\min}(\beta_j) \to e_{\min}(\hat{\beta}) - e_{\min}(\hat{\beta}) = 0$ as well as $|s(\sqrt{\hat{e}(\beta_j)} \sin\theta_j, \beta_j)| \geq \varepsilon_0$ for all $j \in \mathbb{N}$; here it was used that $e_{\min}(\beta) = U_{\rm eff}(r_0(\beta), \beta)$ is continuous in $\beta \in]0, \beta_*[$, cf. Remark A.3. Thus, $\sqrt{\hat{e}(\beta_j)} \sin\theta_j \to 0$ and $s(\sqrt{\hat{e}(\beta_j)} \sin\theta_j, \beta_j) \to s(0, \hat{\beta}) = 0$, which is a contradiction. For the latter convergence, note that $s \mapsto h(s, \beta)$ for $s \in [s_-, s_+]$ is an increasing function that connects $-\sqrt{\hat{e}(\beta)}$ to $\sqrt{\hat{e}(\beta)}$. Since $\hat{e}(\beta_j) \to 0$, we must also have $s_\pm(e_j, \beta_j) \to 0$: for instance, if we had $s_+(e_j, \beta_j) \to \hat{s}_+ > 0$ (along a subsequence), then $h(s, \hat{\beta}) = 0$ for $s \in [0, \hat{s}_+]$, which is impossible. Thus, $s_\pm(e_j, \beta_j) \to 0$, and due to $|s(R, \beta)| \leq \max\{|s_-(e, \beta)|, s_+(e, \beta)\}$, we obtain $s(\sqrt{\hat{e}(\beta_j)} \sin\theta_j, \beta_j) \to 0$ as claimed.

Coming back to (3.22) and using Lemma A.7(d) and (3.24), we estimate

$$\begin{aligned}
&\left| \int_0^1 U''_{\rm eff}(r_0(\beta) + \rho s(\sqrt{\hat{e}(\beta)} \sin\theta, \beta), \beta)\, d\rho - B(r_0(\beta)) \right| \\
&= \left| \int_0^1 [U''_{\rm eff}(r_0(\beta) + \rho s(\sqrt{\hat{e}(\beta)} \sin\theta, \beta), \beta) - U''_{\rm eff}(r_0(\beta), \beta)]\, d\rho \right| \\
&\leq C \int_0^1 |s(\sqrt{\hat{e}(\beta)} \sin\theta, \beta)|\, d\rho \leq C S(e, \beta), \\
&\quad S(e, \beta) = \sup_{\theta \in [-\pi/2, \pi/2]} |s(\sqrt{\hat{e}(\beta)} \sin\theta, \beta)|.
\end{aligned} \tag{3.26}$$

Similarly,

$$\left| \int_0^1 (1 - \rho)\, U''_{\rm eff}(r_0(\beta) + \rho s(\sqrt{\hat{e}(\beta)} \sin\theta, \beta), \beta)\, d\rho - \frac{1}{2}\, B(r_0(\beta)) \right| \leq C S(e, \beta). \tag{3.27}$$

From (3.26), (3.27) and (3.25), in conjunction with Lebesgue's dominated convergence theorem and $B(r_0(\hat{\beta})) > 0$, we deduce (3.21). □

Remark 3.9 Note that $T_1(e, \beta)$ is defined for $e = e_0$ and $\beta \in]0, \beta_*]$; it is the period of the orbit of $\ddot{r} = -U''_{\text{eff}}(r, \beta)$ that has the largest energy $e = e_0$. Therefore, it is straightforward that

$$T_1(e_0, \beta) = 2 \int_{r_-(e_0,\beta)}^{r_+(e_0,\beta)} \frac{dr}{\sqrt{2(e_0 - U_{\text{eff}}(r, \beta))}}$$

extends T_1 continuously to $\{(e, \beta) : e = e_0, \beta \in]0, \beta_*]\}$. ◇

There is yet another way to represent T_1; see [24, Exercise 1, p. 40].

Lemma 3.10 *Define*

$$\chi(r, e, \beta) = \int_0^1 d\tau\,(1-\tau) \int_0^1 d\sigma\, U''_{\text{eff}}(\tau r_+(e, \beta) + \sigma(1-\tau)r + (1-\sigma)(1-\tau)r_-(e, \beta), \beta).$$

Then

$$T_1(e, \beta) = \sqrt{2} \int_{r_-(e,\beta)}^{r_+(e,\beta)} \frac{dr}{\sqrt{(r_+(e, \beta) - r)(r - r_-(e, \beta))\, \chi(r, e, \beta)}}. \tag{3.28}$$

Proof If $r > s$, then

$$U_{\text{eff}}(r, \beta) - U_{\text{eff}}(s, \beta) = (r - s) \int_0^1 U'_{\text{eff}}(\tau r + (1-\tau)s, \beta)\, d\tau,$$

and in particular $U_{\text{eff}}(r_\pm, \beta) = e$ yields $\int_0^1 U'_{\text{eff}}(\tau r_+ + (1-\tau)r_-, \beta)\, d\tau = 0$. Therefore, we can write

$$\begin{aligned}
e - U_{\text{eff}}(r, \beta) &= U_{\text{eff}}(r_+, \beta) - U_{\text{eff}}(r, \beta) \\
&= (r_+ - r) \int_0^1 U'_{\text{eff}}(\tau r_+ + (1-\tau)r, \beta)\, d\tau \\
&= (r_+ - r) \int_0^1 [U'_{\text{eff}}(\tau r_+ + (1-\tau)r, \beta) - U'_{\text{eff}}(\tau r_+ + (1-\tau)r_-, \beta)]\, d\tau \\
&= (r_+ - r)(r - r_-) \int_0^1 d\tau\,(1-\tau) \int_0^1 d\sigma\, U''_{\text{eff}}(\tau r_+ + \sigma(1-\tau)r \\
&\qquad + (1-\sigma)(1-\tau)r_-, \beta),
\end{aligned}$$

which leads to (3.28). □

Lemma 3.11 *We have*

$$T_1(e,\beta) \to \frac{2\pi}{\sqrt{B(0)}} \quad as \quad \mathring{D} \ni (e,\beta) \to (U_Q(0), 0). \tag{3.29}$$

Proof First, we note that, although it won't be used, $e - e_{\min}(\beta) \geq 0$ together with Lemma A.7(f) implies $e - U_Q(0) \geq e_{\min}(\beta) - U_Q(0) \sim \sqrt{U_Q''(0)}\sqrt{\beta}$ as $\beta \to 0$, which means that as $e \to U_Q(0)$, the quantity $e - U_Q(0)$ can't be too small in terms of $\beta \to 0$; due to $U_Q''(r) + \frac{2}{r}U_Q'(r) = 4\pi\rho_Q(r)$, we have $U_Q''(0) = \frac{4\pi}{3}\rho_Q(0) > 0$.

To actually verify (3.29), we are going to write

$$T_1(e,\beta) = \sqrt{2}\int_{r_-}^{r_+} \frac{dr}{\sqrt{(r_+ - r)(r - r_-)\,\chi(r)}} \tag{3.30}$$

as in (3.28) from Lemma 3.10, where

$$\begin{aligned}\chi(r) = \chi(r,e,\beta) = \int_0^1 d\tau\,(1-\tau)\int_0^1 d\sigma\, U_{\mathrm{eff}}''(\tau r_+ + \sigma(1-\tau)r \\ + (1-\sigma)(1-\tau)r_-, \beta).\end{aligned}$$

Due to Lemma A.6(c), we have $U_{\mathrm{eff}}''(r,\beta) = \frac{3\beta}{r^4} + B(r) - 3A(r)$. By explicit integration,

$$\begin{aligned}
&3\beta\int_0^1 d\sigma\int_0^1 d\tau\,(1-\tau)\,\frac{1}{(\tau r_+ + \sigma(1-\tau)r + (1-\sigma)(1-\tau)r_-)^4}\\
&= \frac{\beta}{2r_+^2}\int_0^1 d\sigma\,\frac{2r_+ + \sigma r + (1-\sigma)r_-}{(\sigma r + (1-\sigma)r_-)^3}\\
&= \frac{\beta}{r_+}\int_0^1 d\sigma\,\frac{1}{(\sigma r + (1-\sigma)r_-)^3} + \frac{\beta}{2r_+^2}\int_0^1 d\sigma\,\frac{1}{(\sigma r + (1-\sigma)r_-)^2}\\
&= \frac{\beta}{r_+}\,\frac{r + r_-}{2r^2 r_-^2} + \frac{\beta}{2r_+^2}\,\frac{1}{r r_-}\\
&= \frac{\beta}{2}\,\frac{r(r_- + r_+) + r_- r_+}{r^2 r_-^2 r_+^2}\\
&= \frac{\beta}{2}\,\frac{(r_+ + r)(r_- + r)}{r^2 r_-^2 r_+^2} - \frac{\beta}{2}\,\frac{1}{r_-^2 r_+^2}.
\end{aligned}$$

Hence, we obtain

$$\chi(r) = \frac{\beta}{2} \frac{(r_+ + r)(r_- + r)}{r^2 r_-^2 r_+^2} + \chi_2(r), \tag{3.31}$$

$$\chi_2(r) = \int_0^1 d\tau\,(1-\tau) \int_0^1 d\sigma\,(B - 3A)(\tau r_+ + \sigma(1-\tau) r + (1-\sigma)(1-\tau) r_-) - \frac{\beta}{2} \frac{1}{r_-^2 r_+^2},$$

and $\chi_2(r) = \chi_2(r, e, \beta)$. From Lemma A.6(a) and (b), we get $(B - 3A)(0) = \frac{16\pi}{3}\,\rho_Q(0) - \frac{12\pi}{3}\,\rho_Q(0) = \frac{4\pi}{3}\,\rho_Q(0) = U_Q''(0)$. Since $\tau r_+ + \sigma(1-\tau) r + (1-\sigma)(1-\tau) r_- \in [r_-, r_+] \subset [0, r_+]$ for $\tau, \sigma \in [0, 1]$ and $r \in [r_-, r_+]$, it follows from

$$\chi_2(r) = \int_0^1 d\tau\,(1-\tau) \int_0^1 d\sigma\,[(B - 3A)(\tau r_+ + \sigma(1-\tau) r + (1-\sigma)(1-\tau) r_-) - (B - 3A)(0)] + \frac{1}{2}\,U_Q''(0) - \frac{\beta}{2} \frac{1}{r_-^2 r_+^2}$$

and (A.26) in Lemma A.5 that

$$\sup_{r \in [r_-, r_+]} |\chi_2(r, e, \beta)| \le \frac{1}{2} \sup_{s \in [0, r_+(e,\beta)]} |(B - 3A)(s) - (B - 3A)(0)| + \frac{1}{2} \sup_{s \in [0, r_+(e,\beta)]} |U_Q''(s) - U_Q''(0)| \tag{3.32}$$

for $(e, \beta) \in \mathring{D}$ and $r_\pm = r_\pm(e, \beta)$.

Next, we assert that

$$r_+(e, \beta) \to 0 \quad \text{as} \quad \mathring{D} \ni (e, \beta) \to (U_Q(0), 0). \tag{3.33}$$

To establish this claim, we will use the relation

$$U_Q(r_+) - U_Q(r_0) = e - \frac{\beta}{2 r_+^2} - e_{\min}(\beta) + \frac{\beta}{2 r_0^2} = e - e_{\min}(\beta) + \frac{\beta}{2 r_0^2 r_+^2}(r_+^2 - r_0^2).$$

Hence, (3.3) from Lemma 3.1 yields

$$\frac{\pi}{12}\,\rho_Q\Big(\frac{r_Q}{2}\Big)(r_+^2 - r_0^2) \le e - U_Q(0) + U_Q(0) - e_{\min}(\beta) + \frac{\beta}{2 r_0^2 r_+^2}(r_+^2 - r_0^2). \tag{3.34}$$

Due to Lemma A.7(f), we have $|e_{\min}(\beta) - U_Q(0)| = \mathcal{O}(\beta^{1/2})$ and $r_0 = \mathcal{O}(\beta^{1/4})$ as $\beta \to 0$. Thus, if $r_+(e, \beta) \to \hat{r}_+ > 0$ as $(e, \beta) \to (U_Q(0), 0)$, and along some subsequence, then (3.34) would imply that $\frac{\pi}{12}\,\rho_Q(\frac{r_Q}{2})\,\hat{r}_+^2 \le 0$, which is a contradiction and confirms (3.33). Since both $(B-3A)(s)$ and $U_Q''(s)$ are continuous at $s=0$, (3.32) in turn shows that

$$\lim_{e\to U_Q(0),\,\beta\to 0}\ \sup_{r\in[r_-,r_+]} |\chi_2(r,e,\beta)| = 0. \tag{3.35}$$

A further preparatory step is to rewrite (3.31) as

$$\begin{aligned}\chi(r) &= \frac{\beta}{2}\,\frac{(r_+ + r)(r_- + r)}{r^2 r_-^2 r_+^2}\,(1+\chi_3(r)), \\ \chi_3(r) &= \frac{2}{\beta}\,\frac{r^2 r_-^2 r_+^2}{(r_+ + r)(r_- + r)}\,\chi_2(r),\end{aligned} \tag{3.36}$$

for $\chi_3(r)=\chi_3(r,e,\beta)$. Owing to Lemma 3.1(b), we have

$$r_-^2 r_+^2 \le C\beta. \tag{3.37}$$

Since also $\frac{r^2}{(r_+ + r)(r_- + r)} \le 1$, it follows from (3.35) that

$$\lim_{e\to U_Q(0),\,\beta\to 0}\ \sup_{r\in[r_-,r_+]} |\chi_3(r,e,\beta)| = 0. \tag{3.38}$$

Coming back to (3.29), consider sequences $e_j \to U_Q(0)$ and $\beta_j \to 0$. Let $\varepsilon > 0$ be given. According to (3.38), there is $j_0 \in \mathbb{N}$ such that $\sup_{r\in[r_-(e_j,\beta_j),r_+(e_j,\beta_j)]} |\chi_3(r,e_j,\beta_j)| \le \varepsilon$ for $j \ge j_0$. Due to (3.36), this yields for $j \ge j_0$ and $r \in [r_-(e_j,\beta_j), r_+(e_j,\beta_j)]$

$$\frac{\beta_j}{2}\,\frac{(r_{+,j}+r)(r_{-,j}+r)}{r^2 r_{-,j}^2 r_{+,j}^2}\,(1-\varepsilon) \le \chi(r,e_j,\beta_j) \le \frac{\beta_j}{2}\,\frac{(r_{+,j}+r)(r_{-,j}+r)}{r^2 r_{-,j}^2 r_{+,j}^2}\,(1+\varepsilon),$$

where $r_{\pm,j} = r_\pm(e_j,\beta_j)$. Therefore, (3.30) leads to

$$\frac{r_{-,j}r_{+,j}}{\sqrt{1+\varepsilon}}\,\frac{2}{\beta_j^{1/2}}\,I_j \le T_1(e_j,\beta_j) \le \frac{r_{-,j}r_{+,j}}{\sqrt{1-\varepsilon}}\,\frac{2}{\beta_j^{1/2}}\,I_j$$

for $j \ge j_0$, where

$$I_j = \int_{r_{-,j}}^{r_{+,j}} \frac{r}{(r_{+,j}^2 - r^2)^{1/2}(r^2 - r_{-,j}^2)^{1/2}}\,dr.$$

Setting $s=r^2$, $ds = 2r\,dr$, this integral may be evaluated as $I_j = \pi/2$. Thus, we obtain

$$\frac{1-\varepsilon}{\pi^2}\,\frac{\beta_j}{r_{-,j}^2 r_{+,j}^2} \le \frac{1}{T_1(e_j,\beta_j)^2} \le \frac{1+\varepsilon}{\pi^2}\,\frac{\beta_j}{r_{-,j}^2 r_{+,j}^2} \tag{3.39}$$

for $j \geq j_0$. From (A.26) in Lemma A.5, we know that

$$\left|\frac{\beta_j}{r_{-,j}^2 r_{+,j}^2} - U_Q''(0)\right| \leq \sup_{r\in[0,r_{+,j}]} |U_Q''(r) - U_Q''(0)|.$$

As $r_{+,j} \to 0$ by (3.33), we may assume that j_0 is already taken so large that

$$U_Q''(0) - \varepsilon \leq \frac{\beta_j}{r_{-,j}^2 r_{+,j}^2} \leq U_Q''(0) + \varepsilon$$

for $j \geq j_0$. Therefore, (3.39) implies that

$$\frac{1-\varepsilon}{\pi^2}(U_Q''(0) - \varepsilon) \leq \frac{1}{T_1(e_j, \beta_j)^2} \leq \frac{1+\varepsilon}{\pi^2}(U_Q''(0) + \varepsilon)$$

for $j \geq j_0$. Altogether, this shows that $\lim_{j\to\infty} T_1(e_j, \beta_j) = \frac{\pi}{\sqrt{U_Q''(0)}}$, and it remains to recall that $B(0) = \frac{16\pi}{3}\rho_Q(0) = 4U_Q''(0)$, cf. Lemma A.6(a), (b). □

Lemma 3.12 *Let* $\hat{e} \in]U_Q(0), e_0]$. *Then*

$$T_1(e, \beta) \to 2\int_0^{\hat{r}(\hat{e})} \frac{dr}{\sqrt{2(\hat{e} - U_Q(r))}} \quad \text{as} \quad \mathring{D} \ni (e, \beta) \to (\hat{e}, 0), \tag{3.40}$$

where $\hat{r}(e) \in [0, r_Q]$ *is the unique solution to* $U_Q(\hat{r}(e)) = e$.

Proof First, we are going to show that r_+ stays away from zero in the limiting case that we are considering here. For this, we may assume that $r_+ \leq r_Q/2$. Due to (3.6), we have

$$r_+^2\varphi(r_+) + \frac{\beta}{2r_+^2} = U_Q(r_+) - U_Q(0) + \frac{\beta}{2r_+^2} = e - U_Q(0) \tag{3.41}$$

for

$$\varphi(r_+) = 4\pi\int_0^1 d\tau\tau^2 \int_0^1 dt\, t\, \rho_Q(\tau t r_+).$$

Since ρ_Q is radially decreasing and $0 \leq \tau t r_+ \leq r_Q/2$, it follows that

$$0 < c_1 = \frac{2\pi}{3}\rho_Q\Big(\frac{r_Q}{2}\Big) \leq \varphi(r_+) \leq \frac{2\pi}{3}\rho_Q(0) = C_1. \tag{3.42}$$

In (3.41), solving the resulting quadratic equation for r_+^2, we obtain

$$r_+^2 = \frac{e - U_Q(0) \pm \sqrt{(e - U_Q(0))^2 - 2\varphi(r_+)\beta}}{2\varphi(r_+)}. \tag{3.43}$$

Let us suppose that the sign were '−', along (a subsequence) of $e \to \hat{e}$ and $\beta \to 0$.

Then

$$r_+^2 = \frac{\beta}{e - U_Q(0) + \sqrt{(e - U_Q(0))^2 - 2\varphi(r_+)\beta}}$$

together with $\hat{e} - U_Q(0) > 0$ and (3.42) would yield $c_2\beta \leq r_+^2 \leq C_2\beta$ for suitable constants $C_2 > c_2 > 0$. By Lemma 3.1(b), we have the general estimate

$$c\beta \leq r_-^2 r_+^2.$$

As a consequence,

$$\frac{c}{C_2} \leq r_-^2.$$

However, $r_-^2 \leq r_0^2 = \mathcal{O}(\beta^{1/2})$ as $\beta \to 0$ by Lemma A.7(f), which gives a contradiction. To summarize, we may suppose that the sign is '+' in (3.43). Hence,

$$r_+^2 = \frac{e - U_Q(0) + \sqrt{(e - U_Q(0))^2 - 2\varphi(r_+)\beta}}{2\varphi(r_+)} \geq \frac{1}{2C_1}(e - U_Q(0))$$

for $\beta \leq \frac{1}{2C_1}(e - U_Q(0))^2$ yields the desired lower bound for r_+. Thus, in what follows, we can assume that $r_+(e, \beta) \geq \eta_0 > 0$ for an appropriate constant η_0 and $(e, \beta) \to (\hat{e}, 0)$.

Next, we are going to show that

$$T_1^-(e, \beta) = 2\int_{r_-}^{r_0} \frac{dr}{\sqrt{2(e - U_{\text{eff}}(r, \beta))}} \to 0 \quad \text{as} \quad \mathring{D} \ni (e, \beta) \to (\hat{e}, 0). \tag{3.44}$$

For, owing to (3.10), we get

$$T_1^-(e, \beta) \leq 4\,\frac{\sqrt{r_- r_+}}{\sqrt{\beta}}\,\frac{r_0}{\sqrt{r_+ - r_0}}\,\sqrt{r_0 - r_-} \leq 4\,\frac{\sqrt{r_- r_+}}{\sqrt{\beta}}\,\frac{r_0^{3/2}}{\sqrt{r_+ - r_0}}.$$

Since $r_-^2 r_+^2 \leq C\beta$ by (3.37) and $r_0 = \mathcal{O}(\beta^{1/4})$ by Lemma A.7(f), $r_+ \geq \eta_0$ yields

$$T_1^-(e, \beta) \leq C\beta^{3/8}$$

and completes the argument for (3.44).

Thus, in order to establish (3.40), we need to prove that

$$\int_{r_0}^{r_+} \frac{dr}{\sqrt{e - U_{\text{eff}}(r, \beta)}} \to \int_0^{\hat{r}(\hat{e})} \frac{dr}{\sqrt{\hat{e} - U_Q(r)}} \quad \text{as} \quad \mathring{D} \ni (e, \beta) \to (\hat{e}, 0); \tag{3.45}$$

note that $U_Q(r_+) \leq U_Q(r_+) + \frac{\beta}{2r_+^2} = e = U_Q(\hat{r}(e))$ implies $r_+ \leq \hat{r}(e)$. In addition, using (3.3), we obtain

$$\frac{\pi}{12}\,\rho_Q\Big(\frac{r_Q}{2}\Big)\eta_0\,(\hat r(e)-r_+)\le\frac{\pi}{12}\,\rho_Q\Big(\frac{r_Q}{2}\Big)(\hat r(e)^2-r_+^2)$$
$$\le U_Q(\hat r(e))-U_Q(r_+)=\frac{\beta}{2r_+^2}\le\frac{1}{2\eta_0^2}\,\beta.$$

Similarly, by (3.2),

$$\begin{aligned}\frac{\beta}{2r_Q^2}\le\frac{\beta}{2r_+^2}&=U_Q(\hat r(e))-U_Q(r_+)\\&\le\frac{2\pi}{3}\rho_Q(0)(\hat r(e)^2-r_+^2)\\&\le\frac{4\pi}{3}\rho_Q(0)\,r_Q\,(\hat r(e)-r_+),\end{aligned}$$

so that $c_3\beta\le\hat r(e)-r_+\le C_3\beta$. To validate (3.45), we are going to show

$$\left|\int_{r_0}^{r_+}\frac{dr}{\sqrt{e-U_{\rm eff}(r,\beta)}}-\int_0^{\hat r(e)}\frac{dr}{\sqrt{e-U_Q(r)}}\right|\to 0,\tag{3.46}$$

$$\left|\int_0^{\hat r(e)}\frac{dr}{\sqrt{e-U_Q(r)}}-\int_0^{\hat r(\hat e)}\frac{dr}{\sqrt{\hat e-U_Q(r)}}\right|\to 0,\tag{3.47}$$

both as $\mathring{D}\ni(e,\beta)\to(\hat e,0)$; the second relation is independent of β.

To begin with,

$$\int_0^{r_0}\frac{dr}{\sqrt{e-U_Q(r)}}\to 0\quad\text{and}\quad\int_{r_+}^{\hat r(e)}\frac{dr}{\sqrt{e-U_Q(r)}}\to 0.\tag{3.48}$$

For the first claim, if $0\le r\le r_0=\mathcal{O}(\beta^{1/4})$ and $e\to\hat e>U_Q(0)$, we may suppose that $e-U_Q(r)\ge\eta_1>0$ for the e and r in question; therefore, the first claim in (3.48) follows. Regarding the second assertion, we write

$$e-U_Q(r)=U_Q(\hat r(e))-U_Q(r)=(\hat r(e)-r)\int_0^1 U_Q'(\tau\hat r(e)+(1-\tau)r)\,d\tau$$

for $r\in[\frac{r_+}{2},\hat r(e)]$. If $s\in[\frac{r_+}{2},\hat r(e)]$, then the fact that ρ_Q is radially decreasing yields

$$U_Q'(s)=\frac{4\pi}{s^2}\int_0^s\sigma^2\rho_Q(\sigma)\,d\sigma\ge\frac{4\pi}{r_Q^2}\int_0^{r_+/2}\sigma^2\rho_Q(\sigma)\,d\sigma\ge\frac{\pi r_+^3}{6r_Q^2}\,\rho_Q\Big(\frac{r_+}{2}\Big)\ge\eta_2>0.\tag{3.49}$$

Hence,

$$e-U_Q(r)\ge\eta_2(\hat r(e)-r),\quad r\in\Big[\frac{r_+}{2},\hat r(e)\Big],\tag{3.50}$$

and accordingly,

$$\int_{r_+}^{\hat{r}(e)} \frac{dr}{\sqrt{e-U_Q(r)}} \le \eta_2^{-1/2} \int_{r_+}^{\hat{r}(e)} \frac{dr}{\sqrt{\hat{r}(e)-r}} = 2\eta_2^{-1/2}\sqrt{\hat{r}(e)-r_+} \le C\beta^{1/2} \to 0.$$

Thus, both relations in (3.48) hold, and therefore (3.46) comes down to proving that

$$\left| \int_{r_0}^{r_+} \frac{dr}{\sqrt{e-U_{\rm eff}(r,\beta)}} - \int_{r_0}^{r_+} \frac{dr}{\sqrt{e-U_Q(r)}} \right| \to 0 \quad \text{as} \quad \mathring{D} \ni (e,\beta) \to (\hat{e},0).$$

If $r \in [r_0, (1-\beta^{1/4})r_+]$, then $\frac{\beta}{2r^2} \le \frac{\beta}{2r_0^2} \le C\beta^{1/2}$, as $r_0 = \mathcal{O}(\beta^{1/4})$. Therefore, (3.3) yields

$$\begin{aligned} e - U_Q(r) \ge e - U_{\rm eff}(r,\beta) &= e - U_Q(r) - \frac{\beta}{2r^2} \\ &\ge U_Q(\hat{r}(e)) - U_Q((1-\beta^{1/4})r_+) - C\beta^{1/2} \\ &\ge \frac{\pi}{12}\,\rho_Q\!\left(\frac{r_Q}{2}\right)(\hat{r}(e)^2 - (1-\beta^{1/4})^2 r_+^2) - C\beta^{1/2} \\ &\ge \frac{\pi}{12}\,\rho_Q\!\left(\frac{r_Q}{2}\right)\eta_0\,(\hat{r}(e) - r_+ + \beta^{1/4} r_+) - C\beta^{1/2} \\ &\ge c_4\beta + c_5\beta^{1/4} - C\beta^{1/2} \\ &\ge c_6\beta^{1/4}. \end{aligned}$$

From the estimate $\frac{1}{\sqrt{a}} - \frac{1}{\sqrt{b}} \le \frac{b-a}{a\sqrt{b}}$ for $b \ge a > 0$, we hence infer

$$\begin{aligned} &\left| \int_{r_0}^{(1-\beta^{1/4})r_+} \frac{dr}{\sqrt{e-U_{\rm eff}(r,\beta)}} - \int_{r_0}^{(1-\beta^{1/4})r_+} \frac{dr}{\sqrt{e-U_Q(r)}} \right| \\ &\le \frac{\beta}{2} \int_{r_0}^{(1-\beta^{1/4})r_+} \frac{1}{(e-U_{\rm eff}(r,\beta))\sqrt{e-U_Q(r)}}\,\frac{dr}{r^2} \\ &\le \frac{\beta}{2r_0^2} \int_{r_0}^{(1-\beta^{1/4})r_+} \frac{1}{c_6^{3/2}\beta^{3/8}}\,dr \le C\beta^{1/8} \to 0. \end{aligned} \tag{3.51}$$

For the remaining part, $r \in [(1-\beta^{1/4})r_+, r_+]$, we note that for such r, by (3.50),

$$e - U_Q(r) \ge \eta_2(\hat{r}(e) - r) \ge \eta_2(\hat{r}(e) - r_+) \ge \eta_2 c_3\,\beta.$$

In addition,

$$\begin{aligned} e - U_{\rm eff}(r,\beta) &= U_{\rm eff}(r_+,\beta) - U_{\rm eff}(r,\beta) \\ &= (r_+ - r)\int_0^1 U_{\rm eff}'(\tau r_+ + (1-\tau)r, \beta)\,d\tau. \end{aligned} \tag{3.52}$$

If $r \in [(1-\beta^{1/4})r_+, r_+]$, then $s = \tau r_+ + (1-\tau)r \in [(1-\beta^{1/4})r_+, r_+] \subset [\frac{r_+}{2}, \hat{r}(e)]$ for instance, and

$$U'_{\rm eff}(s,\beta) = U'_Q(s) - \frac{\beta}{s^3} \geq \eta_2 - \frac{8\beta}{r_+^3} \geq \eta_2 - \frac{8\beta}{\eta_0^3} \geq \frac{1}{2}\eta_2$$

by (3.49), if β is small enough. Thus, (3.52) leads to

$$e - U_{\rm eff}(r,\beta) \geq \frac{1}{2}\eta_2\,(r_+ - r), \quad r \in [(1-\beta^{1/4})r_+, r_+].$$

If we now use that $\frac{1}{\sqrt{a}} - \frac{1}{\sqrt{b}} \leq \frac{b-a}{\sqrt{ab}}$ for $b \geq a > 0$, we obtain the bound

$$\begin{aligned} &\left| \int_{(1-\beta^{1/4})r_+}^{r_+} \frac{dr}{\sqrt{e - U_{\rm eff}(r,\beta)}} - \int_{(1-\beta^{1/4})r_+}^{r_+} \frac{dr}{\sqrt{e - U_Q(r)}} \right| \\ &\leq \frac{\beta}{2} \int_{(1-\beta^{1/4})r_+}^{r_+} \frac{1}{\sqrt{e - U_{\rm eff}(r,\beta)(e - U_Q(r))}}\,\frac{dr}{r^2} \\ &\leq \frac{\beta}{2}\,\frac{4}{\eta_0^2}\,\frac{1}{\eta_2 c_3\,\beta}\sqrt{\frac{2}{\eta_2}} \int_{(1-\beta^{1/4})r_+}^{r_+} \frac{1}{\sqrt{r_+ - r}}\,dr \leq C\beta^{1/8} \to 0. \end{aligned} \tag{3.53}$$

By (3.51) and (3.53), the proof of (3.46) is complete.

Therefore, it remains to check that (3.47) is satisfied. This is not worked out, since it is just the continuity of the standard period function in the potential $V(x) = U_Q(x) - U_Q(0)$ for $x \geq 0$ and $V(x) = U_Q(-x) - U_Q(0)$ for $x \leq 0$, for energies $\hat{e} = e - U_Q(0) \in]0, e_0 - U_Q(0)]$. □

If we now summarize Lemma 3.8, Remark 3.9 and Lemmas 3.11 and 3.12, then we have shown the following result (note that $e_{\rm min}(0) = U_Q(0)$ and $r_0(0) = 0$).

Theorem 3.13 *We have $T_1 \in C(D)$. The extensions to ∂D are given by*

$$T_1(e,\beta) = \begin{cases} \frac{2\pi}{\sqrt{B(r_0(\beta))}} & : \ e = e_{\rm min}(\beta),\ \beta \in [0,\beta_*] \\ 2\int_{r_-(e_0,\beta)}^{r_+(e_0,\beta)} \frac{dr}{\sqrt{2(e_0 - U_{\rm eff}(r,\beta))}} & : \ e = e_0,\ \beta \in]0,\beta_*] \\ 2\int_0^{\hat{r}(e)} \frac{dr}{\sqrt{2(e - U_Q(r))}} & : \ e \in]U_Q(0), e_0],\ \beta = 0 \end{cases},$$

where $\hat{r}(e) \in [0, r_Q]$ is the unique solution to $U_Q(\hat{r}(e)) = e$.

In the remaining part of this section, we will discuss some monotonicity properties of T_1.

Lemma 3.14 *The function $[0,\beta_*] \ni \beta \mapsto T_1(e_{\rm min}(\beta),\beta)$ is strictly increasing.*

Proof We know from Lemma A.7(e) that $\beta \mapsto r_0(\beta)$ is strictly increasing, and furthermore $r \mapsto B(r)$ is strictly decreasing by Lemma A.6(b). Hence, the claim follows from $T_1(e_{\min}(\beta), \beta) = \frac{2\pi}{\sqrt{B(r_0(\beta))}}$. □

Lemma 3.15 *The function* $[U_Q(0), e_0] \ni e \mapsto T_1(e, 0)$ *is strictly increasing.*

Proof The argument is analogous to the fact that for a one degree of freedom oscillator $\ddot{x} = -V'(x)$ about a stable center, where $V(0) = V'(0) = 0$ and $V(-x) = V(x)$ for simplicity, the condition $V'(x) > 0$ and $V''(x) > V'(x)/x$ for $x > 0$ guarantees that the period function of the periodic orbits about $x = 0$ is decreasing in the energy $e = \frac{1}{2}\dot{x}^2 + V(x)$. The first reference to point this out seems to be [64] (which we basically follow); related papers are [11, 78, 79]. To see the connection, first observe that, by (1.13), Remark A.1 and (A.32),

$$U_Q''(r) - \frac{U_Q'(r)}{r} = 4\pi\rho_Q(r) - \frac{3}{r}\,U_Q'(r) = \frac{4\pi}{r^3}\int_0^r s^3\rho_Q'(s)\,ds < 0, \quad r > 0.$$

Thus, $(U_Q'(r)/r)' = (rU_Q''(r) - U_Q'(r))/r^2 < 0$ for $r > 0$, and it follows that

$$\frac{1}{p}\,U_Q'(pr) < U_Q'(r), \quad p > 1, \quad r > 0. \tag{3.54}$$

The function $\hat{r}$ is strictly increasing, due to $1 = U_Q'(\hat{r}(e))\,\hat{r}'(e)$ and $U_Q'(r) > 0$ for $r > 0$. Therefore, its inverse $[0, r_Q] \ni \hat{r} \mapsto e(\hat{r}) \in [U_Q(0), e_0]$ is well-defined and strictly increasing too; note that $\hat{r}(U_Q(0)) = 0$ and $\hat{r}(e_0) = r_Q$. Let

$$\hat{T}(\hat{r}) = 2\int_0^{\hat{r}} \frac{dr}{\sqrt{2(U_Q(\hat{r}) - U_Q(r))}}.$$

Then

$$T_1(e, 0) = 2\int_0^{\hat{r}(e)} \frac{dr}{\sqrt{2(U_Q(\hat{r}(e)) - U_Q(r)}} = \hat{T}(\hat{r}(e)),$$

which implies that $e \mapsto T_1(e, 0)$ is increasing if and only if $\hat{r} \mapsto \hat{T}(\hat{r})$ is increasing. If $p > 1$ and $s \in [0, \hat{r}]$, then by (3.54), one has

$$U_Q(p\hat{r}) - U_Q(ps) = p\int_s^{\hat{r}} U_Q'(p\tau)\,d\tau \le p^2\int_s^{\hat{r}} U_Q'(\tau)\,d\tau = p^2(U_Q(\hat{r}) - U_Q(s)).$$

As a consequence,

$$\begin{aligned}
\hat{T}(p\hat{r}) &= 2\int_0^{p\hat{r}} \frac{dr}{\sqrt{2(U_Q(p\hat{r}) - U_Q(r))}} = 2p\int_0^{\hat{r}} \frac{ds}{\sqrt{2(U_Q(p\hat{r}) - U_Q(ps))}} \\
&\ge 2\int_0^{\hat{r}} \frac{ds}{\sqrt{2(U_Q(\hat{r}) - U_Q(s))}} = \hat{T}(\hat{r}),
\end{aligned}$$

which completes the proof. □

Corollary 3.16 *Suppose that* $\delta_1 = \inf_{\mathring{D}} \omega_1 = \min_D \omega_1$ *is attained at some point* $(\hat{e}, \hat{\beta}) \in \partial D$. *Then* $(\hat{e}, \hat{\beta})$ *lies on the 'upper line'* $\{(e, \beta) : e = e_0, \beta \in [0, \beta_*]\}$ *of the boundary.*

Proof This follows from $\omega_1 = \frac{2\pi}{T_1}$ together with Lemmas 3.14 and 3.15. □

Remark 3.17 It will certainly be important to gain a better understanding of the monotonicity properties of ω_1 (or, equivalently, T_1) in D. In particular, we expect that it should be significant to locate those points in D, where ω_1 attains its minimum δ_1. Some relations for $\frac{\partial T_1}{\partial e}$ and $\frac{\partial T_1}{\partial \beta}$ are stated in Lemma A.12(b), (c). For instance, we have

$$\frac{\partial T_1}{\partial \beta}(e, \beta) = -\frac{1}{2}\frac{\partial}{\partial e}\int_0^{T_1(e,\beta)} \frac{ds}{r(s)^2} = -\frac{\partial}{\partial e}\int_{r_-(e,\beta)}^{r_+(e,\beta)} \frac{dr}{r^2 p_r}, \tag{3.55}$$

which could provide a way to approach the monotonicity of T_1 in β. To see this, we apply the transformation $\rho = \sqrt{\beta}\, r^{-1}$, $d\rho = -\sqrt{\beta}\, r^{-2}\, dr$, like for the 'apsidal angle' [77]. Defining

$$\tilde{U}(\rho, \beta) = \frac{1}{2}\rho^2 + U_Q\Big(\frac{\sqrt{\beta}}{\rho}\Big), \quad \rho_\mp(e, \beta) = \frac{\sqrt{\beta}}{r_\pm(e, \beta)},$$

and recalling that $p_r = \sqrt{2(e - U_Q(r) - \frac{\beta}{2r^2})}$, we get

$$\frac{\partial T_1}{\partial \beta}(e, \beta) = -\frac{1}{\sqrt{\beta}}\frac{\partial}{\partial e}\int_{\rho_-(e,\beta)}^{\rho_+(e,\beta)} \frac{d\rho}{\sqrt{2(e - \tilde{U}(\rho, \beta))}}.$$

At fixed β, this has turned the integral on the right-hand side of (3.55) into the period function

$$\tilde{T}(e) = \int_{\rho_-(e,\beta)}^{\rho_+(e,\beta)} \frac{d\rho}{\sqrt{2(e - \tilde{U}(\rho, \beta))}}$$

for the transformed potential $\tilde{U}$; note that $0 < \rho_- < \rho_+$ and $\tilde{U}(\rho_\pm, \beta) = e$. One could study the monotonicity of $\tilde{T}(e)$ in the energy e by checking the criteria that have been listed in the papers we mentioned in the proof of Lemma 3.15 or which can be found in similar works. Let us state a remarkable relation that could be useful in this respect. Writing $\tilde{U}(\rho) = \tilde{U}(\rho, \beta)$, it is calculated that

$$\tilde{U}'(\rho) = -\frac{\sqrt{\beta}}{\rho^2} U_Q'\Big(\frac{\sqrt{\beta}}{\rho}\Big) + \rho, \quad \tilde{U}''(\rho) = \frac{\beta}{\rho^4} U_Q''\Big(\frac{\sqrt{\beta}}{\rho}\Big) + \frac{2\sqrt{\beta}}{\rho^3} U_Q'\Big(\frac{\sqrt{\beta}}{\rho}\Big) + 1,$$

and using (1.13) this yields

$$
\begin{aligned}
\tilde{U}''(\rho) - \frac{\tilde{U}'(\rho)}{\rho} &= \frac{\beta}{\rho^4}\, U_Q''\Big(\frac{\sqrt{\beta}}{\rho}\Big) + \frac{3\sqrt{\beta}}{\rho^3}\, U_Q'\Big(\frac{\sqrt{\beta}}{\rho}\Big) \\
&= \frac{\beta}{\rho^4}\left[4\pi\rho_Q\Big(\frac{\sqrt{\beta}}{\rho}\Big) - \frac{2\rho}{\sqrt{\beta}}\, U_Q'\Big(\frac{\sqrt{\beta}}{\rho}\Big)\right] + \frac{3\sqrt{\beta}}{\rho^3}\, U_Q'\Big(\frac{\sqrt{\beta}}{\rho}\Big) \\
&= \frac{\beta}{\rho^4}\left[4\pi\rho_Q\Big(\frac{\sqrt{\beta}}{\rho}\Big) + \frac{\rho}{\sqrt{\beta}}\, U_Q'\Big(\frac{\sqrt{\beta}}{\rho}\Big)\right] \\
&= \frac{\beta}{\rho^4}\, B\Big(\frac{\sqrt{\beta}}{\rho}\Big).
\end{aligned}
$$

In other words,

$$\Big(\frac{\tilde{U}'(\rho)}{\rho}\Big)' = \frac{\beta}{\rho^5}\, B\Big(\frac{\sqrt{\beta}}{\rho}\Big),$$

and the function B is strictly positive. Comparing to the reasoning in Lemma 3.15, this looks promising for proving that $\tilde{T}(e)$ is increasing in e, i.e., that $\frac{\partial T_1}{\partial \beta} < 0$. However, the argument does not seem to work properly, since the integral defining $\tilde{T}(e)$ is on $[\rho_-, \rho_+]$, instead of it beginning at zero, as is the case in Lemma 3.15. ◇

3.4 $\lambda_* \leq \delta_1^2$

From (1.20), recall the definition of λ_*.

Lemma 3.18 *We have* $\lambda_* \leq \delta_1^2$.

Proof From (1.18), cf. Corollary B.19 and Lemma B.8(c), we deduce that, for $u \in X^2_{\mathrm{odd}}$,

$$
\begin{aligned}
(Lu, u)_Q &= \int_{\mathbb{R}^3}\int_{\mathbb{R}^3} \frac{dx\,dv}{|Q'(e_Q)|}\,|\mathcal{T}u|^2 - \frac{1}{4\pi}\int_{\mathbb{R}^3} |\nabla_x U_{\mathcal{T}u}|^2\,dx \\
&\leq \int_{\mathbb{R}^3}\int_{\mathbb{R}^3} \frac{dx\,dv}{|Q'(e_Q)|}\,|\mathcal{T}u|^2 = \|\mathcal{T}u\|^2_{X^0} = 16\pi^3 \sum_{k\neq 0} k^2 \|\omega_1 u_k\|^2_{L^2_{\frac{1}{|Q'|}}(D)}.
\end{aligned}
$$

Since $u_{-k} = -u_k$ by Lemma B.3(b), this yields

$$\lambda_* \leq 32\pi^3 \sum_{k=1}^{\infty} k^2 \|\omega_1 u_k\|^2_{L^2_{\frac{1}{|Q'|}}(D)} \tag{3.56}$$

for all $u \in X^2_{\mathrm{odd}}$ such that $\|u\|_{X^0} = \|u\|_Q = 1$. Now we specialize (3.56) to $u \cong (\ldots, 0, u_{-1}, 0, u_1, 0, \ldots) = (\ldots, 0, -u_1, 0, u_1, 0, \ldots)$ to find that

$$\lambda_* \le 32\pi^3 \|\omega_1 u_1\|^2_{L^2_{\frac{1}{|Q'|}}(D)} = 32\pi^3 \iint_D dI\, d\ell\, \ell\, \frac{1}{|Q'(e)|}\, \omega_1^2(I,\ell)\, |u_1(I,\ell)|^2$$

$$= 32\pi^3 \iint_D de\, d\ell\, \ell\, \frac{1}{|Q'(e)|}\, \omega_1(e,\ell)\, |u_1(e,\ell)|^2 \tag{3.57}$$

for all $u_1 = u_1(I,\ell) = u_1(e,\ell) \in L^2_{\frac{1}{|Q'|}}(D)$ satisfying

$$1 = 32\pi^3 \iint_D dI\, d\ell\, \ell\, \frac{1}{|Q'(e)|}\, |u_1(I,\ell)|^2$$

$$= 32\pi^3 \iint_D de\, d\ell\, \ell\, \frac{1}{|Q'(e)|}\, \frac{1}{\omega_1(e,\ell)}\, |u_1(e,\ell)|^2;$$

see Definition B.1 and cf. (A.18). Let $\varepsilon > 0$. Since $\delta_1 = \inf_{\mathring{D}} \omega_1$, there is $(\hat{e},\hat{l}) \in \mathring{D}$ such that $\omega_1(\hat{e},\hat{l}) < \delta_1 + \varepsilon/2$. As ω_1 is continuous in $\mathring{D}$ by Theorem 3.6, there is an open neighborhood $U \subset \mathring{D}$ of $(\hat{e},\hat{l})$ with the property that $\omega_1(e,l) < \delta_1 + \varepsilon$ for $(e,l) \in U$; then $\iint_U de\, d\ell\, \ell > 0$. Define

$$\chi(e,\ell) = \begin{cases} 1 & : \ (e,\ell) \in U \\ 0 & : \ (e,\ell) \in D \setminus U \end{cases}$$

and $u_1(e,\ell) = a\,|Q'(e)|^{1/2}\omega_1(e,\ell)^{1/2}\,\chi(e,\ell)$ for $a = (32\pi^3 \iint_U de\, d\ell\, \ell)^{-1/2}$. It follows that

$$32\pi^3 \iint_D de\, d\ell\, \ell\, \frac{1}{|Q'(e)|\omega_1}\, |u_1|^2 = 32\pi^3 a^2 \iint_U de\, d\ell\, \ell = 1.$$

Thus, by (3.57),

$$\lambda_* \le 32\pi^3 \iint_D de\, d\ell\, \ell\, \frac{\omega_1}{|Q'|}\, |u_1|^2 = 32\pi^3 a^2 \iint_U de\, d\ell\, \ell\, \omega_1^2$$

$$\le 32\pi^3 a^2\, (\delta_1 + \varepsilon)^2 \iint_U de\, d\ell\, \ell = (\delta_1 + \varepsilon)^2.$$

As $\varepsilon \to 0^+$, we get $\lambda_* \le \delta_1^2$. □

Chapter 4
A Birman-Schwinger Type Operator

As has been outlined in the introduction, the eigenvalues $\lambda < \delta_1^2$ of $L = -\mathcal{T}^2 - \mathcal{K}\mathcal{T}$ from (1.16) are in one-to-one correspondence with the eigenvalues 1 of a certain Birman-Schwinger type operator $\mathcal{Q}_\lambda$ that acts on functions $\Psi = \Psi(r)$.

4.1 The Operator $\mathcal{Q}_z$

Let L_r^2 denote the L^2-Lebesgue space of radially symmetric functions $\Psi(x) = \Psi(r)$ on $\mathbb{R}^3$, where we take

$$\langle \Psi, \Phi \rangle = \int_{\mathbb{R}^3} \overline{\Psi(x)}\, \Phi(x)\, dx = 4\pi \int_0^\infty r^2\, \overline{\Psi(r)}\, \Phi(r)\, dr$$

as the inner product of $\Psi, \Phi \in L_r^2$. Unless otherwise stated, a generic constant (denoted by C) is allowed to depend only upon Q.

Definition 4.1 For $z \in \Omega = \mathbb{C} \setminus [\delta_1^2, \infty[$, we introduce

$$\begin{aligned} &\mathcal{Q}_z : L_r^2 \to L_r^2, \\ &(\mathcal{Q}_z \Psi)(r) = \frac{16\pi}{r^2} \sum_{k \neq 0} \int_0^\infty d\tilde{r}\, \Psi(\tilde{r}) \iint_D d\ell\, \ell\, de\, \mathbf{1}_{\{r_-(e,\ell) \le r,\, \tilde{r} \le r_+(e,\ell)\}} \\ &\qquad \times \frac{\omega_1(e,\ell)\, |Q'(e)|}{k^2 \omega_1^2(e,\ell) - z} \sin(k\theta(r,e,\ell)) \sin(k\theta(\tilde{r},e,\ell)), \end{aligned} \tag{4.1}$$

M. Kunze, *A Birman-Schwinger Principle in Galactic Dynamics*,
Progress in Mathematical Physics 77,
https://doi.org/10.1007/978-3-030-75186-9_4

where $r_\pm(e,\ell)$ and $\theta(r,e,\ell)$ are as in Appendix I, Sect. A.1, and D is given by (3.1). Along with $\mathcal{Q}_z$, we also introduce the integral kernels

$$K_z(r,\tilde{r}) = \frac{4}{r^2\tilde{r}^2}\sum_{k\neq 0}\iint_D d\ell\,\ell\,de\,\mathbf{1}_{\{r_-(e,\ell)\leq r,\,\tilde{r}\leq r_+(e,\ell)\}} \times \frac{\omega_1(e,\ell)\,|Q'(e)|}{k^2\omega_1^2(e,\ell)-z}\sin(k\theta(r,e,\ell))\sin(k\theta(\tilde{r},e,\ell)). \tag{4.2}$$

Remark 4.2 (a) If $z = a+ib\in\mathbb{C}\setminus\mathbb{R}$, then $|k^2\omega_1^2(e,\ell)-z|\geq|b|>0$. More precisely,

$$|k|\geq\left[\frac{\sqrt{2|a|}}{\delta_1}\right]+1 \implies |k^2\omega_1^2(e,\ell)-z|^2 = (k^2\omega_1^2(e,\ell)-a)^2+b^2 \geq (k^2\delta_1^2-|a|)^2+b^2 \geq \frac{1}{4}k^4\delta_1^4+b^2. \tag{4.3}$$

On the other hand, if $z=\lambda\in]-\infty,\delta_1^2[$, then

$$|k^2\omega_1^2(e,\ell)-z| = k^2\omega_1^2(e,\ell)-\lambda\geq k^2\delta_1^2-\lambda\geq\delta_1^2-\lambda>0,$$

and hence

$$|k|\geq 2\implies |k^2\omega_1^2(e,\ell)-z|\geq k^2\delta_1^2-\lambda\geq(k^2-1)\delta_1^2\geq\frac{1}{2}k^2\delta_1^2. \tag{4.4}$$

In particular, $\frac{1}{k^2\omega_1^2(e,\ell)-z}$ in (4.1) and (4.2) is well-defined for $z\in\Omega$.
(b) In the definitions, we understand the factor $|Q'(e)|$ to be zero outside of K, the support of Q, instead of carrying around another characteristic function all the time. In particular, always $r_+(e,\ell)\leq r_Q$ holds, which means the following: in (4.1), $\int_0^\infty d\tilde{r}\,\Psi(\tilde{r})$ can be replaced by $\int_0^{r_Q} d\tilde{r}\,\Psi(\tilde{r})$; $(\mathcal{Q}_z\Psi)(r)$ can be replaced by $(\mathcal{Q}_z\Psi)(r)\,\mathbf{1}_{\{0\leq r\leq r_Q\}}$ and $K_z(r,\tilde{r})$ can be replaced by $K_z(r,\tilde{r})\,\mathbf{1}_{\{0\leq r,\tilde{r}\leq r_Q\}}$. ◇

Lemma 4.3 *[Properties of $\mathcal{Q}_z$] The following assertions hold.*

(a) For every $z\in\Omega$, we have $\mathcal{Q}_z\in\mathcal{B}(L_r^2)$, the space of linear and bounded operators on L_r^2. In addition, the map

$$\Omega\ni z\mapsto\mathcal{Q}_z\in\mathcal{B}(L_r^2) \tag{4.5}$$

is analytic, and for the derivatives

$$(\mathcal{Q}_z^{(j)}\Psi)(r) = \frac{16\pi j!}{r^2}\sum_{k\neq 0}\int_0^\infty d\tilde{r}\,\Psi(\tilde{r})\iint_D d\ell\,\ell\,de\,\mathbf{1}_{\{r_-(e,\ell)\leq r,\,\tilde{r}\leq r_+(e,\ell)\}}$$
$$\times\frac{\omega_1(e,\ell)\,|Q'(e)|}{(k^2\omega_1^2(e,\ell)-z)^{j+1}}\sin(k\theta(r,e,\ell))\sin(k\theta(\tilde{r},e,\ell))$$

for $\Psi \in L_r^2$.

(*b*) *If* $z \in \Omega$, *then*

$$(\mathcal{Q}_z\Psi)(r) = \langle K_{\bar{z}}(r,\cdot),\Psi\rangle$$

for $\Psi \in L_r^2$. *In particular,*

$$\langle \mathcal{Q}_z\Psi,\Phi\rangle = \langle\Psi,\mathcal{Q}_{\bar{z}}\Phi\rangle$$

for $\Psi, \Phi \in L_r^2$, *so that* $Q_z^* = \mathcal{Q}_{\bar{z}}$. *Thus, if* $\lambda \in]-\infty, \delta_1^2[$, *then* $\mathcal{Q}_\lambda$ *is symmetric.*

(*c*) *If* $z \in \Omega$, *then* $\mathcal{Q}_z$ *is a Hilbert-Schmidt operator on* L_r^2.

(*d*) *If* $z \in \Omega$, *then*

$$\langle \mathcal{Q}_z\Psi,\Psi\rangle$$
$$= 64\pi^2\sum_{k\neq 0}\iint_D d\ell\,\ell\,de\,\frac{\omega_1(e,\ell)\,|Q'(e)|}{k^2\omega_1^2(e,\ell)-\bar{z}}\left|\int_{r_-(e,\ell)}^{r_+(e,\ell)}\Psi(r)\sin(k\theta(r,e,\ell))\,dr\right|^2$$

for $\Psi \in L_r^2$. *In particular, if* $\lambda \in]-\infty, \delta_1^2[$, *then* $\langle\mathcal{Q}_\lambda\Psi,\Psi\rangle \geq 0$ *for* $\Psi \in L_r^2$, *i.e.,* $\mathcal{Q}_\lambda$ *is positive. In addition, for the derivatives*

$$\langle \mathcal{Q}_z^{(j)}\Psi,\Psi\rangle = 64\pi^2 j!\sum_{k\neq 0}\iint_D d\ell\,\ell\,de\,\frac{\omega_1(e,\ell)\,|Q'(e)|}{(k^2\omega_1^2(e,\ell)-\bar{z})^{j+1}}$$
$$\times\left|\int_{r_-(e,\ell)}^{r_+(e,\ell)}\Psi(r)\sin(k\theta(r,e,\ell))\,dr\right|^2 \quad (4.6)$$

for $\Psi \in L_r^2$.

(*e*) *There is a constant* $C > 0$ *such that for* $\lambda, \tilde{\lambda} \in]-\infty, \delta_1^2[$,

$$\|\mathcal{Q}_\lambda - \mathcal{Q}_{\tilde{\lambda}}\|_{\mathrm{HS}} \leq C\left(1+\frac{1}{(\delta_1^2-\lambda)(\delta_1^2-\tilde{\lambda})}\right)|\lambda-\tilde{\lambda}|,$$

where $\|\cdot\|_{\mathrm{HS}}$ *denotes the Hilbert-Schmidt norm.*

(*f*) *If* $\lambda \in]-\infty, \delta_1^2[$, *then the spectrum of* $\mathcal{Q}_\lambda$ *consists of* $\mu_1(\lambda) \geq \mu_2(\lambda) \geq \ldots \to 0$ *(the eigenvalues are listed according to their multiplicities). In addition,*

$$\mu_1(\lambda) = \|\mathcal{Q}_\lambda\| = \sup\{\langle\mathcal{Q}_\lambda\Psi,\Psi\rangle : \|\Psi\|_{L_r^2} \leq 1\}, \quad (4.7)$$

where $\|\cdot\| = \|\cdot\|_{\mathcal{B}(L_r^2)}$, *and every function*

$$\mu_k(\cdot) :]-\infty, \delta_1^2[\to]0, \infty[$$

for $k \in \mathbb{N}$ is monotone increasing and locally Lipschitz continuous (and hence differentiable a.e. by Rademacher's Theorem).

Proof (a) Let $z \in \Omega$ be fixed. By Remark 4.2(a), there is $\alpha_0 > 0$ such that $|k^2\omega_1^2(e, \ell) - z| \geq \alpha_0$ for $|k| \geq 1$ and $(e, \ell) \in D$. In addition, according to (4.3) and (4.4), there is $k_0 \in \mathbb{N}$ so that $|k^2\omega_1^2(e, \ell) - z| \geq \frac{1}{2}k^2\delta_1^2$ for $|k| \geq k_0$ and $(e, \ell) \in D$; if k_0 is taken to be large enough, we can also make sure that $\frac{1}{2}k^2\delta_1^2 \geq k^{3/2}$. First, we observe that

$$r_-(e, \ell) \leq r \leq r_+(e, \ell) \implies \ell^2 \leq 2r^2(e_0 - U_Q(0)). \tag{4.8}$$

To establish this claim, we recall from (3.7) that $\ell^2 = 2r_-^2(e - U_Q(r_-))$ holds, where $r_\pm = r_\pm(e, \ell)$. Since U_Q is increasing and $e \leq e_0$, we get $\ell^2 \leq 2r_-^2(e_0 - U_Q(0)) \leq 2r^2(e_0 - U_Q(0))$.

For $1 \leq |k| \leq k_0$ and $i \in \mathbb{N}_0$, we now apply (4.8) to r and $\tilde{r}$ in order to estimate

$$\begin{aligned} s_{k,i}(r, \tilde{r}, z) = &\iint_D d\ell\, \ell\, de\, \mathbf{1}_{\{r_-(e,\ell)\leq r, \tilde{r}\leq r_+(e,\ell)\}} \\ &\times \frac{\omega_1(e, \ell)\, |Q'(e)|}{(k^2\omega_1^2(e, \ell) - z)^{i+1}} \sin(k\theta(r, e, \ell)) \sin(k\theta(\tilde{r}, e, \ell)) \end{aligned} \tag{4.9}$$

as

$$\begin{aligned} |s_{k,i}(r, \tilde{r}, z)| &\leq \alpha_0^{-(i+1)} \Delta_1 \mathbf{1}_{\{0\leq r,\tilde{r}\leq r_Q\}} \int_0^{l_*} d\ell\, \ell \int_{e_{\min}(\ell)}^{e_0} de \\ &\qquad \times \mathbf{1}_{\{r_-(e,\ell)\leq r, \tilde{r}\leq r_+(e,\ell)\}}\, |Q'(e)| \\ &\leq \alpha_0^{-(i+1)} \Delta_1 \mathbf{1}_{\{0\leq r,\tilde{r}\leq r_Q\}} \int_0^{l_*} d\ell\, \ell \int_{e_{\min}(\ell)}^{e_0} de \\ &\qquad \times \mathbf{1}_{\{\ell^2\leq 2(e_0 - U_Q(0)) \min\{r^2,\tilde{r}^2\}\}}\, |Q'(e)| \\ &\leq \alpha_0^{-(i+1)} \Delta_1 \mathbf{1}_{\{0\leq r,\tilde{r}\leq r_Q\}} (e_0 - U_Q(0)) \left(\int_{U_Q(0)}^{e_0} |Q'(e)|\, de \right) \min\{r^2, \tilde{r}^2\}. \end{aligned}$$

Analogously, for $|k| \geq k_0$ and $i \in \mathbb{N}_0$, we deduce

$$|s_{k,i}(r, \tilde{r}, z)| \leq \frac{1}{k^{3(i+1)/2}} \Delta_1 \mathbf{1}_{\{0\leq r,\tilde{r}\leq r_Q\}} (e_0 - U_Q(0)) \left(\int_{U_Q(0)}^{e_0} |Q'(e)|\, de \right) \min\{r^2, \tilde{r}^2\}.$$

It follows that

$$\begin{aligned} \sum_{k\neq 0} |s_{k,i}(r, \tilde{r}, z)| \leq &\sum_{|k|\leq k_0} \alpha_0^{-(i+1)} \Delta_1 \mathbf{1}_{\{0\leq r,\tilde{r}\leq r_Q\}} (e_0 - U_Q(0)) \\ &\times \left(\int_{U_Q(0)}^{e_0} |Q'(e)|\, de \right) \min\{r^2, \tilde{r}^2\} \end{aligned}$$

$$
\begin{aligned}
&+ \sum_{|k|\geq k_0} \frac{1}{k^{3(i+1)/2}} \Delta_1 \, \mathbf{1}_{\{0\leq r,\tilde{r}\leq r_Q\}} \, (e_0 - U_Q(0)) \\
&\qquad \times \left(\int_{U_Q(0)}^{e_0} |Q'(e)| \, de \right) \min\{r^2, \tilde{r}^2\} \\
&\leq C_{1,i} \, \mathbf{1}_{\{0\leq r,\tilde{r}\leq r_Q\}} \min\{r^2, \tilde{r}^2\}
\end{aligned}
\tag{4.10}
$$

for

$$
C_{1,i} = \left(2k_0 \alpha_0^{-(i+1)} + 2 \sum_{k=1}^{\infty} \frac{1}{k^{3/2}} \right) \Delta_1 \, (e_0 - U_Q(0)) \left(\int_{U_Q(0)}^{e_0} |Q'(e)| \, de \right); \tag{4.11}
$$

this constant depends upon z and Q, but k_0 is independent of i. Therefore,

$$
\begin{aligned}
|(Q_z\Psi)(r)| &= \left| \frac{16\pi}{r^2} \sum_{k\neq 0} \int_0^\infty \Psi(\tilde{r}) \, s_{k,0}(r, \tilde{r}, z) \, d\tilde{r} \right| \\
&\leq \frac{16\pi C_{1,0}}{r^2} \, \mathbf{1}_{\{0\leq r\leq r_Q\}} \int_0^{r_Q} |\Psi(\tilde{r})| \min\{r^2, \tilde{r}^2\} \, d\tilde{r}.
\end{aligned}
$$

Next, note that

$$
\min\{r^2, \tilde{r}^2\} \leq r\tilde{r}. \tag{4.12}
$$

Thus, using Hölder's inequality,

$$
\begin{aligned}
|(Q_z\Psi)(r)|^2 &\leq \frac{256\pi^2 C_{1,0}^2}{r^2} \, \mathbf{1}_{\{0\leq r\leq r_Q\}} \left(\int_0^{r_Q} \tilde{r} \, |\Psi(\tilde{r})| \, d\tilde{r} \right)^2 \\
&\leq \frac{256\pi^2 C_{1,0}^2 \, r_Q}{r^2} \, \mathbf{1}_{\{0\leq r\leq r_Q\}} \int_0^{r_Q} \tilde{r}^2 \, |\Psi(\tilde{r})|^2 \, d\tilde{r} \\
&\leq \frac{64\pi C_{1,0}^2 \, r_Q}{r^2} \, \mathbf{1}_{\{0\leq r\leq r_Q\}} \, \|\Psi\|_{L_r^2}^2,
\end{aligned}
$$

and this in turn leads to

$$
\|Q_z\Psi\|_{L_r^2}^2 = 4\pi \int_0^\infty r^2 \, |(Q_z\Psi)(r)|^2 \, dr \leq 264\pi^2 C_{1,0}^2 \, r_Q^2 \, \|\Psi\|_{L_r^2}^2.
$$

To prove the analyticity of (4.5), we recall that it suffices to show weak analyticity, in the sense that all maps $\Omega \ni z \mapsto \langle \Psi, Q_z\Phi \rangle \in \mathbb{C}$ for $\Psi, \Phi \in L_r^2$ are analytic; see [85, Thm. 3.1.12]. Fix $z_0 \in \Omega$. If $|z - z_0|$ is sufficiently small, then $z \in \Omega$ and we have the series expansion

$$
\frac{1}{k^2\omega_1^2(e,\ell) - z} = \sum_{i=0}^{\infty} \frac{1}{(k^2\omega_1^2(e,\ell) - z_0)^{i+1}} \, (z - z_0)^i
$$

for every $k \neq 0$ and $(e, l) \in D$, which suggests that

$$\begin{aligned}\langle \Psi, Q_z \Phi \rangle &= 64\pi^2 \int_0^{r_Q} \int_0^{r_Q} dr\, d\tilde{r}\, \overline{\Psi(r)}\, \Phi(\tilde{r}) \sum_{k\neq 0} \iint_D d\ell\, \ell\, de\, \mathbf{1}_{\{r_-(e,\ell)\le r,\, \tilde{r}\le r_+(e,\ell)\}} \\ &\qquad \times \frac{\omega_1(e,\ell)\,|Q'(e)|}{k^2\omega_1^2(e,\ell) - z} \sin(k\theta(r,e,\ell)) \sin(k\theta(\tilde{r},e,\ell)) \\ &= \sum_{i=0}^{\infty} a_i (z - z_0)^i \end{aligned} \tag{4.13}$$

for

$$\begin{aligned} a_i = 64\pi^2 \int_0^{r_Q} \int_0^{r_Q} dr\, d\tilde{r}\, \overline{\Psi(r)}\, \Phi(\tilde{r}) \sum_{k\neq 0} \iint_D d\ell\, \ell\, de\, \mathbf{1}_{\{r_-(e,\ell)\le r,\, \tilde{r}\le r_+(e,\ell)\}} \\ \times \frac{\omega_1(e,\ell)\,|Q'(e)|}{(k^2\omega_1^2(e,\ell) - z_0)^{i+1}} \sin(k\theta(r,e,\ell)) \sin(k\theta(\tilde{r},e,\ell)). \end{aligned}$$

We are going to show that the series (4.13) converges near z_0. For this, due to (4.10) and (4.12), we deduce that

$$\begin{aligned} |a_i| &= 64\pi^2 \left| \int_0^{r_Q} \int_0^{r_Q} dr\, d\tilde{r}\, \overline{\Psi(r)}\, \Phi(\tilde{r}) \sum_{k\neq 0} s_{k,i}(r, \tilde{r}, z_0) \right| \\ &\le 64\pi^2\, C_{1,i} \int_0^{r_Q} \int_0^{r_Q} dr\, d\tilde{r}\, |\Psi(r)|\, |\Phi(\tilde{r})| \min\{r^2, \tilde{r}^2\} \\ &\le 64\pi^2\, C_{1,i} \left(\int_0^{r_Q} r\, |\Psi(r)|\, dr \right) \left(\int_0^{r_Q} \tilde{r}\, |\Phi(\tilde{r})|\, d\tilde{r} \right) \\ &\le 16\pi r_Q\, C_{1,i}\, \|\Psi\|_{L_r^2}\, \|\Phi\|_{L_r^2}. \end{aligned}$$

If we write the constant $C_{1,i}$ from (4.11) as $C_{1,i} = \tilde{C}_1 \alpha_0^{-(i+1)} + \hat{C}_1$, with α_0 depending only on z_0, then $|z - z_0| < \min\{\frac{\alpha_0}{2}, \frac{1}{2}\}$ ensures that

$$C_{1,i} |z - z_0|^i \le \tilde{C}_1 \alpha_0^{-1} 2^{-i} + \hat{C}_1 2^{-i},$$

which has a finite $\sum_{i=0}^{\infty}$. It follows that (4.13) converges for $z \in \Omega$ such that $|z - z_0| < \min\{\frac{\alpha_0}{2}, \frac{1}{2}\}$, i.e., on a sufficiently small ball about z_0. The formula for the derivative is gotten from a_1 and those for the higher order derivatives follow from this one inductively.

(b) By the definition of K_z in (4.2), we have

$$K_z(r, \tilde{r}) = \frac{4}{r^2 \tilde{r}^2} \sum_{k\neq 0} s_{k,0}(r, \tilde{r}, z). \tag{4.14}$$

Hence,

$$(\mathcal{Q}_z\Psi)(r) = \frac{16\pi}{r^2}\sum_{k\neq 0}\int_0^\infty \Psi(\tilde{r})\, s_{k,0}(r,\tilde{r},z)\,d\tilde{r} = 4\pi\int_0^\infty \tilde{r}^2\, K_z(r,\tilde{r})\,\Psi(\tilde{r})\,d\tilde{r} = \langle K_{\bar{z}}(r,\cdot),\Psi\rangle, \tag{4.15}$$

observing that $\overline{K_z} = K_{\bar{z}}$. Due to $K_z(r,\tilde{r}) = K_z(\tilde{r},r)$, we hence obtain

$$\begin{aligned}
\langle \mathcal{Q}_z\Psi,\Phi\rangle &= 4\pi\int_0^\infty r^2\,\overline{(\mathcal{Q}_z\Psi)(r)}\,\Phi(r)\,dr = 4\pi\int_0^\infty r^2\,\overline{\langle K_{\bar{z}}(r,\cdot),\Psi\rangle}\,\Phi(r)\,dr\\
&= 16\pi^2\int_0^\infty dr\,r^2\int_0^\infty d\tilde{r}\,\tilde{r}^2\,K_{\bar{z}}(r,\tilde{r})\,\overline{\Psi(\tilde{r})}\,\Phi(r)\\
&= 16\pi^2\int_0^\infty d\tilde{r}\,\tilde{r}^2\,\overline{\Psi(\tilde{r})}\int_0^\infty dr\,r^2\,K_{\bar{z}}(\tilde{r},r)\,\Phi(r)\\
&= 4\pi\int_0^\infty d\tilde{r}\,\tilde{r}^2\,\overline{\Psi(\tilde{r})}\,\langle\overline{K_{\bar{z}}(\tilde{r},\cdot)},\Phi\rangle = \langle\Psi,\mathcal{Q}_{\bar{z}}\Phi\rangle.
\end{aligned}$$

(c) According to (b), the operator $\mathcal{Q}_z$ on L_r^2 has the integral kernel $K_{\bar{z}}$. Hence, in order to verify that $\mathcal{Q}_z$ is Hilbert-Schmidt, we need to verify that

$$\begin{aligned}
\|\mathcal{Q}_z\|_{\mathrm{HS}}^2 &= \int_{\mathbb{R}^3}\int_{\mathbb{R}^3}|K_z(x,\bar{x})|^2\,dx\,d\bar{x}\\
&= 16\pi^2\int_0^\infty\int_0^\infty r^2\,\tilde{r}^2\,|K_z(r,\tilde{r})|^2\,dr\,d\tilde{r}\\
&= 16\pi^2\int_0^{r_Q}\int_0^{r_Q} r^2\,\tilde{r}^2\,|K_z(r,\tilde{r})|^2\,dr\,d\tilde{r} < \infty
\end{aligned} \tag{4.16}$$

for every $z\in\Omega$, where K_z is viewed both as a function of $(x,\bar{x})$ and a function of $(r,\tilde{r})$ and we used Remark 4.2(b); see [35, Prop. 6.36]. From (4.14), (4.10) and (4.12), we get

$$\begin{aligned}
\int_0^{r_Q}\int_0^{r_Q} r^2\,\tilde{r}^2\,|K_z(r,\tilde{r})|^2\,dr\,d\tilde{r} &\le 16\int_0^{r_Q}\int_0^{r_Q}\frac{1}{r^2\tilde{r}^2}\Big(\sum_{k\neq 0}|s_{k,0}(r,\tilde{r},z)|\Big)^2\,dr\,d\tilde{r}\\
&\le 16\,C_1^2\int_0^{r_Q}\int_0^{r_Q}\frac{1}{r^2\tilde{r}^2}\,(\min\{r^2,\tilde{r}^2\})^2\,dr\,d\tilde{r}\\
&\le 16\,C_1^2\int_0^{r_Q}\int_0^{r_Q}dr\,d\tilde{r} = 16\,C_1^2 r_Q^2 < \infty.
\end{aligned}$$

Note that from $\mathcal{Q}_z$ being Hilbert-Schmidt it follows that $\mathcal{Q}_z$ is bounded and $\|\mathcal{Q}_z\|\le\|\mathcal{Q}_z\|_{\mathrm{HS}}$, i.e., once again we see that (a) holds. However, since the key of the argument is (4.10) and (4.12), it needs very little additional work to derive both bounds. (d) Here, we calculate

$$
\begin{aligned}
\langle \mathcal{Q}_z \Psi, \Psi\rangle &= 4\pi \int_0^\infty r^2\, \overline{(\mathcal{Q}_z\Psi)(r)}\, \Psi(r)\, dr \\
&= 64\pi^2 \sum_{k\neq 0} \int_0^\infty dr\, \Psi(r) \int_0^\infty d\tilde{r}\, \overline{\Psi(\tilde{r})} \\
&\qquad \times \iint_D d\ell\, \ell\, de\, \mathbf{1}_{\{r_-(e,\ell)\leq r,\, \tilde{r}\leq r_+(e,\ell)\}}\, \frac{\omega_1(e,\ell)\, |Q'(e)|}{k^2\omega_1^2(e,\ell)-\bar{z}} \\
&= 64\pi^2 \sum_{k\neq 0} \iint_D d\ell\, \ell\, de\, \frac{\omega_1(e,\ell)\, |Q'(e)|}{k^2\omega_1^2(e,\ell)-\bar{z}} \\
&\qquad \left| \int_{r_-(e,\ell)}^{r_+(e,\ell)} \Psi(r) \sin(k\theta(r,e,\ell))\, dr \right|^2 \geq 0.
\end{aligned}
$$

The proof of (4.6) is analogous. (e) For $\lambda, \tilde{\lambda} < \delta_1^2$, we have, cf. (4.16),

$$
\begin{aligned}
\|\mathcal{Q}_\lambda - \mathcal{Q}_{\tilde{\lambda}}\|_{\mathrm{HS}}^2 &= \int_{\mathbb{R}^3}\int_{\mathbb{R}^3} |K_\lambda(x,\tilde{x}) - K_{\tilde{\lambda}}(x,\tilde{x})|^2\, dx\, d\tilde{x} \\
&= 16\pi^2 \int_0^{r_Q}\int_0^{r_Q} r^2\, \tilde{r}^2\, |K_\lambda(r,\tilde{r}) - K_{\tilde{\lambda}}(r,\tilde{r})|^2\, dr\, d\tilde{r} \\
&= 256\,\pi^2 \int_0^{r_Q}\int_0^{r_Q} \frac{dr}{r^2}\frac{d\tilde{r}}{\tilde{r}^2} \Bigg| \sum_{k\neq 0} \iint_D d\ell\, \ell\, de\, \mathbf{1}_{\{r_-(e,\ell)\leq r,\, \tilde{r}\leq r_+(e,\ell)\}} \\
&\qquad \times \omega_1(e,\ell)\, |Q'(e)| \left[\frac{1}{k^2\omega_1^2(e,\ell)-\lambda} - \frac{1}{k^2\omega_1^2(e,\ell)-\tilde{\lambda}} \right] \\
&\qquad \times \sin(k\theta(r,e,\ell)) \sin(k\theta(\tilde{r},e,\ell)) \Bigg|^2 \\
&\leq 512\,\pi^2 \Delta_1^2 \int_0^{r_Q}\int_0^{r_Q} \frac{dr}{r^2}\frac{d\tilde{r}}{\tilde{r}^2} \\
&\qquad \times \Bigg(\sum_{k=1}^\infty \iint_D d\ell\, \ell\, de\, \mathbf{1}_{\{r_-(e,\ell)\leq r,\, \tilde{r}\leq r_+(e,\ell)\}}\, |Q'(e)| \\
&\qquad \left| \frac{1}{k^2\omega_1^2(e,\ell)-\lambda} - \frac{1}{k^2\omega_1^2(e,\ell)-\tilde{\lambda}} \right| \Bigg)^2.
\end{aligned}
$$

Using (4.8) and (4.12), we may continue this estimate for suitable constants $C, \hat{C} > 0$ as

$$
\begin{aligned}
\|\mathcal{Q}_\lambda - \mathcal{Q}_{\tilde{\lambda}}\|_{\mathrm{HS}}^2 &\leq 256\,\pi^2 \Delta_1^2 \int_0^{r_Q}\int_0^{r_Q} \frac{dr}{r^2}\frac{d\tilde{r}}{\tilde{r}^2} \\
&\qquad \times \Bigg(\sum_{k=1}^\infty \iint_D d\beta\, de\, \mathbf{1}_{\{\beta\leq C\min\{r^2,\tilde{r}^2\}\}}\, |Q'(e)|
\end{aligned}
$$

$$
\left. \left| \frac{1}{k^2\omega_1^2(e,\beta)-\lambda} - \frac{1}{k^2\omega_1^2(e,\beta)-\tilde{\lambda}} \right| \right)^2 \tag{4.17}
$$

$$
\le 256\,\pi^2\Delta_1^2 \int_0^{r_Q}\int_0^{r_Q} \frac{dr}{r^2}\frac{d\bar{r}}{\bar{r}^2}
$$

$$
\times \left(\sum_{k=1}^{\infty} \iint_D d\beta\, de\, \mathbf{1}_{\{\beta\le\hat{C}r\bar{r}\}}\, |Q'(e)| \right.
$$

$$
\left. \frac{|\lambda-\tilde{\lambda}|}{(k^2\omega_1^2(e,\beta)-\lambda)(k^2\omega_1^2(e,\beta)-\tilde{\lambda})} \right)^2.
$$

For $k \ge 2$, we know from Remark 4.2(a) that $k^2\omega_1^2(e,\beta)-\lambda \ge k^2\delta_1^2/2$ and $k^2\omega_1^2(e,\beta)-\tilde{\lambda} \ge k^2\delta_1^2/2$ are verified. If $k=1$, then always $\omega_1^2(e,\beta)-\lambda \ge \delta_1^2-\lambda$ and $\omega_1^2(e,\beta)-\tilde{\lambda} \ge \delta_1^2-\tilde{\lambda}$ hold. Thus, we arrive at

$$
\begin{aligned}
\|\mathcal{Q}_\lambda-\mathcal{Q}_{\tilde{\lambda}}\|_{\mathrm{HS}}^2 &\le C|\lambda-\tilde{\lambda}|^2 \int_0^{r_Q}\int_0^{r_Q} \frac{dr}{r^2}\frac{d\bar{r}}{\bar{r}^2}\, r^2\bar{r}^2 \\
&\qquad \times \left(\delta_1^{-4} \sum_{k=2}^{\infty} \frac{1}{k^4} \int_{U_Q(0)}^{e_0} |Q'(e)|\, de \right)^2 \\
&\quad + C\, \frac{|\lambda-\tilde{\lambda}|^2}{(\delta_1^2-\lambda)^2(\delta_1^2-\tilde{\lambda})^2} \int_0^{r_Q}\int_0^{r_Q} \frac{dr}{r^2}\frac{d\bar{r}}{\bar{r}^2}\, r^2\bar{r}^2 \\
&\qquad \times \left(\int_{U_Q(0)}^{e_0} |Q'(e)|\, de \right)^2 \\
&\le C\left(1+\frac{1}{(\delta_1^2-\lambda)^2(\delta_1^2-\tilde{\lambda})^2}\right)|\lambda-\tilde{\lambda}|^2,
\end{aligned}
$$

and this yields the claim. (f) According to (b–d), $\mathcal{Q}_\lambda$ is a symmetric and positive Hilbert-Schmidt operator, which is in particular compact. Thus, the assertions up to and including (4.7) are a consequence of the spectral theory for compact positive self-adjoint operators; see [35, Section 6]. Concerning the $\mu_k(\lambda)$, we have the characterization

$$
\mu_k(\lambda) = \max\left\{ \min_{\Psi\in S,\, \|\Psi\|_{L_r^2}=1} \langle \mathcal{Q}_\lambda\Psi,\Psi\rangle : S\subset L_r^2 \text{ is a subspace of dimension } k \right\} \tag{4.18}
$$

according to the Courant max-min principle. In the present situation, this follows from the spectral decomposition theorem for symmetric and compact operators. By (d), we obtain for $\tilde{\lambda}\ge\lambda$, both in $]-\infty,\delta_1^2[$ and $\Psi\in L_r^2$,

$$
\langle \mathcal{Q}_{\tilde{\lambda}}\Psi,\Psi\rangle = 64\pi^2 \sum_{k\neq 0} \iint_D d\ell\,\ell\, de\, \frac{\omega_1\,|Q'(e)|}{k^2\omega_1^2-\tilde{\lambda}} \left| \int_{r_-}^{r_+} \Psi(r)\sin(k\theta)\, dr \right|^2
$$

$$\geq 64\pi^2 \sum_{k\neq 0} \iint_D d\ell\, \ell\, de\, \frac{\omega_1\, |Q'(e)|}{k^2\omega_1^2 - \lambda} \left| \int_{r_-}^{r_+} \Psi(r) \sin(k\theta)\, dr \right|^2$$

$$= \langle \mathcal{Q}_\lambda \Psi, \Psi \rangle, \tag{4.19}$$

where $r_\pm = r_\pm(e, \ell)$ and $\theta = \theta(r, e, \ell)$. Hence, (4.18) implies that $\mu_k(\tilde{\lambda}) \geq \mu_k(\lambda)$ for all $k \in \mathbb{N}$. To establish the local Lipschitz continuity of $\mu_k(\cdot)$, note that

$$|\langle \mathcal{Q}_\lambda \Psi, \Psi \rangle - \langle \mathcal{Q}_{\tilde{\lambda}} \Psi, \Psi \rangle| \leq \|\mathcal{Q}_\lambda - \mathcal{Q}_{\tilde{\lambda}}\|\, \|\Psi\|_{L_r^2}^2,$$

whence we deduce from (e) and $\|\cdot\| \leq \|\cdot\|_{\mathrm{HS}}$ that for $\Psi \in L_r^2$ satisfying $\|\Psi\|_{L_r^2} \leq 1$, one has

$$|\langle \mathcal{Q}_\lambda \Psi, \Psi \rangle - \langle \mathcal{Q}_{\tilde{\lambda}} \Psi, \Psi \rangle| \leq C \left(1 + \frac{1}{(\delta_1^2 - \lambda)(\delta_1^2 - \tilde{\lambda})}\right) |\lambda - \tilde{\lambda}|.$$

Applying (4.18) once more, we arrive at

$$|\mu_k(\lambda) - \mu_k(\tilde{\lambda})| \leq C \left(1 + \frac{1}{(\delta_1^2 - \lambda)(\delta_1^2 - \tilde{\lambda})}\right) |\lambda - \tilde{\lambda}|,$$

which completes the proof. □

In the following, we are going to derive some more specific properties of the $\mathcal{Q}_z$. See Appendix II, Sect. B.1 below for the function spaces that are being used. Once again, we understand that $|Q'(e_Q)|$ vanishes outside of K.

Lemma 4.4 *If $z \in \Omega$ and $\psi(r, p_r, \ell) = |Q'(e_Q)|\, p_r \Psi(r)$ for $\Psi \in L_r^2$, then $\psi \in X_{\mathrm{odd}}^0$,*

$$\|\psi\|_{X^0} \leq \rho_Q(0)^{1/2}\, \|\Psi\|_{L_r^2} \tag{4.20}$$

and

$$\mathcal{K}\mathcal{T}(-\mathcal{T}^2 - z)^{-1}\psi = |Q'(e_Q)|\, p_r(\mathcal{Q}_z \Psi). \tag{4.21}$$

In particular,

$$\mathcal{Q}_z \Psi = U'_{\mathcal{T}(-\mathcal{T}^2 - z)^{-1}\psi} = 4\pi \int_{\mathbb{R}^3} p_r\, (-\mathcal{T}^2 - z)^{-1}\psi\, dv. \tag{4.22}$$

Moreover, if also $\tilde{\psi}(r, p_r, \ell) = |Q'(e_Q)|\, p_r \tilde{\Psi}(r)$ for some $\tilde{\Psi} \in L_r^2$, then

$$\|\psi - \tilde{\psi}\|_{X^0} \leq \rho_Q(0)^{1/2}\, \|\Psi - \tilde{\Psi}\|_{L_r^2}. \tag{4.23}$$

Proof First, note that ψ is odd in v and has its support in K. Furthermore, due to Remark B.2(a), Lemma 2.5 and (A.32),

$$
\begin{aligned}
\|\psi\|_{X^0}^2 &= \|\psi\|^2_{L^2_{\mathrm{sph},\frac{1}{|Q'|}}(K)} \\
&= \iint_K \frac{1}{|Q'(e_Q)|}\,|\psi(x,v)|^2\,dx\,dv \\
&= \iint_K |Q'(e_Q)|\,p_r^2\,|\Psi(r)|^2\,dx\,dv \\
&= \int_{|x|<r_Q} dx\,|\Psi(r)|^2 \int_{\mathbb{R}^3} dv\,|Q'(e_Q)|\,p_r^2 \\
&= \int_{|x|<r_Q} dx\,|\Psi(r)|^2\,\rho_Q(r) \\
&\leq \rho_Q(0)\int_{|x|<r_Q} dx\,|\Psi(r)|^2 \leq \rho_Q(0)\,\|\Psi\|^2_{L^2_r}.
\end{aligned}
$$

Thus, $\psi \in X^0_{\mathrm{odd}} \subset X^0_0$, and accordingly Corollary B.14 yields

$$
\begin{aligned}
&\mathcal{K}\mathcal{T}(-\mathcal{T}^2 - z)^{-1}\psi \\
&= |Q'(e_Q)|\,p_r\,\frac{16\pi^2 i}{r^2}\sum_{k\neq 0}\iint_D d\ell\,\ell\,de\,\mathbf{1}_{[r_-(e,\ell),\,r_+(e,\ell)]}(r)\,\frac{\sin(k\theta(r,e,\ell))}{k^2\omega_1^2(e,\ell)-z}\,\psi_k(I,\ell).
\end{aligned}
$$

On the other hand,

$$
\psi_k(I,\ell) = -\frac{i}{\pi}\,|Q'(e)|\,\omega_1(e,\ell)\int_{r_-(e,\ell)}^{r_+(e,\ell)} d\tilde{r}\,\Psi(\tilde{r})\sin(k\theta(\tilde{r},e,\ell)) \tag{4.24}
$$

by Lemma B.5. Therefore, we arrive at

$$
\begin{aligned}
\mathcal{K}\mathcal{T}(-\mathcal{T}^2 - z)^{-1}\psi &= |Q'(e_Q)|\,p_r\,\frac{16\pi}{r^2}\sum_{k\neq 0}\iint_D d\ell\,\ell\,de\,\mathbf{1}_{[r_-(e,\ell),\,r_+(e,\ell)]}(r) \\
&\quad\times \frac{\sin(k\theta(r,e,\ell))}{k^2\omega_1^2(e,\ell)-z}\,|Q'(e)|\,\omega_1(e,\ell) \\
&\quad\quad \int_{r_-(e,\ell)}^{r_+(e,\ell)} d\tilde{r}\,\Psi(\tilde{r})\sin(k\theta(\tilde{r},e,\ell)) \\
&= |Q'(e_Q)|\,p_r\,\frac{16\pi}{r^2}\sum_{k\neq 0}\int_0^{r_Q} d\tilde{r}\,\Psi(\tilde{r}) \\
&\quad\quad \iint_D d\ell\,\ell\,de\,\mathbf{1}_{\{r_-(e,\ell)\leq r,\,\tilde{r}\leq r_+(e,\ell)\}}(r) \\
&\quad\times \frac{\omega_1(e,\ell)\,|Q'(e)|}{k^2\omega_1^2(e,\ell)-z}\sin(k\theta(r,e,\ell))\sin(k\theta(\tilde{r},e,\ell)) \\
&= |Q'(e_Q)|\,p_r(\mathcal{Q}_z\Psi),
\end{aligned}
$$

and this completes the proof of (4.21), by the definition of $\mathcal{Q}_z$. Concerning (4.22), the first part follows from $\mathcal{K}g = |Q'(e_Q)|\, p_r\, U_g'(r)$, see (B.37), and for the second part, one just has to use Lemma 2.4. Lastly, (4.23) is a direct consequence of (4.20) and the fact that $(\tilde{\psi} - \psi)(r, p_r, \ell) = |Q'(e_Q)|\, p_r(\tilde{\Psi} - \Psi)(r)$. □

Now, we can make the connection from eigenvalues $\lambda < \delta_1^2$ of the self-adjoint operator

$$L = -\mathcal{T}^2 - \mathcal{K}\mathcal{T} : X^2_{\text{odd}} \to X^0_{\text{odd}},$$

cf. (1.16) and Corollary B.19, to eigenvalues 1 of $\mathcal{Q}_\lambda$.

Theorem 4.5 *Let $\lambda < \delta_1^2$. Then λ is an eigenvalue of L if and only if 1 is an eigenvalue of $\mathcal{Q}_\lambda$. More precisely,*

(a) if $u \in X^2_{\text{odd}}$ is an eigenfunction of L for the eigenvalue λ, then $\Psi = U'_{\mathcal{T}u} \in L^2_r$ for $r \in [0, r_Q]$ is an eigenfunction of $\mathcal{Q}_\lambda$ for the eigenvalue 1;

(b) if $\Psi \in L^2_r$ is an eigenfunction of $\mathcal{Q}_\lambda$ for the eigenvalue 1, then $u = (-\mathcal{T}^2 - \lambda)^{-1}(|Q'(e_Q)|\, p_r \Psi) \in X^2_{\text{odd}}$ is an eigenfunction of L for the eigenvalue λ.

Proof First, suppose that $Lu = \lambda u$ for some $u \in X^2_{\text{odd}}$ and $u \neq 0$. Then $(-\mathcal{T}^2 - \lambda)u = \mathcal{K}\mathcal{T}u$. Defining $\psi = (-\mathcal{T}^2 - \lambda)u \in X^0_{\text{odd}}$, Remark B.18(a) implies that $\psi = \mathcal{K}\mathcal{T}(-\mathcal{T}^2 - \lambda)^{-1}\psi$. Since $\mathcal{K}g = |Q'(e_Q)|\, p_r\, U_g'(r)$ by (B.37), we can put

$$\Psi(r) = U'_{\mathcal{T}(-\mathcal{T}^2-\lambda)^{-1}\psi}(r) = U'_{\mathcal{T}u}(r)$$

for $r \in [0, r_Q]$ to obtain $\psi = |Q'(e_Q)|\, p_r \Psi(r)$. Then $\Psi \neq 0$, as otherwise $\psi = 0$ and $u = 0$. Next, we are going to verify that $\Psi \in L^2_r$. Using (B.40) from Lemma B.15 and Lemma B.8(c), we get

$$\begin{aligned}
\|\Psi\|^2_{L^2_r} &= \int_{\mathbb{R}^3} |U'_{\mathcal{T}(-\mathcal{T}^2-\lambda)^{-1}\psi}(r)|^2\, dx \\
&= 4\pi\Big(\mathcal{K}\mathcal{T}(-\mathcal{T}^2 - \lambda)^{-1}\psi, (-\mathcal{T}^2 - \lambda)^{-1}\psi\Big)_{X^0} \\
&= 4\pi(\psi, (-\mathcal{T}^2 - \lambda)^{-1}\psi)_{X^0} \\
&= 4\pi((-\mathcal{T}^2 - \lambda)u, u)_{X^0} \\
&= 4\pi(\|\mathcal{T}u\|^2_{X^0} - \lambda\, \|u\|^2_{X^0}).
\end{aligned}$$

In particular, Lemma B.8(a) implies $\|\Psi\|^2_{L^2_r} \leq 4\pi\|\mathcal{T}u\|^2_{X^0} \leq 4\pi\Delta_1^2\, \|u\|^2_{X^1} < \infty$, so that indeed $\Psi \in L^2_r$. Thus, we deduce from Lemma 4.4 that

$$|Q'(e_Q)|\, p_r(\mathcal{Q}_\lambda\Psi) = \mathcal{K}\mathcal{T}(-\mathcal{T}^2 - \lambda)^{-1}\psi = \psi = |Q'(e_Q)|\, p_r\, \Psi,$$

and consequently $\mathcal{Q}_\lambda\Psi = \Psi$.

Conversely, suppose that $\mathcal{Q}_\lambda\Psi = \Psi$ is verified for some $\Psi \in L^2_r$ and $\Psi \neq 0$. According to Remark 4.2(b), Ψ has its support in $[0, r_Q]$. Defining $\psi = |Q'(e_Q)$

$|\, p_r \Psi(r)$, we obtain $\psi \in X^0_{\rm odd}$ from Lemma 4.4. As a consequence, $u = (-\mathcal{T}^2 - \lambda)^{-1}\psi \in X^2_{\rm odd}$. Also $u \neq 0$, since otherwise $\psi = 0$ and $\Psi = 0$. From Lemma 4.4, we finally get

$$\begin{aligned}(-\mathcal{T}^2 - \lambda)u = \psi &= |Q'(e_Q)|\, p_r\, \Psi = |Q'(e_Q)|\, p_r(\mathcal{Q}_\lambda \Psi)\\ &= \mathcal{K}\mathcal{T}(-\mathcal{T}^2 - \lambda)^{-1}\psi = \mathcal{K}\mathcal{T}u,\end{aligned}$$

so that $Lu = -\mathcal{T}^2 u - \mathcal{K}\mathcal{T}u = \lambda u$. □

Lemma 4.6 *The following assertions hold.*

(a) To $\Psi \in L^2_r$ we associate the function $\psi(r, p_r, \ell) = |Q'(e_Q)|\, p_r \Psi(r)$. If $z \in \Omega$, then

$$\langle \mathcal{Q}_z \Psi, \Psi\rangle = 64\pi^4 \sum_{k\neq 0} \iint_D d\ell\, \ell\, de\, \frac{1}{\omega_1(e,\ell)\,|Q'(e)|}\, \frac{1}{k^2\omega_1^2(e,\ell) - \bar{z}}\, |\psi_k(I,\ell)|^2. \tag{4.25}$$

(b) Let $\Psi \in L^2_r$ be given and suppose that $F(r) = F(0) + \int_0^r \Psi(s)\, ds$ for $r \in [0, r_Q]$ as well as $g = -|Q'(e_Q)|(F - F_0)$, where F_0 is the zero'th Fourier coefficient of F. Then $\mathcal{Q}_0\Psi = U'_g$ and furthermore

$$\langle \mathcal{Q}_0 \Psi, \Psi\rangle = 4\pi \iint_K \frac{dx\, dv}{|Q'(e_Q)|}\, |g|^2 = 4\pi \iint_K |Q'(e_Q)|\, (F - F_0)^2\, dx\, dv. \tag{4.26}$$

(c) Let $\Psi \in L^2_r$ be given and suppose that $F(r) = F(0) + \int_0^r \Psi(s)\, ds$ for $r \in [0, r_Q]$. Define $u = -\mathcal{T}^{-1}(|Q'(e_Q)|(F - F_0))$. Then $u \in X^2_{\rm odd}$ and

$$(Lu, u)_{X^0} = \frac{1}{4\pi}\Big(\langle \mathcal{Q}_0\Psi, \Psi\rangle - \|\mathcal{Q}_0\Psi\|^2_{L^2_r}\Big). \tag{4.27}$$

Proof (a) The relation (4.25) follows from Lemma 4.3(d) and (4.24).

(b) Owing to Lemma B.9, we have $g \in X^1_{\rm even}$ as well as $\mathcal{T}g = -\psi$ for ψ as in (a). In addition, $g_0 = 0$ by (B.24), so that $g \in X^1_0$. Thus, Lemma B.13(c) yields $-\mathcal{T}^{-1}\psi = g - g_0 = g$.

Next, recall that ψ is odd in v and $\|\psi\|_{X^0} \le \rho_Q(0)^{1/2}\, \|\Psi\|_{L^2_r} < \infty$ by (4.20), which means that $\psi \in X^0_{\rm odd} \subset X^0_0$. As a consequence, $\mathcal{T}(-\mathcal{T}^2)^{-1}\psi = -\mathcal{T}^{-1}\psi = g$ by Lemma B.13(e). Hence, if we take $z = 0 \in \Omega$ in (4.22) of Lemma 4.4, then we get

$$\mathcal{Q}_0\Psi = U'_{\mathcal{T}(-\mathcal{T}^2)^{-1}\psi} = U'_g.$$

To verify (4.26), note first that $ik\omega_1 g_k = -\psi_k$ for $k \in \mathbb{Z}$. Applying (B.4) from Remark B.2(a), we obtain

$$\iint_K \frac{dx\,dv}{|Q'(e_Q)|}|g|^2 = 16\pi^3 \sum_{k\neq 0} \iint_D dI\,d\ell\,\ell\,\frac{1}{|Q'(e_Q)|}|g_k|^2$$
$$= 16\pi^3 \sum_{k\neq 0} \iint_D dI\,d\ell\,\ell\,\frac{1}{|Q'(e_Q)|}\,\frac{1}{k^2\omega_1^2}|\psi_k|^2$$
$$= 16\pi^3 \sum_{k\neq 0} \iint_D de\,d\ell\,\ell\,\frac{1}{|Q'(e_Q)|}\,\frac{1}{k^2\omega_1^3}|\psi_k|^2,$$

where we have used that $\frac{\partial e}{\partial I} = \omega_1$ owing to (A.18). Thus, the claim follow from (a) for $z = 0$. (c) We continue to use the notation and the observations from (b). Since $g \in X_0^1$, we have $u = \mathcal{T}^{-1}g \in X_0^2$. As also $g \in X^1_{\text{even}}$ and $\mathcal{T}^{-1}$ reverses the parity by Remark B.18, we get $u \in X^2_{\text{odd}}$. Accordingly, we deduce from (B.44) in Corollary B.19 that

$$(Lu, u)_{X^0} = \|\mathcal{T}u\|^2_{X^0} - (\mathcal{K}\mathcal{T}u, u)_{X^0}.$$

Now $\mathcal{T}u = \mathcal{T}\mathcal{T}^{-1}g = g$ due to Lemma B.13(d), so that

$$\|\mathcal{T}u\|^2_{X^0} = \|g\|^2_{X^0} = \|g\|^2_{L^2_{\text{sph},\frac{1}{|Q'|}}(K)} = \frac{1}{4\pi}\langle \mathcal{Q}_0\Psi, \Psi\rangle$$

by Remark B.2(a) and (4.26). Furthermore, using (B.40) from Lemma B.15 in conjunction with (b), it follows that

$$(\mathcal{K}\mathcal{T}u, u)_{X^0} = \frac{1}{4\pi}\int_{\mathbb{R}^3} |U'_{\mathcal{T}u}|^2\,dx = \frac{1}{4\pi}\int_{\mathbb{R}^3} |U'_g|^2\,dx$$
$$= \frac{1}{4\pi}\int_{\mathbb{R}^3} |\mathcal{Q}_0\Psi|^2\,dx = \frac{1}{4\pi}\|\mathcal{Q}_0\Psi\|^2_{L^2_r},$$

Altogether, this yields (4.27). □

Lemma 4.7 *Let $\mu_1 :]-\infty, \delta_1^2[\to]0, \infty[$ be defined as in Lemma 4.3(f). Then*

(a) $0 < \mu_1(0) < 1$.

(b) If $\lambda_* < \delta_1^2$ and $\lambda \in [0, \lambda_*]$, or $\lambda_* = \delta_1^2$ and $\lambda \in [0, \lambda_*[= [0, \delta_1^2[$, then $\mu_1(\lambda) \leq 1$.

(c) For $\lambda \in [0, \delta_1^2[$, let $\Psi_\lambda \in L_r^2$ denote a normalized eigenfunction of $\mathcal{Q}_\lambda$ for $\mu_1(\lambda)$. Define $\psi_\lambda(r, p_r, \ell) = |Q'(e_Q)|\,p_r\Psi_\lambda(r) \in X^0_{\text{odd}}$ and $g_\lambda = (-\mathcal{T}^2 - \lambda)^{-1}\psi_\lambda \in X^2_{\text{odd}}$. Then

$$\mu_1(\lambda) = 4\pi\,(\psi_\lambda, g_\lambda)_{X^0}$$

and

$$Lg_\lambda = (1 - \mu_1(\lambda))\psi_\lambda + \lambda g_\lambda,$$

as well as

$$(Lg_\lambda, g_\lambda)_Q = \frac{1}{4\pi}\mu_1(\lambda)(1 - \mu_1(\lambda)) + \lambda\|g_\lambda\|^2_{X^0}.$$

(d) *The function* $\mu_1 :]-\infty, \delta_1^2[\to]0, \infty[$ *is convex.*
(e) *We have*

$$\mu_1(\lambda) \le 16\pi \Bigg(\int_0^{r_Q} \int_0^{r_Q} \frac{dr}{r^2} \frac{d\tilde{r}}{\tilde{r}^2} \Bigg| \sum_{k\neq 0} \iint_D d\ell\, \ell\, de\, \mathbf{1}_{\{r_-(e,\ell)\le r,\, \tilde{r}\le r_+(e,\ell)\}} \times \frac{\omega_1(e,\ell)\,|Q'(e)|}{k^2\omega_1^2(e,\ell)-\lambda} \sin(k\theta(r,e,\ell)) \sin(k\theta(\tilde{r},e,\ell)) \Bigg|^2 \Bigg)^{1/2}.$$

Proof (a) Clearly $\mu_1(0) > 0$, since otherwise $\|\mathcal{Q}_0\| = 0$, and thus $\mathcal{Q}_0 = 0$. To show that $\mu_1(0) < 1$, let $\Psi \in L_r^2$ be given. Define $F(r) = \int_0^r \Psi(s)\,ds$ as well as $u = -\mathcal{T}^{-1}(|Q'(e_Q)|(F - F_0))$. Then $u \in X_{\text{odd}}^2$ and

$$0 \le \lambda_* \|u\|_{X^0}^2 \le (Lu, u)_{X^0} = \frac{1}{4\pi}\Big(\langle \mathcal{Q}_0\Psi, \Psi\rangle - \|\mathcal{Q}_0\Psi\|_{L_r^2}^2\Big) \tag{4.28}$$

by (1.20) and (4.27) from Lemma 4.6. As a consequence,

$$\|\mathcal{Q}_0\Psi\|_{L_r^2}^2 \le \langle \mathcal{Q}_0\Psi, \Psi\rangle \le \|\mathcal{Q}_0\Psi\|_{L_r^2}\|\Psi\|_{L_r^2}$$

implies that $\mu_1(0) = \|\mathcal{Q}_0\| \le 1$. Lastly, suppose that $\mu_1(0) = 1$. Since $\mu_1(0)$ is an eigenvalue, we have $\mathcal{Q}_0\Psi = \Psi$ for some $\Psi = \Psi(r) \neq 0$ such that $\Psi \in L_r^2$; Remark 4.2(b) implies that Ψ has its support in $[0, r_Q]$. For the corresponding u, we deduce $u = 0$ from (4.28). Therefore, (B.24), Lemma B.13(d) and Lemma B.9(b) lead to

$$\begin{aligned} 0 = \mathcal{T}^2 u &= -\mathcal{T}^2\mathcal{T}^{-1}(|Q'(e_Q)|(F - F_0)) \\ &= -\mathcal{T}(|Q'(e_Q)|(F - F_0)) = -|Q'(e)|\, p_r\, \Psi, \end{aligned}$$

which is impossible. (b) Recall from Lemma 3.18 that $\lambda_* \le \delta_1^2$. Thus, if we fix λ in one of the two cases: (i) $\lambda_* < \delta_1^2$ and $\lambda \in [0, \lambda_*]$; or (ii) $\lambda_* = \delta_1^2$ and $\lambda \in [0, \lambda_*[= [0, \delta_1^2[$, then $\lambda \in [0, \delta_1^2[$. Let $\Psi_\lambda \in L_r^2$ denote a normalized eigenfunction for $\mu_1(\lambda)$, i.e., we have $\mathcal{Q}_\lambda\Psi_\lambda = \mu_1(\lambda)\Psi_\lambda$ and $\|\Psi_\lambda\|_{L_r^2} = 1$. For $\psi_\lambda(r, p_r, \ell) = |Q'(e_Q)|\, p_r \Psi_\lambda(r)$, we get $\psi_\lambda \in X_{\text{odd}}^0$, cf. the proof of Lemma 4.6(a). Thus, $g_\lambda = (-\mathcal{T}^2 - \lambda)^{-1}\psi_\lambda \in X_{\text{odd}}^2$. Using (4.21) from Lemma 4.4, we calculate

$$\begin{aligned} \mathcal{K}\mathcal{T} g_\lambda &= \mathcal{K}\mathcal{T}(-\mathcal{T}^2 - \lambda)^{-1}\psi_\lambda = |Q'(e_Q)|\, p_r(\mathcal{Q}_\lambda\Psi_\lambda) \\ &= \mu_1(\lambda)\,|Q'(e_Q)|\, p_r\Psi_\lambda = \mu_1(\lambda)\psi_\lambda. \end{aligned}$$

In addition,

$$\mathcal{T}^2 g_\lambda = (\mathcal{T}^2 + \lambda)g_\lambda - \lambda g_\lambda = -\psi_\lambda - \lambda g_\lambda.$$

This yields

$$L g_\lambda = -\mathcal{T}^2 g_\lambda - \mathcal{K}\mathcal{T} g_\lambda = (1 - \mu_1(\lambda))\psi_\lambda + \lambda g_\lambda \tag{4.29}$$

hence in particular

$$(Lg_\lambda, g_\lambda)_Q = (Lg_\lambda, g_\lambda)_{X^0} = (1 - \mu_1(\lambda))\,(\psi_\lambda, g_\lambda)_{X^0} + \lambda\|g_\lambda\|_{X^0}^2. \tag{4.30}$$

Thus, by the Antonov stability estimate, Theorem 1.2,

$$\lambda_*\|g_\lambda\|_{X^0}^2 \le (1 - \mu_1(\lambda))\,(\psi_\lambda, g_\lambda)_{X^0} + \lambda\|g_\lambda\|_{X^0}^2,$$

so that

$$0 \le (\lambda_* - \lambda)\|g_\lambda\|_{X^0}^2 \le (1 - \mu_1(\lambda))\,(\psi_\lambda, g_\lambda)_{X^0}. \tag{4.31}$$

Now, (B.26) in Corollary B.10 yields

$$\begin{aligned}(\psi_\lambda, g_\lambda)_{X^0} &= (\psi_\lambda, (-\mathcal{T}^2 - \lambda)^{-1}\psi_\lambda)_{X^0} = ((-\mathcal{T}^2 - \lambda)^{-1}\psi_\lambda, \psi_\lambda)_{X^0} \\ &= 16\pi^3 \sum_{k\neq 0} \iint_D dI\, d\ell\, \ell\, \frac{1}{|Q'(e)|}\, \frac{|(\psi_\lambda)_k(I,\ell)|^2}{k^2\omega_1^2(I,\ell) - \lambda},\end{aligned} \tag{4.32}$$

and in particular $(\psi_\lambda, g_\lambda)_{X^0} > 0$, as otherwise $\psi_\lambda = 0$ and consequently $\Psi_\lambda = 0$, which is impossible. Hence, (4.31) shows that $\mu_1(\lambda) \le 1$.
(c) Note that due to Lemma 4.6(a),

$$\begin{aligned}\mu_1(\lambda) &= \mu_1(\lambda)\|\Psi_\lambda\|_{L_r^2}^2 = \langle \mu_1(\lambda)\Psi_\lambda, \Psi_\lambda\rangle = \langle \mathcal{Q}_\lambda\Psi_\lambda, \Psi_\lambda\rangle \\ &= 64\pi^4 \sum_{k\neq 0} \iint_D d\ell\, \ell\, de\, \frac{1}{\omega_1(e,\ell)\,|Q'(e)|}\, \frac{|(\psi_\lambda)_k(I,\ell)|^2}{k^2\omega_1^2(e,\ell) - \lambda} \\ &= 64\pi^4 \sum_{k\neq 0} \iint_D d\ell\, \ell\, dI\, \frac{1}{|Q'(e)|}\, \frac{|(\psi_\lambda)_k(I,\ell)|^2}{k^2\omega_1^2(e,\ell) - \lambda},\end{aligned}$$

and therefore the first claim follows by comparing to (4.32). The other relations are due to (4.29) and (4.30). (d) If $\lambda \in]-\infty, \delta_1^2[$ and $\Psi \in L_r^2$, then

$$\begin{aligned}\frac{d^2}{d\lambda^2}\langle \mathcal{Q}_\lambda\Psi, \Psi\rangle &= \langle \mathcal{Q}_\lambda''\Psi, \Psi\rangle \\ &= 128\pi^2 \sum_{k\neq 0} \iint_D d\ell\, \ell\, de\, \frac{\omega_1(e,\ell)\,|Q'(e)|}{(k^2\omega_1^2(e,\ell) - \lambda)^3} \\ &\qquad \times \left| \int_{r_-(e,\ell)}^{r_+(e,\ell)} \Psi(r)\sin(k\theta(r,e,\ell))\, dr \right|^2 \\ &\ge 0\end{aligned}$$

by (4.6) from Lemma 4.3(d). Thus, every function $]-\infty, \delta_1^2[\ \ni \lambda \mapsto \langle \mathcal{Q}_\lambda\Psi, \Psi\rangle$ is convex. As a consequence of (4.7), also $\mu_1(\lambda) = \sup\{\langle \mathcal{Q}_\lambda\Psi, \Psi\rangle : \|\Psi\|_{L_r^2} \le 1\}$ is convex. (e) Here, we use

$$\mu_1(\lambda) = \|\mathcal{Q}_\lambda\|_{\mathcal{B}(L_r^2)} \le \|\mathcal{Q}_\lambda\|_{\mathrm{HS}}$$

and the fact that

$$\begin{aligned}\|\mathcal{Q}_\lambda\|_{\mathrm{HS}}^2 &= 16\pi^2 \int_0^{r_Q} \int_0^{r_Q} r^2 \tilde{r}^2 \, |K_\lambda(r, \tilde{r})|^2 \, dr \, d\tilde{r} \\ &= 256\pi^2 \int_0^{r_Q} \int_0^{r_Q} \frac{dr}{r^2} \frac{d\tilde{r}}{\tilde{r}^2} \Bigg| \sum_{k \neq 0} \iint_D d\ell \, \ell \, de \, \mathbf{1}_{\{r_-(e,\ell) \le r, \tilde{r} \le r_+(e,\ell)\}} \\ &\qquad \times \frac{\omega_1(e,\ell)\, |Q'(e)|}{k^2 \omega_1^2(e,\ell) - \lambda} \sin(k\theta(r,e,\ell)) \sin(k\theta(\tilde{r},e,\ell)) \Bigg|^2,\end{aligned}$$

cf. [35, Prop. 6.36] and (4.16). □

According to Lemma 4.3(f), the monotone limits

$$\mu_{*,k} = \lim_{\lambda \to \delta_1^2 -} \mu_k(\lambda) = \sup\{\mu_k(\lambda) : \lambda \in [0, \delta_1^2[\} \in [\mu_k(0), \infty]$$

do exist. Of particular importance to us will be the number

$$\mu_* = \mu_{*,1} = \lim_{\lambda \to \delta_1^2 -} \mu_1(\lambda) = \sup\{\mu_1(\lambda) : \lambda \in [0, \delta_1^2[\} \in [\mu_1(0), \infty]. \tag{4.33}$$

Remark 4.8 If $\lambda_* = \delta_1^2$, then $\mu_* \le 1$. This follows from Lemma 4.7(b). ◇

The next result will use assumption (ω_1-3). If ω_1 attains its minimum at an interior point $(\hat{e}, \hat{\beta}) \in \mathring{D}$, then we are in the situation of (ω_1-2), and Corollary 4.16 below applies. Otherwise, since ω_1 is continuous on D, its minimum is attained on the boundary ∂D, which consists of three parts: the left side

$$\{(e, 0) : e \in [U_Q(0), e_0]\},$$

the lower boundary curve

$$\{(e, \beta) : e = e_{\min}(\beta), \beta \in [0, \beta_*]\}$$

and the upper line

$$\{(e_0, \beta) : \beta \in [0, \beta_*]\}. \tag{4.34}$$

Corollary 3.16 shows that the minimum can only be attained on this upper line (4.34) at a point $(e_0, \hat{\beta})$, and (ω_1-3) roughly concerns the case where both $\frac{\partial \omega_1}{\partial e}(e_0, \hat{\beta}) \neq 0$ and $\frac{\partial \omega_1}{\partial \beta}(e_0, \hat{\beta}) \neq 0$, which is reasonable to expect for a minimum on the boundary.

Lemma 4.9 *Suppose that (ω_1-3) is satisfied. Then*

$$\mathcal{Q}_{\delta_1^2} = \lim_{\lambda \to \delta_1^2 -} \mathcal{Q}_\lambda \tag{4.35}$$

does exist in the Hilbert-Schmidt norm $\|\cdot\|_{\mathrm{HS}}$ of L_r^2. In particular, the kernel of the symmetric and positive Hilbert-Schmidt operator $\mathcal{Q}_{\delta_1^2}$ is given by

$$K_{\delta_1^2}(r,\tilde{r}) = \frac{4}{r^2\tilde{r}^2}\sum_{k\neq 0}\iint_D d\ell\,\ell\,de\,\mathbf{1}_{\{r_-(e,\ell)\le r,\,\tilde{r}\le r_+(e,\ell)\}} \\ \times \frac{\omega_1(e,\ell)\,|Q'(e)|}{k^2\omega_1^2(e,\ell)-\delta_1^2}\,\sin(k\theta(r,e,\ell))\sin(k\theta(\tilde{r},e,\ell)),$$

and $\mu_* = \|\mathcal{Q}_{\delta_1^2}\| < \infty$. More generally, the k'th eigenvalue of $\mathcal{Q}_{\delta_1^2}$ is $\mu_{*,k}$. For $k \in \mathbb{N}$, the functions

$$\mu_k(\cdot) :]-\infty, \delta_1^2] \to]0, \infty[$$

are monotone increasing, locally Lipschitz continuous on $]-\infty,\delta_1^2[$ and continuous on $]-\infty,\delta_1^2]$, if we set $\mu_k(\delta_1^2) = \mu_{*,k}$. In particular, the μ_k are differentiable a.e. Furthermore, $\mu_1 :]-\infty,\delta_1^2] \to]0,\infty[$ is a convex function.

Proof We need to refine (4.17), from where we know that

$$\|\mathcal{Q}_\lambda - \mathcal{Q}_{\tilde{\lambda}}\|_{\mathrm{HS}}^2 \le 256\,\pi^2\Delta_1^2 \int_0^{r_Q}\int_0^{r_Q}\frac{dr}{r^2}\frac{d\tilde{r}}{\tilde{r}^2}\left(\sum_{k=1}^{\infty}\iint_D d\beta\,de\,\mathbf{1}_{\{\beta\le\hat{C}r\tilde{r}\}}\,|Q'(e)| \right.\\ \left. \times\left|\frac{1}{k^2\omega_1^2(e,\beta)-\lambda}-\frac{1}{k^2\omega_1^2(e,\beta)-\tilde{\lambda}}\right|\right)^2$$

for $\lambda, \tilde{\lambda} < \delta_1^2$ and a suitable constant $\hat{C} > 0$. Thus

$$\|\mathcal{Q}_\lambda - \mathcal{Q}_{\tilde{\lambda}}\|_{\mathrm{HS}}^2 \\ \le 512\,\pi^2\Delta_1^2 \int_0^{r_Q}\int_0^{r_Q}\frac{dr}{r^2}\frac{d\tilde{r}}{\tilde{r}^2}\left(\sum_{k=2}^{\infty}\iint_D d\beta\,de\,\mathbf{1}_{\{\beta\le\hat{C}r\tilde{r}\}}\,|Q'(e)| \right.\\ \left. \times\frac{|\lambda-\tilde{\lambda}|}{(k^2\omega_1^2(e,\beta)-\lambda)(k^2\omega_1^2(e,\beta)-\tilde{\lambda})}\right)^2 \\ + 512\,\pi^2\Delta_1^2 \int_0^{r_Q}\int_0^{r_Q}\frac{dr}{r^2}\frac{d\tilde{r}}{\tilde{r}^2}\left(\iint_D d\beta\,de\,\mathbf{1}_{\{\beta\le\hat{C}r\tilde{r}\}}\,|Q'(e)| \right.\\ \left. \times\left|\frac{1}{\omega_1^2(e,\beta)-\lambda}-\frac{1}{\omega_1^2(e,\beta)-\tilde{\lambda}}\right|\right)^2$$

$$
\begin{aligned}
&\le 8192\,\pi^2\Delta_1^2\,\delta_1^{-8}\,|\lambda-\tilde{\lambda}|^2\int_0^{r_Q}\int_0^{r_Q}\frac{dr}{r^2}\,\frac{d\bar{r}}{\bar{r}^2}\left(\sum_{k=2}^{\infty}\frac{1}{k^4}\iint_D d\beta\,de\,\mathbf{1}_{\{\beta\le\hat{C}r\bar{r}\}}\,|Q'(e)|\right)^2\\
&\quad+1024\,\pi^2\Delta_1^2\int_0^{r_Q}\int_0^{r_Q}\frac{dr}{r^2}\,\frac{d\bar{r}}{\bar{r}^2}\left(\iint_D d\beta\,de\,\mathbf{1}_{\{\beta\le\hat{C}r\bar{r}\}}\,|Q'(e)|\right.\\
&\qquad\qquad\left.\times\left|\frac{1}{\omega_1^2(e,\beta)-\lambda}-\frac{1}{\omega_1^2(e,\beta)-\delta_1^2}\right|\right)^2\\
&\quad+1024\,\pi^2\Delta_1^2\int_0^{r_Q}\int_0^{r_Q}\frac{dr}{r^2}\,\frac{d\bar{r}}{\bar{r}^2}\left(\iint_D d\beta\,de\,\mathbf{1}_{\{\beta\le\hat{C}r\bar{r}\}}\,|Q'(e)|\right.\\
&\qquad\qquad\left.\times\left|\frac{1}{\omega_1^2(e,\beta)-\tilde{\lambda}}-\frac{1}{\omega_1^2(e,\beta)-\delta_1^2}\right|\right)^2\\
&\le C|\lambda-\tilde{\lambda}|^2\int_0^{r_Q}\int_0^{r_Q}dr\,d\bar{r}\left(\int_{U_Q(0)}^{e_0}|Q'(e)|\,de\right)^2+CT(\lambda)+CT(\tilde{\lambda})\\
&\le C|\lambda-\tilde{\lambda}|^2+CT(\lambda)+CT(\tilde{\lambda}),
\end{aligned}
\tag{4.36}
$$

where

$$
T(\lambda)=\int_0^{r_Q}\int_0^{r_Q}\frac{dr}{r^2}\,\frac{d\bar{r}}{\bar{r}^2}\left(\iint_D d\beta\,de\,\mathbf{1}_{\{\beta\le\hat{C}r\bar{r}\}}\left|\frac{1}{\omega_1^2(e,\beta)-\lambda}-\frac{1}{\omega_1^2(e,\beta)-\delta_1^2}\right|\right)^2.
$$

We assert that

$$
\lim_{\lambda\to\delta_1^2-}T(\lambda)=0,\tag{4.37}
$$

and to establish this claim, we are going to use Lebesgue's dominated convergence in $\int_0^{r_Q}\int_0^{r_Q}dr\,d\bar{r}$ together with the condition

$$
|\omega_1(e,\beta)-\delta_1|\ge c_1|(e,\beta)-(e_0,\hat{\beta})|,\quad (e,\beta)\in D,\tag{4.38}
$$

from (ω_1-3), where $(e_0,\hat{\beta})\in D$ satisfies $\omega_1(e_0,\hat{\beta})=\delta_1$. Let $r,\bar{r}>0$ be fixed and define

$$
\tau(r,\bar{r})=\iint_D d\beta\,de\,\mathbf{1}_{\{\beta\le\hat{C}r\bar{r}\}}\left|\frac{1}{\omega_1^2(e,\beta)-\lambda}-\frac{1}{\omega_1^2(e,\beta)-\delta_1^2}\right|.
$$

If $(e,\beta)\in D$ are such that $\beta\le\hat{C}r\bar{r}$ and $(e,\beta)\neq(\hat{e},\hat{\beta})$, then $\omega_1(e,\beta)-\delta_1\ge\alpha>0$ for $\alpha=\alpha(e,\beta)$ by (4.38), and accordingly

$$
\begin{aligned}
\left|\frac{1}{\omega_1^2(e,\beta)-\lambda}-\frac{1}{\omega_1^2(e,\beta)-\delta_1^2}\right|&=\frac{\delta_1^2-\lambda}{(\omega_1^2(e,\beta)-\lambda)(\omega_1^2(e,\beta)-\delta_1^2)}\\
&\le\delta_1^{-2}\alpha^{-2}(\delta_1^2-\lambda)\to 0,\quad \lambda\to\delta_1^2-,
\end{aligned}
\tag{4.39}
$$

for this (e, β). On the other hand,

$$\left|\frac{1}{\omega_1^2(e,\beta)-\lambda}-\frac{1}{\omega_1^2(e,\beta)-\delta_1^2}\right| \le 2\delta_1^{-1}\frac{1}{\omega_1(e,\beta)-\delta_1} \le 2\delta_1^{-1}c_1^{-1}\frac{1}{|(e,\beta)-(e_0,\hat{\beta})|} \tag{4.40}$$

by (4.38). Next, we are going to bound

$$I(R)=\iint_D d\beta\, de\, \mathbf{1}_{\{\beta\le R\}}\frac{1}{|(e,\beta)-(e_0,\hat{\beta})|},\quad R>0. \tag{4.41}$$

<u>Case 1:</u> $\hat{\beta}>0$. If $\beta\le R\le\hat{\beta}/2$, then $|(e,\beta)-(e_0,\hat{\beta})|\ge|\beta-\hat{\beta}|\ge\hat{\beta}/2$ and hence

$$I(R)\le 2\hat{\beta}^{-1}(e_0-U_Q(0))\,R,\quad R\le\hat{\beta}/2. \tag{4.42}$$

If $R\ge\hat{\beta}/2$, then we always have

$$\begin{aligned} I(R) &\le \int_0^{\beta_*}d\beta\int_{U_Q(0)}^{e_0}de\,\frac{1}{|(e-e_0,\beta-\hat{\beta})|}\le\int_{-\hat{\beta}}^{\beta_*-\hat{\beta}}dx_2\int_0^{e_0-U_Q(0)}dx_1\,\frac{1}{|(x_1,x_2)|}\\ &\le\int_{-\beta_*}^{\beta_*}dx_2\int_0^{e_0-U_Q(0)}dx_1\,\frac{1}{|(x_1,x_2)|}\le C. \end{aligned} \tag{4.43}$$

<u>Case 2:</u> $\hat{\beta}=0$. Then

$$\begin{aligned} I(R) &\le \int_0^{R}d\beta\int_{U_Q(0)}^{e_0}de\,\frac{1}{|(e-e_0,\beta)|}\le\int_0^{R}dx_2\int_0^{e_0-U_Q(0)}dx_1\,\frac{1}{|(x_1,x_2)|}\\ &=\int_0^R dx_2\ln\left(x_1+\sqrt{x_1^2+x_2^2}\right)\Big|_{x_1=0}^{x_1=e_0-U_Q(0)}\\ &=\int_0^R dx_2\left[\ln\left(e_0-U_Q(0)+\sqrt{(e_0-U_Q(0))^2+x_2^2}\right)-\ln x_2\right]\\ &\le CR-R(\ln R-1)\le CR-R\ln R. \end{aligned} \tag{4.44}$$

Thus, if we summarize (4.39) and (4.42)–(4.44) for $R=\hat{C}r\bar{r}$, it follows that $\tau(r,\bar{r})\to 0$ as $\lambda\to\delta_1^2-$ for all $r,\bar{r}>0$. Hence, to complete the proof of (4.37), we need to obtain an integrable majorant. For, using (4.40), we can bound

$$\begin{aligned} \mathcal{I}(\lambda) &= \int_0^{r_Q}\int_0^{r_Q}\frac{dr}{r^2}\frac{d\bar{r}}{\bar{r}^2}\left(\iint_D d\beta\, de\,\mathbf{1}_{\{\beta\le\hat{C}r\bar{r}\}}\left|\frac{1}{\omega_1^2(e,\beta)-\lambda}-\frac{1}{\omega_1^2(e,\beta)-\delta_1^2}\right|\right)^2\\ &=\int_0^{r_Q}\int_0^{r_Q}\frac{dr}{r^2}\frac{d\bar{r}}{\bar{r}^2}\tau(r,\bar{r})^2\le C\int_0^{r_Q}\int_0^{r_Q}\frac{dr}{r^2}\frac{d\bar{r}}{\bar{r}^2}I(\hat{C}r\bar{r})^2. \end{aligned}$$

Case 1: $\hat{\beta} > 0$. Let $\hat{\varepsilon} = \min\{r_Q, \frac{\hat{\beta}}{2\hat{C}r_Q}\}$. If $r \leq \hat{\varepsilon}$ or $\hat{r} \leq \hat{\varepsilon}$, then $\hat{C}r\bar{r} \leq \hat{C}\hat{\varepsilon}r_Q \leq \hat{\beta}/2$, and thus we can apply (4.42) in this case, as well as (4.43) in the opposite case. Therefore, we split the integral to obtain

$$\begin{aligned}\mathcal{I}(\lambda) &\leq C \int_0^{r_Q} \int_0^{r_Q} dr\, d\bar{r}\, \mathbf{1}_{\{r \leq \hat{\varepsilon} \text{ or } \hat{r} \leq \hat{\varepsilon}\}} \frac{1}{r^2 \bar{r}^2} I(\hat{C}r\bar{r})^2 \\ &\quad + C \int_0^{r_Q} \int_0^{r_Q} dr\, d\bar{r}\, \mathbf{1}_{\{r > \hat{\varepsilon} \text{ and } \hat{r} > \hat{\varepsilon}\}} \frac{1}{r^2 \bar{r}^2} I(\hat{C}r\bar{r})^2 \\ &\leq C \int_0^{r_Q} \int_0^{r_Q} dr\, d\bar{r}\, \mathbf{1}_{\{r \leq \hat{\varepsilon} \text{ or } \hat{r} \leq \hat{\varepsilon}\}} \frac{1}{r^2 \bar{r}^2} r^2 \bar{r}^2 \\ &\quad + C \int_0^{r_Q} \int_0^{r_Q} dr\, d\bar{r}\, \mathbf{1}_{\{r > \hat{\varepsilon} \text{ and } \hat{r} > \hat{\varepsilon}\}} \frac{1}{r^2 \bar{r}^2} \\ &\leq C \int_0^{r_Q} \int_0^{r_Q} dr\, d\bar{r},\end{aligned}$$

which shows that a suitably large constant provides an integrable majorant. Case 2: $\hat{\beta} = 0$. By (4.44), we get

$$\begin{aligned}\mathcal{I}(\lambda) &\leq C \int_0^{r_Q} \int_0^{r_Q} \frac{dr}{r^2} \frac{d\bar{r}}{\bar{r}^2} I(\hat{C}r\bar{r})^2 \\ &\leq C \int_0^{r_Q} \int_0^{r_Q} \frac{dr}{r^2} \frac{d\bar{r}}{\bar{r}^2} (C\hat{C}r\bar{r} - \hat{C}r\bar{r} \ln(\hat{C}r\bar{r}))^2 \\ &\leq C \int_0^{r_Q} \int_0^{r_Q} (1 - \ln(\hat{C}r\bar{r}))^2\, dr\, d\bar{r} \\ &\leq C \int_0^{r_Q} \int_0^{r_Q} (1 + |\ln r|^2 + |\ln \bar{r}|^2)\, dr\, d\bar{r}.\end{aligned}$$

Since $1 + |\ln r|^2 + |\ln \bar{r}|^2$ is integrable on $[0, r_Q] \times [0, r_Q]$, we have found an integrable majorant also in this case. Altogether, we have shown that (4.37) is verified. At the same time, this yields $\lim_{\tilde{\lambda} \to \delta_1^2 -} T(\tilde{\lambda}) = 0$, and going back to (4.36), we deduce that (4.35) holds for an appropriate Hilbert-Schmidt operator $\mathcal{Q}_{\delta_1^2}$ on L_r^2. Since $\|\cdot\|_{\mathcal{B}(L_r^2)} \leq \|\cdot\|_{\mathrm{HS}}$, (4.35) in particular implies that $\mathcal{Q}_{\delta_1^2} = \lim_{\lambda \to \delta_1^2 -} \mathcal{Q}_\lambda$ in $\mathcal{B}(L_r^2)$. Recalling from (4.7) that $\mu_1(\lambda) = \|\mathcal{Q}_\lambda\|_{\mathcal{B}(L_r^2)}$, we can use (4.33) to get

$$\mu_* = \lim_{\lambda \to \delta_1^2 -} \mu_1(\lambda) = \lim_{\lambda \to \delta_1^2 -} \|\mathcal{Q}_\lambda\|_{\mathcal{B}(L_r^2)} = \|\mathcal{Q}_{\delta_1^2}\|_{\mathcal{B}(L_r^2)},$$

as claimed.

Let $\kappa_1 \geq \kappa_2 \geq \ldots \to 0$ denote the eigenvalues (listed according to their multiplicities) of the symmetric and positive Hilbert-Schmidt operator $\mathcal{Q}_{\delta_1^2}$. Then

$$\kappa_k = \max\left\{ \min_{\Psi \in S,\, \|\Psi\|_{L_r^2} = 1} \langle \mathcal{Q}_{\delta_1^2} \Psi, \Psi \rangle : S \subset L_r^2 \text{ is a subspace of dimension } k \right\}$$

by the Courant max-min principle. Passing to the limit $\lim_{\tilde{\lambda}\to\delta_1^2-}$ in (4.36), we derive

$$\|\mathcal{Q}_\lambda - \mathcal{Q}_{\delta_1^2}\|_{\mathrm{HS}} \le C|\lambda - \delta_1^2| + CT(\lambda)^{1/2},$$

where $\lim_{\lambda\to\delta_1^2-} T(\lambda) = 0$. Thus, if $\Psi \in L_r^2$ is such that $\|\Psi\|_{L_r^2} = 1$, then we have

$$|\langle \mathcal{Q}_\lambda \Psi, \Psi\rangle - \langle \mathcal{Q}_{\delta_1^2}\Psi, \Psi\rangle| \le \|\mathcal{Q}_\lambda - \mathcal{Q}_{\delta_1^2}\|_{\mathrm{HS}} \le C|\lambda - \delta_1^2| + CT(\lambda)^{1/2}.$$

Since the $\mu_k(\lambda)$ are also characterized by the Courant max-min principle, see (4.18), it follows that

$$|\mu_k(\lambda) - \kappa_k| \le C|\lambda - \delta_1^2| + CT(\lambda)^{1/2},$$

and accordingly $\mu_{*,k} = \lim_{\lambda\to\delta_1^2-} \mu_k(\lambda) = \kappa_k$.

The next assertion is due to the definition of $\mu_{*,k}$ and Lemma 4.3(f), whereas the convexity of μ_1 on $]-\infty, \delta_1^2]$ is a consequence of Lemma 4.7(d). □

Corollary 4.10 *Suppose that (ω_1-3) is satisfied.*

(a) There is a constant $C > 0$ such that for every $\lambda \in [0, \delta_1^2]$ and $r, \tilde{r} \in]0, r_Q]$, we have

$$|K_\lambda(r, \tilde{r})| \le \frac{C}{\tilde{r}^2}(1 + |\ln r|).$$

(b) For $\lambda \in [0, \delta_1^2[$, let $\Psi_\lambda \in L_r^2$ denote a normalized eigenfunction of $\mathcal{Q}_\lambda$ for $\mu_1(\lambda)$. Then there is a constant $C > 0$ such that for every $\lambda \in [0, \delta_1^2[$ and $r \in]0, r_Q]$, we have

$$|\Psi_\lambda(r)| \le C(1 + |\ln r|)\, \|\Psi_\lambda\|_{L_r^2}.$$

(c) For Ψ_λ as in (b), define $\psi_\lambda(r, p_r, \ell) = |Q'(e_Q)|\, p_r \Psi_\lambda(r) \in X^0_{\mathrm{odd}}$. Then there is a constant $C > 0$ such that for every $\lambda \in [0, \delta_1^2[$ and $k \in \mathbb{Z}$, we have

$$|(\psi_\lambda)_k(I, \ell)| \le C\, |Q'(e)|\, \|\Psi_\lambda\|_{L_r^2}, \quad (I, \ell) \in D,$$

where $(\psi_\lambda)_k$ are the Fourier coefficients of ψ_λ.

Proof (a) From (4.14) and similar to the argument following (4.9), we obtain with $\min\{r^2, \tilde{r}^2\} \le r^2$ and using (ω_1-3)

$$\begin{aligned}
|K_\lambda(r, \tilde{r})| &= \frac{4}{r^2\tilde{r}^2}\left|\sum_{k\ne 0} s_{k,0}(r, \tilde{r}, \lambda)\right| \\
&\le \frac{C}{r^2\tilde{r}^2}\,\mathbf{1}_{\{0\le r,\tilde{r}\le r_Q\}} \sum_{k\ne 0} \iint_D d\beta\, de\, \mathbf{1}_{\{\beta\le Cr^2\}} \frac{1}{k^2\omega_1^2(e,\beta) - \lambda} \\
&\le \frac{C}{r^2\tilde{r}^2}\,\mathbf{1}_{\{0\le r,\tilde{r}\le r_Q\}} \sum_{|k|\ge 2} \iint_D d\beta\, de\, \mathbf{1}_{\{\beta\le Cr^2\}} \frac{2}{\delta_1^2 k^2}
\end{aligned}$$

$$
\begin{aligned}
&+ \frac{C}{r^2\tilde{r}^2}\,\mathbf{1}_{\{0\le r,\tilde{r}\le r_Q\}} \iint_D d\beta\, de\, \mathbf{1}_{\{\beta\le Cr^2\}}\, \frac{1}{\omega_1^2(e,\beta)-\lambda}\\
&\le \frac{C}{r^2\tilde{r}^2}\,\mathbf{1}_{\{0\le r,\tilde{r}\le r_Q\}}\, r^2 + \frac{C}{r^2\tilde{r}^2}\,\mathbf{1}_{\{0\le r,\tilde{r}\le r_Q\}} \iint_D d\beta\, de\, \mathbf{1}_{\{\beta\le Cr^2\}}\, \frac{1}{\omega_1^2(e,\beta)-\delta_1^2}\\
&\le \frac{C}{\tilde{r}^2}\,\mathbf{1}_{\{0\le r,\tilde{r}\le r_Q\}} + \frac{C}{r^2\tilde{r}^2}\,\mathbf{1}_{\{0\le r,\tilde{r}\le r_Q\}} \iint_D d\beta\, de\, \mathbf{1}_{\{\beta\le Cr^2\}}\, \frac{1}{|(e,\beta)-(e_0,\hat{\beta})|}.
\end{aligned}
$$

By means of the function I from (4.41), this can be expressed as

$$
|K_\lambda(r,\tilde{r})| \le \frac{C}{\tilde{r}^2}\,\mathbf{1}_{\{0\le r,\tilde{r}\le r_Q\}} + \frac{C}{r^2\tilde{r}^2}\,\mathbf{1}_{\{0\le r,\tilde{r}\le r_Q\}}\, I(\hat{C}r^2)
$$

for certain constants $C, \hat{C} > 0$ that only depend on Q. Once again, we distinguish two cases. <u>Case 1:</u> $\hat{\beta} > 0$. If $r^2 \le \frac{\hat{\beta}}{2\hat{C}}$, then we can apply (4.42) to get

$$
|K_\lambda(r,\tilde{r})| \le \frac{C}{\tilde{r}^2}\,\mathbf{1}_{\{0\le r,\tilde{r}\le r_Q\}}.
$$

On the other hand, if $r^2 \ge \frac{\hat{\beta}}{2\hat{C}}$, then (4.43) leads to

$$
|K_\lambda(r,\tilde{r})| \le \frac{C}{\tilde{r}^2}\,\mathbf{1}_{\{0\le r,\tilde{r}\le r_Q\}} + \frac{C}{r^2\tilde{r}^2}\,\mathbf{1}_{\{(\frac{\hat{\beta}}{2\hat{C}})^{1/2}\le r\le r_Q,\,0\le\tilde{r}\le r_Q\}} \le \frac{C}{\tilde{r}^2}\,\mathbf{1}_{\{0\le r,\tilde{r}\le r_Q\}}.
$$

<u>Case 2:</u> $\hat{\beta} = 0$. Here, we invoke (4.44) to deduce that

$$
|K_\lambda(r,\tilde{r})| \le \frac{C}{\tilde{r}^2}\,\mathbf{1}_{\{0\le r,\tilde{r}\le r_Q\}} + \frac{C}{r^2\tilde{r}^2}\,\mathbf{1}_{\{0\le r,\tilde{r}\le r_Q\}}\,|\hat{C}r^2\ln(\hat{C}r^2)| \le \frac{C}{\tilde{r}^2}\,\mathbf{1}_{\{0\le r,\tilde{r}\le r_Q\}}\,(1+|\ln r|).
$$

Hence, in any case, we arrive at the bound

$$
|K_\lambda(r,\tilde{r})| \le \frac{C}{\tilde{r}^2}\,(1+|\ln r|),
$$

as desired. (b) Using (a), we obtain from (4.15) and Remark 4.2(b)

$$
\begin{aligned}
\mu_1(0)|\Psi_\lambda(r)| \le \mu_1(\lambda)|\Psi_\lambda(r)| = |(\mathcal{Q}_\lambda\Psi_\lambda)(r)| &= 4\pi \left| \int_0^{r_Q} \tilde{r}^2\, K_\lambda(r,\tilde{r})\, \Psi_\lambda(\tilde{r})\, d\tilde{r} \right|\\
&\le C(1+|\ln r|) \int_0^{r_Q} |\Psi_\lambda(\tilde{r})|\, d\tilde{r},
\end{aligned}
$$

so that

$$
|\Psi_\lambda(r)| \le C_*(1+|\ln r|) \int_0^{r_Q} |\Psi_\lambda(\tilde{r})|\, d\tilde{r} \tag{4.45}
$$

for a certain constant $C_* > 0$ that only depends on Q. Fix $a_* \in]0, r_Q[$ such that $\int_0^{a_*}(1 + |\ln r|)\, dr \le \frac{1}{2C_*}$. Then

$$\int_0^{a_*} |\Psi_\lambda(r)|\, dr \le C_* \int_0^{a_*} (1 + |\ln r|)\, dr \int_0^{r_Q} |\Psi_\lambda(\tilde{r})|\, d\tilde{r} \le \frac{1}{2} \int_0^{r_Q} |\Psi_\lambda(\tilde{r})|\, d\tilde{r}$$
$$= \frac{1}{2} \int_0^{a_*} |\Psi_\lambda(\tilde{r})|\, d\tilde{r} + \frac{1}{2} \int_{a_*}^{r_Q} |\Psi_\lambda(\tilde{r})|\, d\tilde{r}$$

entails $\int_0^{a_*} |\Psi_\lambda(\tilde{r})|\, d\tilde{r} \le \int_{a_*}^{r_Q} |\Psi_\lambda(\tilde{r})|\, d\tilde{r}$. Going back to (4.45), it follows by means of Hölder's inequality that

$$|\Psi_\lambda(r)| \le C_*(1 + |\ln r|)\Big[\int_0^{a_*} |\Psi_\lambda(\tilde{r})|\, d\tilde{r} + \int_{a_*}^{r_Q} |\Psi_\lambda(\tilde{r})|\, d\tilde{r}\Big] \le 2C_*(1 + |\ln r|) \int_{a_*}^{r_Q} |\Psi_\lambda(\tilde{r})|\, d\tilde{r}$$
$$\le \frac{2C_*}{a_*}(1 + |\ln r|) \int_{a_*}^{r_Q} \tilde{r}|\Psi_\lambda(\tilde{r})|\, d\tilde{r} \le \frac{2C_* r_Q^{1/2}}{\sqrt{4\pi}\, a_*}(1 + |\ln r|)\, \|\Psi_\lambda\|_{L_r^2},$$

from where a suitable $C > 0$ can be read off. (c) Owing to (4.24), Theorem 3.5 and (b), we have

$$|(\psi_\lambda)_k(I, \ell)| = \frac{1}{\pi} |Q'(e)|\, \omega_1(e, \ell) \left| \int_{r_-(e,\ell)}^{r_+(e,\ell)} \Psi_\lambda(\tilde{r}) \sin(k\theta(\tilde{r}, e, \ell))\, d\tilde{r} \right|$$
$$\le C\, |Q'(e)| \int_0^{r_Q} |\Psi_\lambda(\tilde{r})|\, d\tilde{r}$$
$$\le C\, |Q'(e)| \Big(\int_0^{r_Q} (1 + |\ln \tilde{r}|)\, d\tilde{r} \Big) \|\Psi_\lambda\|_{L_r^2} \le C\, |Q'(e)|\, \|\Psi_\lambda\|_{L_r^2},$$

which completes the proof. □

Corollary 4.11 *Suppose that (ω_1-3) is satisfied. Let $(\lambda_j) \subset [0, \delta_1^2[$ be such that $\lim_{j\to\infty} \lambda_j = \delta_1^2$. For $j \in \mathbb{N}$, let $\Psi_j \in L_r^2$ denote a normalized eigenfunction of $\mathcal{Q}_{\lambda_j}$ for $\mu_1(\lambda_j)$. Furthermore, define $\psi_j(r, p_r, \ell) = |Q'(e_Q)|\, p_r \Psi_j(r) \in X^0_{\text{odd}}$. Then there is a subsequence $j' \to \infty$ so that*

$$\Psi_* = \lim_{j'\to\infty} \Psi_{j'}$$

does exist in L_r^2 and

$$\psi_* = \lim_{j'\to\infty} \psi_{j'}$$

does exist in X^0, where $\psi_*(r, p_r, \ell) = |Q'(e_Q)|\, p_r \Psi_*(r)$. In addition, $\|\Psi_*\|_{L_r^2} = 1$ and $\mathcal{Q}_{\delta_1^2} \Psi_* = \mu_* \Psi_*$ as well as $\mu_* = \|\mathcal{Q}_{\delta_1^2}\|$.

Proof Recall from (4.33) and Lemma 4.9 that $\mu_* \in [\mu_1(0), \infty[$. For $j, k \in \mathbb{N}$, we can estimate

$$
\begin{aligned}
\mu_* \|\Psi_j - \Psi_k\|_{L_r^2} &\le (\mu_* - \mu_1(\lambda_j))\|\Psi_j\|_{L_r^2} + \|\mathcal{Q}_{\lambda_j}\Psi_j - \mathcal{Q}_{\lambda_k}\Psi_k\|_{L_r^2} \\
&\quad + (\mu_* - \mu_1(\lambda_k))\|\Psi_k\|_{L_r^2} \\
&\le (\mu_* - \mu_1(\lambda_j)) + (\mu_* - \mu_1(\lambda_k)) + \|(\mathcal{Q}_{\lambda_j} - \mathcal{Q}_{\delta_1^2})\Psi_j\|_{L_r^2} \\
&\quad + \|\mathcal{Q}_{\delta_1^2}\Psi_j - \mathcal{Q}_{\delta_1^2}\Psi_k\|_{L_r^2} + \|(\mathcal{Q}_{\delta_1^2} - \mathcal{Q}_{\lambda_k})\Psi_k\|_{L_r^2} \\
&\le (\mu_* - \mu_1(\lambda_j)) + (\mu_* - \mu_1(\lambda_k)) + \|\mathcal{Q}_{\lambda_j} - \mathcal{Q}_{\delta_1^2}\|_{\mathcal{B}(L_r^2)} \\
&\quad + \|\mathcal{Q}_{\delta_1^2}\Psi_j - \mathcal{Q}_{\delta_1^2}\Psi_k\|_{L_r^2} + \|\mathcal{Q}_{\delta_1^2} - \mathcal{Q}_{\lambda_k}\|_{\mathcal{B}(L_r^2)} \\
&\le (\mu_* - \mu_1(\lambda_j)) + (\mu_* - \mu_1(\lambda_k)) + \|\mathcal{Q}_{\delta_1^2} - \mathcal{Q}_{\lambda_j}\|_{\mathrm{HS}} \\
&\quad + \|\mathcal{Q}_{\delta_1^2} - \mathcal{Q}_{\lambda_k}\|_{\mathrm{HS}} + \|\mathcal{Q}_{\delta_1^2}\Psi_j - \mathcal{Q}_{\delta_1^2}\Psi_k\|_{L_r^2}.
\end{aligned}
\tag{4.46}
$$

According to Lemma 4.9, we have $\lim_{\lambda\to\delta_1^2-} \|\mathcal{Q}_{\delta_1^2} - \mathcal{Q}_\lambda\|_{\mathrm{HS}} = 0$ and $\mathcal{Q}_{\delta_1^2} : L_r^2 \to L_r^2$ is a Hilbert-Schmidt operator, and hence compact. Thus, since $\|\Psi_j\|_{L_r^2} = 1$, the set $\{\mathcal{Q}_{\delta_1^2}\Psi_j : j \in \mathbb{N}\} \subset L_r^2$ is relatively compact. Therefore, there is a subsequence $j' \to \infty$ and a function $\hat{\Psi} \in L_r^2$ so that $\lim_{j'\to\infty} \mathcal{Q}_{\delta_1^2}\Psi_{j'} = \hat{\Psi}$ in L_r^2. From (4.46), we deduce that along the subsequence

$$
\begin{aligned}
\mu_* \|\Psi_{j'} - \Psi_{k'}\|_{L_r^2} &\le (\mu_* - \mu_1(\lambda_{j'})) + (\mu_* - \mu_1(\lambda_{k'})) \\
&\quad + \|\mathcal{Q}_{\delta_1^2} - \mathcal{Q}_{\lambda_{j'}}\|_{\mathrm{HS}} + \|\mathcal{Q}_{\delta_1^2} - \mathcal{Q}_{\lambda_{k'}}\|_{\mathrm{HS}} \\
&\quad + \|\mathcal{Q}_{\delta_1^2}\Psi_{j'} - \mathcal{Q}_{\delta_1^2}\Psi_{k'}\|_{L_r^2} \to 0, \quad j', k' \to \infty.
\end{aligned}
$$

As a consequence, $\Psi_* = \lim_{j'\to\infty} \Psi_{j'}$ does exist in L_r^2. Since

$$
\|\psi_{j'} - \psi_{k'}\|_{X^0} \le \rho_Q(0)^{1/2} \|\Psi_{j'} - \Psi_{k'}\|_{L_r^2}
$$

by (4.23), also $\psi_* = \lim_{j'\to\infty} \psi_{j'}$ does exist in X^0, where $\psi_*(r, p_r, \ell) = |Q'(e_Q)|\, p_r \Psi_*(r)$ a.e. Lastly,

$$
\begin{aligned}
\|\mathcal{Q}_{\delta_1^2}\Psi_* - \mu_*\Psi_*\|_{L_r^2} &\le \|\mathcal{Q}_{\delta_1^2}(\Psi_* - \Psi_{j'})\|_{L_r^2} + \|(\mathcal{Q}_{\delta_1^2} - \mathcal{Q}_{\lambda_{j'}})\Psi_{j'}\|_{L_r^2} \\
&\quad + (\mu_* - \mu_1(\lambda_{j'}))\, \|\Psi_{j'}\|_{L_r^2} + \mu_* \|\Psi_{j'} - \Psi_*\|_{L_r^2} \\
&\le 2\mu_* \|\Psi_* - \Psi_{j'}\|_{L_r^2} + \|\mathcal{Q}_{\delta_1^2} - \mathcal{Q}_{\lambda_{j'}}\|_{\mathcal{B}(L_r^2)} \\
&\quad + (\mu_* - \mu_1(\lambda_{j'})) \to 0, \quad j' \to \infty,
\end{aligned}
$$

implies that $\mathcal{Q}_{\delta_1^2}\Psi_* = \mu_*\Psi_*$. □

The following criterion is useful for proving that δ_1^2 is an eigenvalue of L in the case where $\mu_* = 1$.

Lemma 4.12 *Suppose that (ω_1-3) is satisfied and that $\mu_* = 1$. Let $(\lambda_j) \subset [0, \delta_1^2[$ be such that $\lim_{j\to\infty} \lambda_j = \delta_1^2$. For $j \in \mathbb{N}$, let $\Psi_j \in L_r^2$ denote a normalized eigenfunction of $\mathcal{Q}_{\lambda_j}$ for $\mu_1(\lambda_j)$. Furthermore, define $\psi_j(r, p_r, \ell) = |Q'(e_Q)|\, p_r \Psi_j(r) \in X^0_{\mathrm{odd}}$*

and $g_j = (-\mathcal{T}^2 - \lambda_j)^{-1}\psi_j \in X^2_{\text{odd}}$. *If* $(g_j) \subset X^0 = L^2_{\text{sph}, \frac{1}{|Q'|}}(K)$ *is bounded, then* δ_1^2 *is an eigenvalue of* L.

Proof From (4.21), we deduce

$$\begin{aligned} \mathcal{K}\mathcal{T}g_j &= \mathcal{K}\mathcal{T}(-\mathcal{T}^2 - \lambda_j)^{-1}\psi_j = |Q'(e_Q)|\, p_r(\mathcal{Q}_{\lambda_j}\Psi_j) \\ &= \mu_1(\lambda_j)\,|Q'(e_Q)|\, p_r\Psi_j = \mu_1(\lambda_j)\psi_j. \end{aligned} \tag{4.47}$$

Since $-\mathcal{T}^2 g_j = \psi_j + \lambda_j g_j$, using Corollary B.19, this implies that for every odd function $h \in X^{00}$, we have

$$\begin{aligned} (g_j, Lh)_{X^0} &= (Lg_j, h)_{X^0} = (-\mathcal{T}^2 g_j, h)_{X^0} - (\mathcal{K}\mathcal{T}g_j, h)_{X^0} \\ &= (\psi_j + \lambda_j g_j, h)_{X^0} - \mu_1(\lambda_j)(\psi_j, h)_{X^0} \\ &= \lambda_j(g_j, h)_{X^0} + (1 - \mu_1(\lambda_j))(\psi_j, h)_{X^0}. \end{aligned} \tag{4.48}$$

Next, from (4.20), we get $\|\psi_j\|_{X^0} \le \rho_Q(0)^{1/2}\, \|\Psi_j\|_{L^2_r} \le \rho_Q(0)^{1/2}$. Since $\lim_{j\to\infty} \mu_1(\lambda_j) = \mu_* = 1$, this yields in particular that

$$\lim_{j\to\infty} [(1 - \mu_1(\lambda_j))(\psi_j, h)_{X^0}] = 0. \tag{4.49}$$

By assumption, $(g_j) \subset X^0$ is bounded. Hence, passing to a subsequence (that is not relabeled), we may assume that $g_j \rightharpoonup g_*$ weakly in X^0 as $j \to \infty$ for some function $g_* \in X^0_{\text{odd}}$. Suppose that $g_* = 0$. Then $g_j \rightharpoonup 0$ weakly in X^0 implies that $\mathcal{K}\mathcal{T}g_j \rightharpoonup 0$ weakly in X^0 as $j \to \infty$, by Lemma B.15(d). Due to (4.47), this yields $\psi_j \rightharpoonup 0$ weakly in X^0 as $j \to \infty$. On the other hand, by Corollary 4.11, we may pass to a subsequence $j' \to \infty$ so that $\Psi_* = \lim_{j'\to\infty} \Psi_{j'}$ does exist in L^2_r and $\psi_* = \lim_{j'\to\infty} \psi_{j'}$ does exist in X^0 as strong limits; the functions are linked via $\psi_*(r, p_r, \ell) = |Q'(e_Q)|\, p_r\Psi_*(r)$. But then we must have $\psi_* = 0$ and accordingly $\Psi_* = 0$, which however contradicts $\|\Psi_*\|_{L^2_r} = 1$, cf. Corollary 4.11. As a consequence, it follows that $g_* \in X^0_{\text{odd}}$ satisfies $g_* \ne 0$. Passing to the limit $j \to \infty$ in (4.48) and using (4.49), we moreover infer that $(g_*, Lh)_{X^0} = \delta_1^2(g_*, h)_{X^0}$ for every odd function $h \in X^{00}$. From Lemma C.11, we conclude that $g_* \in X^2_{\text{odd}}$ and $Lg_* = \delta_1^2 g_*$, which completes the proof. □

4.2 Relating μ_* to the Fact That λ_* is an Eigenvalue of L

Theorem 4.13 *We have*

$$\mu_* > 1 \iff \lambda_* < \delta_1^2.$$

In this case, $\mu_1(\lambda_*) = 1$ *and* λ_* *is an eigenvalue of* L.

Proof If $\mu_* > 1$, then $\lambda_* = \delta_1^2$ is impossible by Remark 4.8, so that we must have $\lambda_* < \delta_1^2$. Conversely, suppose that $\lambda_* < \delta_1^2$ holds. Then, according to Theorem C.8, λ_* is an eigenvalue of L. Let $u_* \in X^2_{\text{odd}}$ be an eigenfunction of L for the eigenvalue λ_*. Using Theorem 4.5(a), it follows that $\Psi_* = U'_{\mathcal{T}u_*} \in L^2_r$ for $r \in [0, r_Q]$ is an eigenfunction of $\mathcal{Q}_{\lambda_*}$ for the eigenvalue 1. Since $\mu_1(\lambda_*)$ is the largest eigenvalue of $\mathcal{Q}_{\lambda_*}$, we get $\mu_1(\lambda_*) \geq 1$. On the other hand, $\mu_1(\lambda_*) \leq 1$ by Lemma 4.7(b), and hence $\mu_1(\lambda_*) = 1$. It remains to show that $\mu_* > 1$. Suppose that on the contrary $\mu_* \leq 1$ is satisfied. For $\lambda \in [\lambda_*, \delta_1^2[$, the monotonicity of μ_1 then yields $1 = \mu_1(\lambda_*) \leq \mu_1(\lambda) \leq \mu_* \leq 1$, which means that $\mu_1(\lambda) = 1$ is constant for $\lambda \in [\lambda_*, \delta_1^2[$. Take $\lambda_* \leq \tilde{\lambda} < \lambda < \delta_1^2$. and let $\Psi_{\tilde{\lambda}}$ denote a normalized eigenfunction for $\mu_1(\tilde{\lambda})$. Then, by (4.19) and (4.7),

$$1 = \mu_1(\tilde{\lambda}) = \langle \mathcal{Q}_{\tilde{\lambda}}\Psi_{\tilde{\lambda}}, \Psi_{\tilde{\lambda}}\rangle \leq \langle \mathcal{Q}_{\lambda}\Psi_{\tilde{\lambda}}, \Psi_{\tilde{\lambda}}\rangle \leq \|\mathcal{Q}_{\lambda}\| \, \|\Psi_{\tilde{\lambda}}\|^2_{L^2_r} = \mu_1(\lambda) = 1,$$

which means that $\langle \mathcal{Q}_{\lambda}\Psi_{\tilde{\lambda}}, \Psi_{\tilde{\lambda}}\rangle = 1$ for all $\lambda_* \leq \tilde{\lambda} < \lambda < \delta_1^2$. Differentiating this relation w.r.t. to λ at a fixed $\lambda_0 \in]\tilde{\lambda}, \delta_1^2[$, it follows from (4.6) that

$$\begin{aligned} 0 &= \langle \mathcal{Q}'_{\lambda_0}\Psi_{\tilde{\lambda}}, \Psi_{\tilde{\lambda}}\rangle \\ &= 64\pi^2 \sum_{k\neq 0} \iint_D d\ell\,\ell\,de\, \frac{\omega_1(e,\ell)\,|Q'(e)|}{(k^2\omega_1^2(e,\ell)-\lambda_0)^2} \left| \int_{r_-(e,\ell)}^{r_+(e,\ell)} \Psi_{\tilde{\lambda}}(r)\sin(k\theta(r,e,\ell))\,dr \right|^2 \end{aligned}$$

for all $\tilde{\lambda} \in [\lambda_*, \lambda_0[$. Defining $\psi_{\tilde{\lambda}}(r, p_r, \ell) = |Q'(e_Q)|\, p_r \Psi_{\tilde{\lambda}}(r) \in X^0_{\text{odd}}$, then (4.24) implies that $(\psi_{\tilde{\lambda}})_k = 0$ for $k \in \mathbb{Z}$, so that $\psi_{\tilde{\lambda}} = 0$ and in turn $\Psi_{\tilde{\lambda}} = 0$, which however is impossible. □

Theorem 4.14 *Suppose that (ω_1-1) is satisfied. If $\mu_* < 1$, then $\lambda_* = \delta_1^2$ and this is not an eigenvalue of L.*

Proof The approach is inspired by [20, Section 2]. Since $\lambda_* \leq \delta_1^2$ by Lemma 3.18, $\mu_* < 1$ together with Theorem 4.13 implies $\lambda_* = \delta_1^2$. Now suppose on the contrary that there is a function $u_* \in X^2_{\text{odd}}$ such that $\|u_*\|_{X^0} = 1$ and $Lu_* = \delta_1^2 u_*$. If we define $\Psi_*(r) = U'_{\mathcal{T}u_*}(r)$ for $r \in [0, r_Q]$, then $\Psi_* \in L^2_r$ and (B.37) yields $\mathcal{KT}u_* = |Q'(e_Q)|\, p_r\, U'_{\mathcal{T}u_*}(r) = |Q'(e_Q)|\, p_r\, \Psi_*(r)$. Hence, for $a > 0$ and $b \in \mathbb{R}$, we get

$$(-\mathcal{T}^2 - (\delta_1^2 - a + ib))u_* = \mathcal{KT}u_* + (a - ib)u_*.$$

Since $z = \delta_1^2 - a + ib \in \Omega$, it follows from (4.21) that

$$\begin{aligned} |Q'(e_Q)|\, p_r(\mathcal{Q}_{\delta_1^2-a+ib}\Psi_*) &= \mathcal{KT}(-\mathcal{T}^2 - (\delta_1^2 - a + ib))^{-1}(\mathcal{KT}u_*) \\ &= \mathcal{KT}u_* - (a - ib)\,\mathcal{KT}(-\mathcal{T}^2 - (\delta_1^2 - a + ib))^{-1}u_* \\ &= |Q'(e_Q)|\, p_r\, \Psi_* \\ &\quad -(a - ib)\,|Q'(e_Q)|\, p_r\, U'_{\mathcal{T}(-\mathcal{T}^2-(\delta_1^2-a+ib))^{-1}u_*}, \end{aligned}$$

and therefore,

$$\mathcal{Q}_{\delta_1^2-a+ib}\Psi_* = \Psi_* - (a-ib)\, U'_{\mathcal{T}(-\mathcal{T}^2-(\delta_1^2-a+ib))^{-1}u_*}. \tag{4.50}$$

We claim that if $a = a(\varepsilon) \to 0^+$ and $b = b(\varepsilon) \to 0$ as $\varepsilon \to 0$, then

$$(a-ib)\, U'_{\mathcal{T}(-\mathcal{T}^2-(\delta_1^2-a+ib))^{-1}u_*} \to 0, \quad \varepsilon \to 0^+, \tag{4.51}$$

in L^2_r. For, we can invoke Corollary B.16 as well as (B.25) to deduce

$$\begin{aligned}
&\|(a-ib)\, U'_{\mathcal{T}(-\mathcal{T}^2-(\delta_1^2-a+ib))^{-1}u_*}\|^2_{L^2_r}\\
&\le 16\pi^2 \rho_Q(0)\,(a^2+b^2)\,\|(-\mathcal{T}^2-(\delta_1^2-a+ib))^{-1}u_*\|^2_{X^0}\\
&= 256\pi^5 \rho_Q(0)\,(a^2+b^2) \sum_{k\neq 0} \iint_D dI\, d\ell\, \ell \,\frac{1}{|Q'(e)|}\, \frac{|(u_*)_k(I,\ell)|^2}{|k^2\omega_1^2(I,\ell)-(\delta_1^2-a+ib)|^2}\\
&= 256\pi^5 \rho_Q(0)\,(a^2+b^2) \sum_{k\neq 0} \iint_D dI\, d\ell\, \ell \,\frac{1}{|Q'(e)|}\, \frac{|(u_*)_k(I,\ell)|^2}{(k^2\omega_1^2(I,\ell)-\delta_1^2+a)^2+b^2}.
\end{aligned}$$

If $|k| \ge 2$, then $k^2\omega_1^2(I,\ell) - \delta_1^2 + a \ge (k^2-1)\delta_1^2 \ge 3\delta_1^2$. Thus,

$$\begin{aligned}
&\|(a-ib)\, U'_{\mathcal{T}(-\mathcal{T}^2-(\delta_1^2-a+ib))^{-1}u_*}\|^2_{L^2_r}\\
&\le 2\pi^2 \delta_1^{-4} \rho_Q(0)\,(a^2+b^2)\,\|u_*\|^2_{X^0}\\
&+ 512\pi^5 \rho_Q(0) \iint_D dI\, d\ell\, \ell\, \frac{a^2+b^2}{(\omega_1^2(I,\ell)-\delta_1^2+a)^2+b^2}\, \phi_1(I,\ell)
\end{aligned}$$

for $\phi_1(I,\ell) = \frac{|(u_*)_1(I,\ell)|^2}{|Q'(e)|} \in L^1(D)$. For almost all $(I,\ell) \in D$, we know from hypothesis (ω_1-1) that $\omega_1(I,\ell) \neq \delta_1$, i.e., $\omega_1(I,\ell) > \delta_1$. For such an (I,ℓ), we have

$$\frac{a^2+b^2}{(\omega_1^2(I,\ell)-\delta_1^2+a)^2+b^2} \le \frac{a^2+b^2}{(\omega_1^2(I,\ell)-\delta_1^2)^2+b^2} \to 0, \quad \varepsilon \to 0.$$

Since always

$$\frac{a^2+b^2}{(\omega_1^2(I,\ell)-\delta_1^2+a)^2+b^2} \le 1,$$

it follows by using Lebesgue's dominated convergence theorem that indeed (4.51) is verified. Going back to (4.50), this entails that

$$\lim_{\varepsilon\to 0} \mathcal{Q}_{\delta_1^2-a+ib}\Psi_* = \Psi_* \quad \text{in} \quad L^2_r. \tag{4.52}$$

Next, we are going to compare $\mathcal{Q}_{\delta_1^2-a+ib}$ to $\mathcal{Q}_{\delta_1^2-a}$. Here, we find

$$
\begin{aligned}
|\langle \mathcal{Q}_{\delta_1^2-a+ib}\Psi_*, \Psi_*\rangle - \langle \mathcal{Q}_{\delta_1^2-a}\Psi_*, \Psi_*\rangle| &= |\langle (\mathcal{Q}_{\delta_1^2-a+ib} - \mathcal{Q}_{\delta_1^2-a})\Psi_*, \Psi_*\rangle| \\
&= 64\pi^2 \Bigg| \sum_{k\neq 0} \iint_D d\ell\, \ell\, de
\end{aligned}
$$

$$
\times \left[\frac{\omega_1(e,\ell)\,|Q'(e)|}{k^2\omega_1^2(e,\ell) - (\delta_1^2 - a - ib)} - \frac{\omega_1(e,\ell)\,|Q'(e)|}{k^2\omega_1^2(e,\ell) - (\delta_1^2 - a)} \right]
$$

$$
\times \int_0^{r_Q}\int_0^{r_Q} dr\, d\tilde{r}\, \Psi_*(r)\,\overline{\Psi_*(\tilde{r})}\, \mathbf{1}_{\{r_-(e,\ell)\le r,\, \tilde{r}\le r_+(e,\ell)\}} \sin(k\theta(r,e,\ell)) \sin(k\theta(\tilde{r},e,\ell)) \Bigg|,
$$

cf. Lemma 4.3(d) and the definition of $\mathcal{Q}_z$. Using (4.8), (4.12) and similar arguments as in the proof of Lemma 4.3(a), we obtain

$$
\begin{aligned}
&|\langle \mathcal{Q}_{\delta_1^2-a+ib}\Psi_*, \Psi_*\rangle - \langle \mathcal{Q}_{\delta_1^2-a}\Psi_*, \Psi_*\rangle| \\
&\le C|b| \sum_{k\neq 0} \iint_D d\ell\, \ell\, de\, |Q'(e)|\, \frac{1}{|k^2\omega_1^2(e,\ell) - (\delta_1^2 - a - ib)||k^2\omega_1^2(e,\ell) - (\delta_1^2 - a)|} \\
&\qquad \times \int_0^{r_Q}\int_0^{r_Q} dr\, d\tilde{r}\, |\Psi_*(r)|\, |\Psi_*(\tilde{r})|\, \mathbf{1}_{\{\beta \le C r\tilde{r}\}}.
\end{aligned}
$$

Now

$$
\begin{aligned}
|k^2\omega_1^2(e,\ell) - (\delta_1^2 - a - ib)|^2 &= (k^2\omega_1^2(e,\ell) - \delta_1^2 + a)^2 + b^2 \ge a^2, \\
|k^2\omega_1^2(e,\ell) - (\delta_1^2 - a)|^2 &= (k^2\omega_1^2(e,\ell) - \delta_1^2 + a)^2 \ge a^2,
\end{aligned}
$$

so that

$$
\begin{aligned}
&|\langle \mathcal{Q}_{\delta_1^2-a+ib}\Psi_*, \Psi_*\rangle - \langle \mathcal{Q}_{\delta_1^2-a}\Psi_*, \Psi_*\rangle| \\
&\le C\,\frac{|b|}{a^2} \sum_{k\neq 0} \int_0^{r_Q}\int_0^{r_Q} dr\, d\tilde{r}\, |\Psi_*(r)|\, |\Psi_*(\tilde{r})|\, r\tilde{r} \left(\int_{U_Q(0)}^{e_0} |Q'(e)|\, de \right) \\
&\le C\,\frac{|b|}{a^2}\, \|\Psi_*\|_{L_r^2}^2.
\end{aligned}
$$

So if we take for instance $b(\varepsilon) = \varepsilon^3$ and $a(\varepsilon) = \varepsilon$, it follows that

$$
\lim_{\varepsilon\to 0} |\langle \mathcal{Q}_{\delta_1^2-\varepsilon+i\varepsilon^3}\Psi_*, \Psi_*\rangle - \langle \mathcal{Q}_{\delta_1^2-\varepsilon}\Psi_*, \Psi_*\rangle| = 0.
$$

Using also (4.52), we conclude that

$$
\lim_{\varepsilon\to 0} \langle \mathcal{Q}_{\delta_1^2-\varepsilon}\Psi_*, \Psi_*\rangle = \|\Psi_*\|_{L_r^2}^2.
$$

As a consequence,

$$\begin{aligned}\|\Psi_*\|_{L_r^2}^2 &= \lim_{\varepsilon\to 0} \langle \mathcal{Q}_{\delta_1^2-\varepsilon}\Psi_*, \Psi_*\rangle \le \limsup_{\varepsilon\to 0} \|\mathcal{Q}_{\delta_1^2-\varepsilon}\| \, \|\Psi_*\|_{L_r^2}^2 \\ &= \limsup_{\varepsilon\to 0} \mu_1(\delta_1^2-\varepsilon)\, \|\Psi_*\|_{L_r^2}^2 \le \mu_* \, \|\Psi_*\|_{L_r^2}^2.\end{aligned}$$

Since $\mu_* < 1$, this enforces $\Psi_* = 0$ and hence $\mathcal{K}\mathcal{T}u_* = 0$. Therefore, $-\mathcal{T}^2 u_* = -\mathcal{T}^2 u_* - \mathcal{K}\mathcal{T}u_* = Lu_* = \delta_1^2 u_*$, i.e., δ_1^2 is an eigenvalue of $-\mathcal{T}^2$ with eigenfunction u_*. However, this contradicts Lemma B.12. □

The next result clarifies the case where $\mu_* = 1$.

Theorem 4.15 *Suppose that (ω_1-3) is satisfied and that $\mu_* = 1$. Then $\lambda_* = \delta_1^2$, and this is an eigenvalue of L if and only if*

$$\|\mu_1'\|_{L^\infty(]-\infty,\delta_1^2[)} < \infty \tag{4.53}$$

holds.

Proof Since $\lambda_* \le \delta_1^2$ by Lemma 3.18, $\mu_* = 1$ together with Theorem 4.13 imply $\lambda_* = \delta_1^2$. For the actual proof, recall from Lemma 4.3(f) that $\mu_1(\cdot) :]-\infty, \delta_1^2[\to]0, \infty[$ is differentiable a.e., so (4.53) makes sense.

First, we consider the case where δ_1^2 is an eigenvalue of L. Let $u_* \in X^2_{\text{odd}}$ be such that $\|u_*\|_{X^0} = 1$ and $Lu_* = \delta_1^2 u_*$. If we define $\Psi_*(r) = U'_{\mathcal{T}u_*}(r)$ for $r \in [0, r_Q]$, then $\Psi_* \in L_r^2$ and (B.37) implies that $\mathcal{K}\mathcal{T}u_* = |Q'(e_Q)|\, p_r\, U'_{\mathcal{T}u_*}(r) = |Q'(e_Q)|\, p_r\, \Psi_*(r) =: \psi_* \in X^0_{\text{odd}}$. For $\lambda < \delta_1^2$, we have

$$(-\mathcal{T}^2 - \lambda)u_* = Lu_* + \mathcal{K}\mathcal{T}u_* - \lambda u_* = \psi_* + (\delta_1^2 - \lambda)u_*, \tag{4.54}$$

and hence

$$(k^2\omega_1^2 - \lambda)(u_*)_k = (\psi_*)_k + (\delta_1^2 - \lambda)(u_*)_k, \quad k \in \mathbb{Z}, \tag{4.55}$$

for the Fourier coefficients. Since

$$(\psi_*, u_*)_{X^0} = (\mathcal{K}\mathcal{T}u_*, u_*)_{X^0} = \frac{1}{4\pi}\int_{\mathbb{R}^3} |U'_{\mathcal{T}u_*}(r)|^2\, dx = \frac{1}{4\pi}\, \|\Psi_*\|_{L_r^2}^2$$

by (B.40) from Lemma B.15(b), taking the inner product in X^0 of (4.54) with u_*, we deduce

$$((-\mathcal{T}^2 - \lambda)u_*, u_*)_{X^0} = \frac{1}{4\pi}\, \|\Psi_*\|_{L_r^2}^2 + (\delta_1^2 - \lambda)\|u_*\|_{X^0}^2. \tag{4.56}$$

Next, due to (4.25) from Lemma 4.6, we have

$$\begin{aligned}\langle \mathcal{Q}_\lambda \Psi_*, \Psi_*\rangle = 64\pi^4 \sum_{k\neq 0} \iint_D d\ell\, \ell\, de\, \frac{1}{\omega_1(e,\ell)\,|Q'(e)|}\, \frac{1}{k^2\omega_1^2(e,\ell) - \lambda} \\ \times |(\psi_*)_k(I,\ell)|^2.\end{aligned}$$

Thus, by (B.4), (A.18), Lemma B.8(b) and (4.55) applied twice,

$$
\begin{aligned}
&((-\mathcal{T}^2 - \lambda)u_*, u_*)_{X^0} \\
&= 16\pi^3 \sum_{k\neq 0} \int_0^\infty dI \int_0^\infty d\ell\, \ell\, \frac{1}{|Q'(e)|}\, \overline{[((-\mathcal{T}^2 - \lambda)u_*]_k(I,\ell)}\, (u_*)_k(I,\ell) \\
&= 16\pi^3 \sum_{k\neq 0} \iint_D d\ell\, \ell\, de\, \frac{1}{\omega_1(e,\ell)\,|Q'(e)|}\, (k^2\omega_1^2(e,\ell) - \lambda)\, |(u_*)_k(I,\ell)|^2 \\
&= 16\pi^3 \sum_{k\neq 0} \iint_D d\ell\, \ell\, de\, \frac{1}{\omega_1(e,\ell)\,|Q'(e)|} (\psi_*)_k(I,\ell)\, \overline{(u_*)_k(I,\ell)} \\
&\quad + 16\pi^3(\delta_1^2 - \lambda) \sum_{k\neq 0} \iint_D d\ell\, \ell\, de\, \frac{1}{\omega_1(e,\ell)\,|Q'(e)|}\, |(u_*)_k(I,\ell)|^2 \\
&= 16\pi^3 \sum_{k\neq 0} \iint_D d\ell\, \ell\, de\, \frac{1}{\omega_1(e,\ell)\,|Q'(e)|} (\psi_*)_k(I,\ell) \\
&\qquad \times \left(\frac{\overline{(\psi_*)_k(I,\ell)}}{k^2\omega_1^2(e,\ell) - \lambda} + (\delta_1^2 - \lambda) \frac{\overline{(u_*)_k(I,\ell)}}{k^2\omega_1^2(e,\ell) - \lambda} \right) \\
&\quad + (\delta_1^2 - \lambda)\, \|u_*\|_{X^0}^2 \\
&= \frac{1}{4\pi} \langle \mathcal{Q}_\lambda \Psi_*, \Psi_* \rangle \\
&\quad + 16\pi^3(\delta_1^2 - \lambda) \sum_{k\neq 0} \iint_D d\ell\, \ell\, de\, \frac{1}{\omega_1(e,\ell)\,|Q'(e)|} (\psi_*)_k(I,\ell)\, \frac{\overline{(u_*)_k(I,\ell)}}{k^2\omega_1^2(e,\ell) - \lambda} \\
&\quad + (\delta_1^2 - \lambda)\, \|u_*\|_{X^0}^2.
\end{aligned}
$$

Comparing to (4.56), this yields

$$
\begin{aligned}
&\frac{1}{4\pi}\, \|\Psi_*\|_{L_r^2}^2 \\
&= \frac{1}{4\pi} \langle \mathcal{Q}_\lambda \Psi_*, \Psi_* \rangle \\
&\quad + 16\pi^3(\delta_1^2 - \lambda) \sum_{k\neq 0} \iint_D d\ell\, \ell\, de\, \frac{1}{\omega_1(e,\ell)\,|Q'(e)|} (\psi_*)_k(I,\ell)\, \frac{\overline{(u_*)_k(I,\ell)}}{k^2\omega_1^2(e,\ell) - \lambda}.
\end{aligned}
\tag{4.57}
$$

If we had $\Psi_* = 0$, then also $\psi_* = 0$ and consequently $(k^2\omega_1^2 - \delta_1^2)(u_*)_k = 0$ in D for $k \neq 0$ by (4.55). This implies that $(u_*)_k = 0$ for $|k| \geq 2$ and $(\omega_1 - \delta_1)(u_*)_1 = 0$ in D. Owing to (ω_1-1), this enforces $(u_*)_1 = 0$ a.e. and therefore $u_* = 0$, which is a contradiction. In other words, we do know that $\Psi_* \neq 0$. Hence, by (4.7) and (4.57),

$$\begin{aligned}\mu_1(\lambda) &= \sup\{\langle \mathcal{Q}_\lambda \Psi, \Psi\rangle : \|\Psi\|_{L^2_r} \le 1\}\\ &\ge \frac{1}{\|\Psi_*\|^2_{L^2_r}}\, \langle \mathcal{Q}_\lambda \Psi_*, \Psi_*\rangle\\ &= 1 - \frac{64\pi^4}{\|\Psi_*\|^2_{L^2_r}}(\delta_1^2 - \lambda)\sum_{k\neq 0}\iint_D d\ell\, \ell\, de\, \frac{1}{\omega_1(e,\ell)\,|Q'(e)|}\\ &\qquad\times (\psi_*)_k(I,\ell)\, \frac{\overline{(u_*)_k(I,\ell)}}{k^2\omega_1^2(e,\ell)-\lambda}.\end{aligned}$$

Thus,

$$\begin{aligned}\frac{1-\mu_1(\lambda)}{\delta_1^2-\lambda} &\le \frac{64\pi^4}{\|\Psi_*\|^2_{L^2_r}}\sum_{k\neq 0}\iint_D d\ell\, \ell\, de\, \frac{1}{\omega_1(e,\ell)\,|Q'(e)|}\\ &\qquad\times (\psi_*)_k(I,\ell)\, \frac{\overline{(u_*)_k(I,\ell)}}{k^2\omega_1^2(e,\ell)-\lambda},\end{aligned}$$

and upon using (4.55) one more time, we conclude that

$$\begin{aligned}\frac{1-\mu_1(\lambda)}{\delta_1^2-\lambda} &\le \frac{64\pi^4}{\|\Psi_*\|^2_{L^2_r}}\sum_{k\neq 0}\iint_D d\ell\, \ell\, de\, \frac{1}{\omega_1(e,\ell)\,|Q'(e)|}\, \frac{k^2\omega_1^2(e,\ell)-\delta_1^2}{k^2\omega_1^2(e,\ell)-\lambda}\,|(u_*)_k(I,\ell)|^2\\ &\le \frac{64\pi^4}{\|\Psi_*\|^2_{L^2_r}}\sum_{k\neq 0}\iint_D d\ell\, \ell\, de\, \frac{1}{\omega_1(e,\ell)\,|Q'(e)|}\,|(u_*)_k(I,\ell)|^2\\ &= \frac{4\pi}{\|\Psi_*\|^2_{L^2_r}}\,\|u_*\|^2_{X^0}\end{aligned} \tag{4.58}$$

for all $\lambda < \delta_1^2$. Since μ_1 is convex on $]-\infty, \delta_1^2]$ by Lemma 4.9(d), the difference quotients

$$\frac{\mu_1(\lambda+h)-\mu_1(\lambda)}{h}$$

for $h > 0$ are monotone increasing in λ (and also in h); see [14, p. 13/14]. Let $\lambda_0 \in]-\infty, \delta_1^2[$ be a point where μ_1 is differentiable and let $h > 0$. For $\lambda_1 = \lambda_0 - h$ and $\lambda_2 = \delta_1^2 - h$, we have $\lambda_1 < \lambda_2$, whence $\mu_1(\delta_1^2) = \mu_* = 1$ in conjunction with (4.58) for $\lambda = \delta_1^2 - h$ leads to

$$\begin{aligned}\frac{\mu_1(\lambda_0)-\mu_1(\lambda_0-h)}{h} &= \frac{\mu_1(\lambda_1+h)-\mu_1(\lambda_1)}{h}\\ &\le \frac{\mu_1(\lambda_2+h)-\mu_1(\lambda_2)}{h}\\ &= \frac{\mu_1(\delta_1^2)-\mu_1(\delta_1^2-h)}{h} \le \frac{4\pi}{\|\Psi_*\|^2_{L^2_r}}\,\|u_*\|^2_{X^0}.\end{aligned}$$

It follows that $\|\mu_1'\|_{L^\infty(]-\infty,\delta_1^2[)} \le \frac{4\pi}{\|\Psi_*\|^2_{L^2_r}} \|u_*\|^2_{X^0}$, which proves (4.53).

To establish the converse, we assume (4.53) to hold, and we are going to verify that δ_1^2 is an eigenvalue of L. For this, we are going to use Lemma 4.12. The operator family Q_z for $z \in \Omega = \mathbb{C} \setminus [\delta_1^2, \infty[$ satisfies the assumptions of Lemma D.1 with $\lambda_0 = \delta_1^2$ and $H = L^2_r$, by Lemmas 4.3 and 4.9. Hence, there are sequences $\lambda_j \nearrow \delta_1^2$, $\varepsilon_j > 0$ and $\Phi_{j,\lambda} \in L^2_r$ for $\lambda \in]\lambda_j - \varepsilon_j, \lambda_j + \varepsilon_j[$ such that $\|\Phi_{j,\lambda}\|_{L^2_r} = 1$,

$$]\lambda_j - \varepsilon_j, \lambda_j + \varepsilon_j[\ni \lambda \mapsto \Phi_{j,\lambda} \in L^2_r$$

is real analytic for $j \in \mathbb{N}$, and $Q_\lambda \Phi_{j,\lambda} = \mu_1(\lambda)\Phi_{j,\lambda}$ for $j \in \mathbb{N}$ and $\lambda \in]\lambda_j - \varepsilon_j, \lambda_j + \varepsilon_j[$. Furthermore, μ_1 is real analytic in $]\lambda_j - \varepsilon_j, \lambda_j + \varepsilon_j[$ and satisfies

$$\mu_1'(\lambda) = \langle Q_\lambda' \Phi_{j,\lambda}, \Phi_{j,\lambda} \rangle \tag{4.59}$$

for $\lambda \in]\lambda_j - \varepsilon_j, \lambda_j + \varepsilon_j[$. By decreasing ε_j further, if necessary, we may assume that $\varepsilon_j \to 0$ as $j \to \infty$. Due to (4.53), there exists a set $N \subset] - \infty, \delta_1^2[$ of measure zero such that $S = \sup_{\lambda \in]-\infty,\delta_1^2[\setminus N} |\mu_1'(\lambda)| < \infty$. For each $j \in \mathbb{N}$, pick $\hat\lambda_j \in]\lambda_j - \varepsilon_j, \lambda_j + \varepsilon_j[\setminus N$ and define $\Psi_j = \Phi_{j,\hat\lambda_j}$. It follows that $\lim_{j\to\infty} \hat\lambda_j = \delta_1^2$ and $\|\Psi_j\|_{L^2_r} = 1$. In addition, $Q_{\hat\lambda_j}\Psi_j = Q_{\hat\lambda_j}\Phi_{j,\hat\lambda_j} = \mu_1(\hat\lambda_j)\Phi_{j,\hat\lambda_j} = \mu_1(\hat\lambda_j)\Psi_j$, i.e., Ψ_j is a normalized eigenfunction for the eigenvalue $\mu_1(\hat\lambda_j)$ of $Q_{\hat\lambda_j}$ such that

$$\sup_{j\in\mathbb{N}} \langle Q'_{\hat\lambda_j} \Psi_j, \Psi_j \rangle \le S, \tag{4.60}$$

the latter due (4.59); recall that generally $\langle Q_\lambda' \Psi, \Psi \rangle \ge 0$ by (4.6). Now define $\psi_j(r, p_r, \ell) = |Q'(e_Q)|\, p_r \Psi_j(r) \in X^0_{\text{odd}}$ and $g_j = (-\mathcal{T}^2 - \hat\lambda_j)^{-1}\psi_j \in X^2_{\text{odd}}$. To complete the proof, we need to show that $(g_j) \subset X^0$ is bounded. From (B.4), (A.18), (B.25), (4.24) and (4.6), we obtain

$$\begin{aligned}
\|g_j\|^2_{X^0} &= 16\pi^3 \sum_{k\neq 0} \int_0^\infty dI \int_0^\infty d\ell\, \ell\, \frac{1}{|Q'(e)|}\, |(g_j)_k(I,\ell)|^2 \\
&= 16\pi^3 \sum_{k\neq 0} \iint_D d\ell\, \ell\, de\, \frac{1}{\omega_1(e,\ell)\,|Q'(e)|}\, \frac{|(\psi_j)_k(I,\ell)|^2}{(k^2\omega_1^2(e,\ell) - \hat\lambda_j)^2} \\
&= 16\pi \sum_{k\neq 0} \iint_D d\ell\, \ell\, de\, \frac{\omega_1(e,\ell)\,|Q'(e)|}{(k^2\omega_1^2(e,\ell) - \hat\lambda_j)^2} \left| \int_{r_-(e,\ell)}^{r_+(e,\ell)} \Psi_j(r) \sin(k\theta(r,e,\ell))\, dr \right|^2 \\
&= \frac{1}{4\pi} \langle Q'_{\hat\lambda_j} \Psi_j, \Psi_j \rangle.
\end{aligned}$$

Thus, the claim follows from (4.60). □

4.3 Some Further Results

The following observation corresponds to the situation where ω_1 is differentiable and attains its minimum at an interior point $(\hat{e}, \hat{\beta})$ of D; cf. assumption (ω_1-2).

Corollary 4.16 *Suppose that* (ω_1-2) *is satisfied. Then* $\mu_* = \infty$, $\lambda_* < \delta_1^2$, $\mu_1(\lambda_*) = 1$ *and* λ_* *is an eigenvalue of* L.

Proof We only need to show that $\mu_* = \infty$, then the remaining assertions do follow from Theorem 4.13. The lower boundary curve $(\partial D)_3 = \{(e, \beta) \in D : e = e_{\min}(\beta)\}$ of D characterizes the (e, β) where $r_-(e, \beta) = r_0(\beta) = r_+(e, \beta)$. Since $(\hat{e}, \hat{\beta}) \in \operatorname{int} D = \{(e, \beta) : \beta \in]0, \beta_*[, e \in]e_{\min}(\beta), e_0[\} \subset D \setminus (\partial D)_3$ by hypothesis, we have that $r_+(\hat{e}, \hat{\beta}) - r_-(\hat{e}, \hat{\beta}) = 6\eta > 0$. The functions $r_\pm$ are known to be continuous (even C^1) on int D; see [30, 50] and [88, Def./Thm. 2.4(b)]. Thus, by shrinking the neighborhood U of $(\hat{e}, \hat{\beta})$ if necessary, we may assume that

$$|r_-(e, \beta) - r_-(\hat{e}, \hat{\beta})| \leq \eta, \quad |r_+(e, \beta) - r_+(\hat{e}, \hat{\beta})| \leq \eta, \quad (e, \beta) \in U,$$

is verified, along with

$$|\omega_1(e, \beta) - \delta_1| \leq C_1 \, |(e, \beta) - (\hat{e}, \hat{\beta})|^2, \quad (e, \beta) \in U, \tag{4.61}$$

from (1.31). Next, we have $\theta(r_-(\hat{e}, \hat{\beta}), \hat{e}, \hat{\beta}) = 0$ and $\theta(r_+(\hat{e}, \hat{\beta}), \hat{e}, \hat{\beta}) = \pi$. Since $\frac{\partial \theta}{\partial r} = \frac{\omega_1}{p_r}$ due to (A.21) and $p_r > 0$ along the half-orbit, $\theta(\cdot, \hat{e}, \hat{\beta})$ is strictly increasing. In particular, we obtain

$$\sin \theta(\hat{r}_m, \hat{e}, \hat{\beta}) = 2\sigma > 0 \quad \text{for} \quad \hat{r}_m = \frac{1}{2} \, (r_-(\hat{e}, \hat{\beta}) + r_+(\hat{e}, \hat{\beta})).$$

As also

$$\theta : \{(r, e, \beta) : (e, \beta) \in \operatorname{int} D, r_-(e, \beta) < r < r_+(e, \beta)\} \to \mathbb{R}$$

is continuous, there is $\varepsilon \in]0, \eta]$ such that $\sin \theta(r, e, \beta) \geq \sigma$ for $(e, \beta) \in U$ so that $|e - \hat{e}| \leq \varepsilon$, $|\beta - \hat{\beta}| \leq \varepsilon$ and $r \in [\hat{r}_m - \varepsilon, \hat{r}_m + \varepsilon] \cap]r_-(e, \beta), r_+(e, \beta)[= [\hat{r}_m - \varepsilon, \hat{r}_m + \varepsilon]$. If $\varepsilon > 0$ is small enough, we may assume that $[\hat{e} - \varepsilon, \hat{e} + \varepsilon] \times [\hat{\beta} - \varepsilon, \hat{\beta} + \varepsilon] \subset U \subset \operatorname{int} D$ as well as $[\hat{r}_m - \varepsilon, \hat{r}_m + \varepsilon] \subset [0, r_Q]$. Furthermore, note that in general $\sin \theta(r, e, \beta) \geq 0$ for $(e, \beta) \in \operatorname{int} D$ and $r_-(e, \beta) < r < r_+(e, \beta)$. Next, owing to $(\hat{e}, \hat{\beta}) \in \operatorname{int} D$, we have $e \in]U_Q(0), e_0[$. Using (Q2), we can thus make sure that $\inf\{|Q'(e)| : e \in [\hat{e} - \varepsilon, \hat{e} + \varepsilon]\} = \alpha > 0$. Now, we consider the function

$$\Psi_0(r) = \gamma^{-1} \mathbf{1}_{[\hat{r}_m - \varepsilon, \hat{r}_m + \varepsilon]}(r), \quad \gamma = \Big(\frac{4\pi}{3}([\hat{r}_m + \varepsilon]^3 - [\hat{r}_m - \varepsilon]^3)\Big)^{1/2},$$

for which $\|\Psi_0\|_{L_r^2} = 1$. Hence, for $\lambda < \delta_1^2$ by Lemma 4.3(d),

$$
\begin{aligned}
\mu_* \geq \mu_1(\lambda) &= \|\mathcal{Q}_\lambda\| = \sup\{\langle \mathcal{Q}_\lambda \Psi, \Psi\rangle : \|\Psi\| \leq 1\} \geq \langle \mathcal{Q}_\lambda \Psi_0, \Psi_0\rangle \\
&= 32\pi^2 \sum_{k\neq 0} \iint_D d\beta\, de\, \frac{\omega_1(e,\beta)\,|Q'(e)|}{k^2\omega_1^2(e,\beta)-\lambda} \left| \int_{r_-(e,\beta)}^{r_+(e,\beta)} \Psi_0(r)\sin(k\theta(r,e,\beta))\,dr \right|^2 \\
&\geq 32\pi^2 \iint_D d\beta\, de\, \frac{\omega_1(e,\beta)\,|Q'(e)|}{\omega_1^2(e,\beta)-\lambda} \left(\int_{r_-(e,\beta)}^{r_+(e,\beta)} \Psi_0(r)\sin(\theta(r,e,\beta))\,dr \right)^2 \\
&\geq 32\pi^2\delta_1\gamma^{-2} \int_{\hat\beta-\varepsilon}^{\hat\beta+\varepsilon} d\beta \int_{\hat e-\varepsilon}^{\hat e+\varepsilon} de\, \frac{|Q'(e)|}{\omega_1^2(e,\beta)-\lambda} \left(\int_{\hat r_m-\varepsilon}^{\hat r_m+\varepsilon} \sin(\theta(r,e,\beta))\,dr \right)^2 \\
&\geq 128\pi^2\delta_1\gamma^{-2}\alpha\,\sigma^2\varepsilon^2 \int_{\hat\beta-\varepsilon}^{\hat\beta+\varepsilon} d\beta \int_{\hat e-\varepsilon}^{\hat e+\varepsilon} de\, \frac{1}{\omega_1^2(e,\beta)-\delta_1^2+a},
\end{aligned}
\tag{4.62}
$$

where $a = \delta_1^2 - \lambda > 0$. From Theorem 3.5 and (4.61), we deduce that

$$
\omega_1^2(e,\beta) - \delta_1^2 + a \leq 2\Delta_1 C_1\, |\xi - \hat\xi|^2 + a, \quad \xi = (e,\beta),\ \hat\xi = (\hat e, \hat\beta).
$$

As a consequence,

$$
\begin{aligned}
\int_{\hat\beta-\varepsilon}^{\hat\beta+\varepsilon} d\beta \int_{\hat e-\varepsilon}^{\hat e+\varepsilon} de\, \frac{1}{\omega_1^2(e,\beta)-\delta_1^2+a} &\geq \int_{|\xi-\hat\xi|\leq\varepsilon} \frac{d^2\xi}{2\Delta_1 C_1\,|\xi-\hat\xi|^2+a} \\
&= 2\pi \int_0^\varepsilon \frac{\rho}{2\Delta_1 C_1\,\rho^2 + a}\,d\rho \\
&= \frac{\pi}{2\Delta_1 C_1} \ln \frac{2\Delta_1 C_1\,\varepsilon^2 + a}{a} \to \infty, \quad a \to 0^+.
\end{aligned}
$$

Thus, if we pass to the limit $\lambda \to \delta_1^2-$, i.e., $a \to 0^+$, in (4.62), it follows that $\mu_* = \infty$.
□

Regarding Theorem 4.15, if (ω_1-3) holds and if $\mu_* = 1$, then one can show that $\lambda = \delta_1^2$ is an eigenvalue of L, provided one is able to gain a little bit from the term $|Q'(e)|$, in the sense that $Q'(e_0) = 0$ in a controlled way, as expressed by (Q5); then the inherent logarithmic singularity can be dealt with. To simplify the presentation, we additionally assume that μ_* is simple as an eigenvalue of $Q_{\delta_1^2}$, but with some more technical efforts, this assumption could be disposed of.

Corollary 4.17 *Suppose that (ω_1-3) and (Q5) are satisfied, and assume that $\mu_* = 1$ is a simple eigenvalue of $Q_{\delta_1^2}$. Then $\lambda_* = \delta_1^2$, and this is an eigenvalue of L.*

Proof We already know that $\lambda_* = \delta_1^2$; see the proof of Theorem 4.15. To verify that δ_1^2 is an eigenvalue of L, we are going to use Theorem 4.15. According to Lemma D.2, there is $\varepsilon > 0$ such that $]\delta_1^2 - \varepsilon, \delta_1^2[\ni \lambda \mapsto \mu_1(\lambda)$ is real analytic. In addition, there are $\Psi_\lambda \in L_r^2$ satisfying $\|\Psi_\lambda\|_{L_r^2} = 1$, $Q_\lambda \Psi_\lambda = \mu_1(\lambda)\Psi_\lambda$, and $]\delta_1^2 - \varepsilon, \delta_1^2[\ni \lambda \mapsto \Psi_\lambda$

is real analytic. Also $\mu_1'(\lambda) = \langle Q_\lambda' \Psi_\lambda, \Psi_\lambda \rangle$ holds for $\lambda \in]\delta_1^2 - \varepsilon, \delta_1^2[$. By Lemma 4.9, the function μ_1 is convex, so that $\mu_1'' \geq 0$ and μ_1' is increasing. In other words,

$$\|\mu_1'\|_{L^\infty(]-\infty,\delta_1^2[)} = \lim_{\lambda \to \delta_1^2 -} \mu_1'(\lambda) =: \mu_*'$$

does exist in $]0, \infty]$, and the issue is to show that $\mu_*' < \infty$. Defining $\psi_\lambda(r, p_r, \ell) = |Q'(e_Q)|\, p_r \Psi_\lambda(r) \in X^0_{\text{odd}}$ as before, we get, from Lemma 4.3(d), (4.24) and Corollary 4.10(c),

$$\begin{aligned}
\mu_1'(\lambda) &= \langle Q_\lambda' \Psi_\lambda, \Psi_\lambda \rangle \\
&= 64\pi^2 \sum_{k\neq 0} \iint_D d\ell\, \ell\, de\, \frac{\omega_1(e,\ell)\,|Q'(e)|}{(k^2\omega_1^2(e,\ell) - \lambda)^2} \\
&\qquad \times \left| \int_{r_-(e,\ell)}^{r_+(e,\ell)} \Psi_\lambda(r) \sin(k\theta(r,e,\ell))\, dr \right|^2 \\
&= 64\pi^4 \sum_{k\neq 0} \iint_D d\ell\, \ell\, de\, \frac{1}{(k^2\omega_1^2(e,\ell) - \lambda)^2}\, \frac{|(\psi_\lambda)_k(I,\ell)|^2}{\omega_1(e,\ell)\,|Q'(e)|} \\
&\leq C \sum_{k\neq 0} \iint_D d\ell\, \ell\, de\, \frac{1}{(k^2\omega_1^2(e,\ell) - \lambda)^2}\, |Q'(e)| \\
&\leq C \sum_{k=2}^{\infty} \iint_D d\ell\, \ell\, de\, \frac{4}{\delta_1^4 k^4} + C \iint_D d\ell\, \ell\, de\, \frac{|Q'(e)|}{(\omega_1(e,\ell) - \delta_1)^2}.
\end{aligned}$$

Thus, using (ω_1-3) and (Q5),

$$\begin{aligned}
\mu_1'(\lambda) &\leq C + C \iint_D d\ell\, \ell\, de\, \frac{(e - e_0)^\alpha}{\left|(e,\beta) - (e_0, \hat{\beta})\right|^2} \\
&\leq C + C \int_0^{\beta_*} d\beta \int_{e_{\min}(\beta)}^{e_0} de\, \frac{1}{\left|(e,\beta) - (e_0,\hat{\beta})\right|^{2-\alpha}} \\
&\leq C + C \int_{-\hat{\beta}}^{\beta_* - \hat{\beta}} dx_2 \int_0^{e_0 - U_Q(0)} dx_1\, \frac{1}{|x|^{2-\alpha}} \leq C,
\end{aligned}$$

where $x = (x_1, x_2)$. Therefore $\mu_*' \leq C$ and the proof is complete. □

Chapter 5
Relation to the Guo-Lin Operator

In [29], the operator

$$A_{\rm GL}: H_r^2(\mathbb{R}^3) \to L_r^2(\mathbb{R}^3), \quad A_{\rm GL} = -\Delta - 4\pi \int_{\mathbb{R}^3} |Q'(e_Q)|(I-P)\,dv,$$

has been introduced (there called A_0), where $H_r^2(\mathbb{R}^3)$ is the Sobolev space of second order of radial functions $\phi(x) = \phi(r)$. Here, P is the projection onto the kernel of $\mathcal{T}$ of such a radial function, cf. Remark B.6 for an explicit expression. The associated quadratic form is

$$\begin{aligned}\langle A_{\rm GL}\phi, \phi\rangle_{2,|Q'|} &= \|\nabla\phi\|^2_{L^2(\mathbb{R}^3)} - 4\pi\left(\|\phi\|^2_{L^2_{|Q'|}(\mathbb{R}^6)} - \|P\phi\|^2_{L^2_{|Q'|}(\mathbb{R}^6)}\right)\\ &= \|\nabla\phi\|^2_{L^2(\mathbb{R}^3)} - 4\pi\,\|\phi - P\phi\|^2_{L^2_{|Q'|}(\mathbb{R}^6)},\end{aligned} \tag{5.1}$$

where we let

$$\|g\|^2_{L^2_{|Q'|}(\mathbb{R}^6)} = \int_{\mathbb{R}^3}\int_{\mathbb{R}^3} |Q'(e_Q)|\,|g(x,v)|^2\,dx\,dv$$

for suitable $g = g(x, v)$. In [29] an important property that was needed for the proofs was that

$$\hat{\lambda}_{\rm GL} = \inf\left\{\frac{\langle A_{\rm GL}\phi, \phi\rangle_{2,|Q'|}}{\|\phi\|^2_{L^2(\mathbb{R}^3)}} : \phi \in H_r^2(\mathbb{R}^3), \phi \neq 0\right\}$$

be positive. However, in fact it holds that $\hat{\lambda}_{\rm GL} = 0$, as has been noted in Theorem 4.6 and the subsequent Remark b) of [88]. In [88], it was also observed that the valid (and for the intended stability proof appropriate) replacement is

M. Kunze, *A Birman-Schwinger Principle in Galactic Dynamics*,
Progress in Mathematical Physics 77,
https://doi.org/10.1007/978-3-030-75186-9_5

$$\lambda_{\mathrm{GL}} = \inf\left\{\frac{\langle A_{\mathrm{GL}}\phi, \phi\rangle_{2,|Q'|}}{\|\nabla\phi\|^2_{L^2(\mathbb{R}^3)}} : \phi \in \dot{H}^1_r(\mathbb{R}^3), \nabla\phi \neq 0\right\} > 0, \tag{5.2}$$

where $\dot{H}^1(\mathbb{R}^3) = \{\phi \in L^2_{\mathrm{loc}}(\mathbb{R}^3) : \nabla\phi \in L^2(\mathbb{R}^3)\}$ denotes the first order homogeneous Sobolev space and $\dot{H}^1_r(\mathbb{R}^3)$ are the radial functions in $\dot{H}^1(\mathbb{R}^3)$.

The next result establishes the connection between λ_{GL} and $\mu_1(0)$.

Lemma 5.1 *We have* $\lambda_{\mathrm{GL}} + \mu_1(0) = 1$.

Proof Let $\phi \in \dot{H}^1_r = \dot{H}^1_r(\mathbb{R}^3)$ be such that $\nabla\phi \neq 0$. Abusing notation, we can write $\nabla\phi(x) = \frac{x}{|x|}\,\Psi(r)$ for $\Psi = \phi'$, and in particular $\|\Psi\|_{L^2_r} = \|\nabla\phi\|_{L^2(\mathbb{R}^3)}$ and $\Psi \in L^2_r$. Since

$$\begin{aligned}\langle \mathcal{Q}_0\Psi, \Psi\rangle &= 4\pi \iint_K |Q'(e_Q)|\,(\phi - P\phi)^2\,dx\,dv\\ &= 4\pi \int_{\mathbb{R}^3}\int_{\mathbb{R}^3} |Q'(e_Q)|\,(\phi - P\phi)^2\,dx\,dv = 4\pi\,\|\phi - P\phi\|^2_{L^2_{|Q'|}(\mathbb{R}^6)}\end{aligned}$$

by (4.26) in Lemma 4.6 and our convention concerning $|Q'(e_Q)|$, (5.1) yields

$$\frac{\langle A_{\mathrm{GL}}\phi, \phi\rangle_{2,|Q'|}}{\|\nabla\phi\|^2_{L^2(\mathbb{R}^3)}} + \frac{\langle \mathcal{Q}_0\Psi, \Psi\rangle}{\|\Psi\|^2_{L^2_r}} = 1. \tag{5.3}$$

Recall from (4.7) that $\mu_1(0) = \sup\{\langle \mathcal{Q}_0\Psi, \Psi\rangle : \|\Psi\|_{L^2_r} \le 1\}$. Therefore, (5.3) leads to

$$1 \le \frac{\langle A_{\mathrm{GL}}\phi, \phi\rangle_{2,|Q'|}}{\|\nabla\phi\|^2_{L^2(\mathbb{R}^3)}} + \mu_1(0)$$

for all such ϕ so that $1 \le \lambda_{\mathrm{GL}} + \mu_1(0)$ is found. Conversely, if $\Psi \in L^2_r$ and $\Psi \neq 0$, we define $\phi(r) = \int_0^r \Psi(s)\,ds$ to obtain $\phi \in \dot{H}^1_r$, and once again (5.3) holds. We deduce that

$$1 = \frac{\langle A_{\mathrm{GL}}\phi, \phi\rangle_{2,|Q'|}}{\|\nabla\phi\|^2_{L^2(\mathbb{R}^3)}} + \frac{\langle \mathcal{Q}_0\Psi, \Psi\rangle}{\|\Psi\|^2_{L^2_r}} \ge \lambda_{\mathrm{GL}} + \frac{\langle \mathcal{Q}_0\Psi, \Psi\rangle}{\|\Psi\|^2_{L^2_r}}$$

for all such Ψ, which entails $1 \ge \lambda_{\mathrm{GL}} + \mu_1(0)$. □

Corollary 5.2 *The infimum* λ_{GL} *in (5.2) is attained, i.e., there is a function* $\phi_* \in \dot{H}^1_r(\mathbb{R}^3)$ *such that* $\|\nabla\phi_*\|_{L^2(\mathbb{R}^3)} = 1$ *and* $\langle A_{\mathrm{GL}}\phi_*, \phi_*\rangle_{2,|Q'|} = \lambda_{\mathrm{GL}}$.

Proof It is implicit in [29], and explicit in [88, Prop. 4.8], that the assertion will follow from $\lambda_{\mathrm{GL}} \neq 1$. However, Lemma 4.7(a) says that $0 < \mu_1(0) < 1$, and hence we also have $0 < \lambda_{\mathrm{GL}} < 1$ by Lemma 5.1. □

There are some further relations of $\mathcal{Q}_\lambda$ and L to A_{GL}. We will argue only formally, without specifying the spaces, etc.

Remark 5.3 (a) The following observation has been used in [29] to prove that $\lambda_{\rm GL} > 0$. Let $\phi = \phi(r)$ be given and write $\phi_0 = P\phi$ for its projection onto the kernel of $\mathcal{T}$. Let $h = h(x, v)$ be such that $\mathcal{T}h = \phi - \phi_0$. Denote $g = -|Q'|h$ and $\psi = U_{-|Q'|(\phi-\phi_0)}$. Then

$$\langle A_{\rm GL}\phi, \phi\rangle_{2,|Q'|} = 4\pi\,(Lg, g)_Q + \int_{\mathbb{R}^3} |\nabla\phi - \nabla\psi|^2\,dx. \tag{5.4}$$

In fact, since $\mathcal{T}|Q'| = 0$ we get

$$\begin{aligned}\int_{\mathbb{R}^3}\int_{\mathbb{R}^3} |Q'|\,P\phi\,(\phi - P\phi)\,dx\,dv &= \int_{\mathbb{R}^3}\int_{\mathbb{R}^3} \phi\,P(|Q'|\,(\phi - P\phi))\,dx\,dv\\ &= \int_{\mathbb{R}^3}\int_{\mathbb{R}^3} \phi\,(|Q'|\,(P\phi - P^2\phi))\,dx\,dv = 0.\end{aligned}$$

Hence from $\mathcal{T}g = -|Q'|\mathcal{T}h = -|Q'|(\phi - \phi_0)$ and (1.18), we obtain

$$\begin{aligned}&\langle A_{\rm GL}\phi, \phi\rangle_{2,|Q'|}\\ &= \int_{\mathbb{R}^3} |\nabla\phi|^2\,dx - 4\pi\int_{\mathbb{R}^3}\int_{\mathbb{R}^3} |Q'|\,|\phi - P\phi|^2\,dx\,dv\\ &= \int_{\mathbb{R}^3} |\nabla\phi|^2\,dx + 4\pi\int_{\mathbb{R}^3}\int_{\mathbb{R}^3} |Q'|\,|\phi - P\phi|^2\,dx\,dv\\ &\quad -8\pi\int_{\mathbb{R}^3}\int_{\mathbb{R}^3} |Q'|\,\phi\,(\phi - P\phi)\,dx\,dv\\ &= \int_{\mathbb{R}^3} |\nabla\phi|^2\,dx + 4\pi\int_{\mathbb{R}^3}\int_{\mathbb{R}^3} \frac{1}{|Q'|}\,|\mathcal{T}g|^2\,dx\,dv + 8\pi\int_{\mathbb{R}^3}\int_{\mathbb{R}^3} \phi\,(\mathcal{T}g)\,dx\,dv\\ &= 4\pi\int_{\mathbb{R}^3}\int_{\mathbb{R}^3} \frac{1}{|Q'|}\,|\mathcal{T}g|^2\,dx\,dv + \int_{\mathbb{R}^3} |\nabla\phi|^2\,dx + 2\int_{\mathbb{R}^3} \phi\,\Delta U_{\mathcal{T}g}\,dx\\ &= 4\pi\int_{\mathbb{R}^3}\int_{\mathbb{R}^3} \frac{1}{|Q'|}\,|\mathcal{T}g|^2\,dx\,dv + \int_{\mathbb{R}^3} |\nabla\phi|^2\,dx - 2\int_{\mathbb{R}^3} \nabla\phi\cdot\nabla U_{\mathcal{T}g}\,dx\\ &= 4\pi\int_{\mathbb{R}^3}\int_{\mathbb{R}^3} \frac{1}{|Q'|}\,|\mathcal{T}g|^2\,dx\,dv - \int_{\mathbb{R}^3} |\nabla U_{\mathcal{T}g}|^2\,dx + \int_{\mathbb{R}^3} |\nabla\phi - \nabla U_{\mathcal{T}g}|^2\,dx\\ &= 4\pi\,(Lg, g)_Q + \int_{\mathbb{R}^3} |\nabla\phi - \nabla U_{\mathcal{T}g}|^2\,dx,\end{aligned}$$

as claimed. There is also a kind of converse statement to (5.4): if $h = h(x, v)$ is given, then $4\pi\,(Lh, h)_Q$ can be written as the sum of $\langle A_{\rm GL}U_{\mathcal{T}h}, U_{\mathcal{T}h}\rangle_{2,|Q'|}$ and a positive term.
(b) We consider the eigenvalue equation $\mathcal{Q}_\lambda\Psi_\lambda = \mu_1(\lambda)\Psi_\lambda$. Let F_λ be such that $F_\lambda' = \Psi_\lambda$. Then, (4.22) yields

$$\mu_1(\lambda)F_\lambda' = \mu_1(\lambda)\Psi_\lambda = \mathcal{Q}_\lambda\Psi_\lambda = U'_{\mathcal{T}(-\mathcal{T}^2-\lambda)^{-1}\psi_\lambda}.$$

for $\psi_\lambda = |Q'|\, p_r \Psi_\lambda$. Therefore $\mu_1(\lambda) F_\lambda = U_{\mathcal{T}(-\mathcal{T}^2-\lambda)^{-1}\psi_\lambda} + \text{const.}$ together with $\mathcal{T}(|Q'| F_\lambda) = |Q'|\,(\mathcal{T} F_\lambda) = |Q'|\, p_r F_\lambda' = \psi_\lambda$ leads to

$$\mu_1(\lambda)\Delta F_\lambda = 4\pi \int_{\mathbb{R}^3} |Q'|\,\mathcal{T}(-\mathcal{T}^2-\lambda)^{-1}\mathcal{T} F_\lambda\, dv. \tag{5.5}$$

Since the operator

$$F \mapsto \mathcal{M}_\lambda F = 4\pi\Delta^{-1} \int_{\mathbb{R}^3} |Q'|\,\mathcal{T}(-\mathcal{T}^2-\lambda)^{-1}\mathcal{T} F\, dv$$

(most likely) will be compact on a suitable space of functions, the eigenvalues of $\mathcal{M}_\lambda$ will correspond to the eigenvalues of $\mathcal{Q}_\lambda$. Observing

$$\mathcal{T}(-\mathcal{T}^2-\lambda)^{-1}\mathcal{T} F = -(F - PF) - \lambda(-\mathcal{T}^2-\lambda)^{-1} F,$$

(5.5) may be rewritten as

$$\mu_1(\lambda)\Delta F_\lambda = -4\pi \int_{\mathbb{R}^3} |Q'|\,(I-P)F_\lambda\, dv - 4\pi\lambda \int_{\mathbb{R}^3} |Q'|\,(-\mathcal{T}^2-\lambda)^{-1} F_\lambda\, dv,$$

so that

$$A_{\mathrm{GL}} F_\lambda + (1-\mu_1(\lambda))\Delta F_\lambda = 4\pi\lambda \int_{\mathbb{R}^3} |Q'|\,(-\mathcal{T}^2-\lambda)^{-1} F_\lambda\, dv,$$

which makes a connection to A_{GL}. ◇

Chapter 6
Invariances

It is well-known that the Vlasov-Poisson system (1.5), (1.3) has many invariances, see [49, p. 427], for instance: if $f = f(t, x, v)$ is a solution, so is

$$\tilde{f}(\tilde{t}, \tilde{x}, \tilde{v}) = \frac{\mu}{\lambda^2} f\Big(\frac{\tilde{t} + t_0}{\mu\lambda}, \frac{\tilde{x} + x_0}{\lambda}, \mu\tilde{v}\Big), \tag{6.1}$$

where $\mu, \lambda > 0$, $t_0 \in \mathbb{R}$ and $x_0 \in \mathbb{R}^3$. The associated potential and density are

$$U_{\tilde{f}}(\tilde{t}, \tilde{x}) = \frac{1}{\mu^2} U_f\Big(\frac{\tilde{t} + t_0}{\mu\lambda}, \frac{\tilde{x} + x_0}{\lambda}\Big), \quad \rho_{\tilde{f}}(\tilde{t}, \tilde{x}) = \frac{1}{\mu^2\lambda^2} \rho_f\Big(\frac{\tilde{t} + t_0}{\mu\lambda}, \frac{\tilde{x} + x_0}{\lambda}\Big). \tag{6.2}$$

It can be expected that quantities that are invariant will play a particularly important role. It is the purpose of this section to determine several such quantities.

Let $Q = Q(x, v)$ be a steady state solution. According to (6.1) and (6.2), then

$$\tilde{Q}(\tilde{x}, \tilde{v}) = \frac{\mu}{\lambda^2} Q\Big(\frac{\tilde{x}}{\lambda}, \mu\tilde{v}\Big) \tag{6.3}$$

is a steady state solution for every $\mu, \lambda > 0$. The associated potential and density are

$$U_{\tilde{Q}}(\tilde{x}) = \frac{1}{\mu^2} U_Q\Big(\frac{\tilde{x}}{\lambda}\Big), \quad \rho_{\tilde{Q}}(\tilde{x}) = \frac{1}{\mu^2\lambda^2} \rho_Q\Big(\frac{\tilde{x}}{\lambda}\Big).$$

The variables transform as $x = \frac{\tilde{x}}{\lambda}$ and $v = \mu\tilde{v}$ so that in particular $r = \frac{\tilde{r}}{\lambda}$ for $r = |x|$ and $\tilde{r} = |\tilde{x}|$.

Next let $Q = Q(e_Q)$ depend only upon $e_Q(x, v) = \frac{1}{2}|v|^2 + U_Q(x)$. Then,

M. Kunze, *A Birman-Schwinger Principle in Galactic Dynamics*,
Progress in Mathematical Physics 77,
https://doi.org/10.1007/978-3-030-75186-9_6

$$\begin{aligned} e_Q(x, v) &= \frac{1}{2}\,|v|^2 + U_Q(x) = \frac{1}{2}\,\mu^2|\tilde{v}|^2 + U_Q(\lambda^{-1}\tilde{x}) \\ &= \frac{1}{2}\,\mu^2|\tilde{v}|^2 + \mu^2 U_{\tilde{Q}}(\tilde{x}) = \mu^2 e_{\tilde{Q}}(\tilde{x}, \tilde{v}), \end{aligned} \tag{6.4}$$

and (6.3) leads to

$$\tilde{Q}(e_{\tilde{Q}}) = \frac{\mu}{\lambda^2}\,Q(e_Q) = \frac{\mu}{\lambda^2}\,Q(\mu^2 e_{\tilde{Q}}).$$

Thus, if $Q = Q(e_Q)$ and $\tilde{Q} = \tilde{Q}(e_{\tilde{Q}})$ are understood as functions of one variable, then

$$\tilde{Q}'(e_{\tilde{Q}}) = \frac{\mu^3}{\lambda^2}\,Q'(e_Q). \tag{6.5}$$

For radial potentials and densities, we have

$$U_{\tilde{Q}}(\tilde{r}) = \frac{1}{\mu^2}\,U_Q(r), \quad \rho_{\tilde{Q}}(\tilde{r}) = \frac{1}{\lambda^2\mu^2}\,\rho_Q(r), \tag{6.6}$$

which leads to

$$U'_{\tilde{Q}}(\tilde{r}) = \frac{1}{\lambda\mu^2}\,U'_Q(r). \tag{6.7}$$

The central densities are related by

$$\rho_{\tilde{Q}}(0) = \frac{1}{\lambda^2\mu^2}\,\rho_Q(0). \tag{6.8}$$

The effective potential from (7.4) is $U_{\mathrm{eff}}(r, \ell) = U_Q(r) + \frac{\ell^2}{2r^2}$, which we also write as $U_{\mathrm{eff}}(r, \beta) = U_Q(r) + \frac{\beta}{2r^2}$ for $\beta = \ell^2$. Let

$$\tilde{\beta} = \frac{\lambda^2}{\mu^2}\,\beta.$$

Then,

$$\tilde{U}_{\mathrm{eff}}(\tilde{r}, \tilde{\beta}) := U_{\tilde{Q}}(\tilde{r}) + \frac{\tilde{\beta}}{2\tilde{r}^2} = \frac{1}{\mu^2}\,U_Q(r) + \frac{\lambda^2}{\mu^2}\,\beta\,\frac{1}{2\lambda^2 r^2} = \frac{1}{\mu^2}\,U_{\mathrm{eff}}(r, \beta)$$

is the corresponding transformation rule. The points $r_\pm = r_\pm(e, \beta)$ are determined by the relation $U_{\mathrm{eff}}(r_\pm(e, \beta), \beta) = e$. Owing to

$$\tilde{U}_{\mathrm{eff}}(\tilde{r}_\pm(\tilde{e}, \tilde{\beta}), \tilde{\beta}) = \tilde{e} \quad\Longleftrightarrow\quad \frac{1}{\mu^2}\,U_{\mathrm{eff}}(\lambda^{-1}\tilde{r}_\pm(\tilde{e}, \tilde{\beta}), \beta) = \frac{1}{\mu^2}\,e$$

we obtain

$$\tilde{r}_\pm(\tilde{e}, \tilde{\beta}) = \lambda\, r_\pm(e, \beta).$$

Next, $r_0 = r_0(\beta)$ is the point where $U_{\mathrm{eff}}(\cdot, \beta)$ attains its minimum. Since $\tilde{U}_{\mathrm{eff}}(\tilde{r}, \tilde{\beta}) = \mu^{-2} U_{\mathrm{eff}}(r, \beta) = \mu^{-2} U_{\mathrm{eff}}(\lambda^{-1}\tilde{r}, \beta)$, we get

$$\tilde{U}'_{\mathrm{eff}}(\tilde{r}, \tilde{\beta}) = \lambda^{-1}\mu^{-2}\, U'_{\mathrm{eff}}(\lambda^{-1}\tilde{r}, \beta),$$

and this implies that

$$\tilde{r}_0(\tilde{\beta}) = \lambda\, r_0(\beta).$$

In terms of the variables e and β, the period function from (A.20) is

$$T_1(e, \beta) = 2\int_{r_-(e,\beta)}^{r_+(e,\beta)} \frac{dr}{\sqrt{2(e - U_{\mathrm{eff}}(r, \beta))}}.$$

Using the transformation $\tilde{r} = \lambda r$, $d\tilde{r} = \lambda dr$, it follows that

$$\begin{aligned}
\tilde{T}_1(\tilde{e}, \tilde{\beta}) &= 2\int_{\tilde{r}_-(\tilde{e},\tilde{\beta})}^{\tilde{r}_+(\tilde{e},\tilde{\beta})} \frac{d\tilde{r}}{\sqrt{2(\tilde{e} - \tilde{U}_{\mathrm{eff}}(\tilde{r}, \tilde{\beta}))}} \\
&= 2\lambda \int_{\lambda^{-1}\tilde{r}_-(\tilde{e},\tilde{\beta})}^{\lambda^{-1}\tilde{r}_+(\tilde{e},\tilde{\beta})} \frac{dr}{\sqrt{2(\tilde{e} - \tilde{U}_{\mathrm{eff}}(\lambda r, \tilde{\beta}))}} \\
&= 2\lambda \int_{r_-(e,\beta)}^{r_+(e,\beta)} \frac{dr}{\sqrt{2(\mu^{-2}e - \mu^{-2}U_{\mathrm{eff}}(r, \beta))}} \\
&= \lambda\mu\, T_1(e, \beta).
\end{aligned}$$

In particular, $\tilde{\omega}_1(\tilde{e}, \tilde{\beta}) = \frac{1}{\lambda\mu}\,\omega_1(e, \beta)$ for $\tilde{\omega}_1 = \frac{2\pi}{\tilde{T}_1}$, and if we denote $\delta_1 = \inf \omega_1$, then also

$$\tilde{\delta}_1 = \frac{1}{\lambda\mu}\,\delta_1. \tag{6.9}$$

Next we consider the space $L^2_{\mathrm{sph},\, \frac{1}{|Q'|}}(K) = X^0$ of spherically symmetric functions with the Q-dependent inner product

$$(u_1, u_2)_Q = \iint_K \frac{1}{|Q'(e_Q)|}\, \overline{u_1(x, v)}\, u_2(x, v)\, dx\, dv,$$

as in Remark B.2. Defining

$$\tilde{u}(\tilde{x}, \tilde{v}) = \frac{\mu}{\lambda^2}\, u\Big(\frac{\tilde{x}}{\lambda}, \mu\tilde{v}\Big) \tag{6.10}$$

in accordance with (6.3), we calculate, using $dx = \lambda^{-3} d\tilde{x}$ and $dv = \mu^3 d\tilde{v}$ as well as (6.5):

$$\begin{aligned} \|\tilde{u}\|_{\tilde{Q}}^2 &= \int\int \frac{d\tilde{x}\, d\tilde{v}}{|\tilde{Q}'(e_{\tilde{Q}})|} |\tilde{u}(\tilde{x}, \tilde{v})|^2 \\ &= \frac{\lambda^2}{\mu^3} \frac{\mu^2}{\lambda^4} \int\int \frac{\lambda^3 dx\, \mu^{-3} dv}{|Q'(e_Q)|} |u(x, v)|^2 = \frac{\lambda}{\mu^4} \|u\|_Q^2. \end{aligned} \tag{6.11}$$

Let the operator $(\mathcal{T} g)(x, v) = v \cdot \nabla_x g(x, v) - \nabla_v g(x, v) \cdot \nabla_x U_Q(x)$ be as in (1.11). From the above relations, it follows that

$$\begin{aligned} (\mathcal{T}\tilde{u})(\tilde{x}, \tilde{v}) &= \tilde{v} \cdot \nabla_{\tilde{x}} \tilde{u} - \nabla_{\tilde{v}} \tilde{u} \cdot \nabla_{\tilde{x}} U_{\tilde{Q}} \\ &= \mu^{-1} \mu \lambda^{-2} \lambda^{-1}\, v \cdot \nabla_x u - \mu \lambda^{-2} \mu \mu^{-2} \lambda^{-1}\, \nabla_v u \cdot \nabla_x U_Q \\ &= \lambda^{-3} (\mathcal{T} u)(x, v). \end{aligned} \tag{6.12}$$

Alternatively, $\mathcal{T} u = \{u, e_Q\}$ can be used. From (6.4) and (6.10), we get

$$\begin{aligned} (\mathcal{T}\tilde{u})(\tilde{x}, \tilde{v}) &= \{\tilde{u}, e_{\tilde{Q}}\} = \nabla_{\tilde{x}}\, \tilde{u} \cdot \nabla_{\tilde{v}}\, e_{\tilde{Q}} - \nabla_{\tilde{x}}\, e_{\tilde{Q}} \cdot \nabla_{\tilde{v}}\, \tilde{u} \\ &= \mu \lambda^{-3} \nabla_x u \cdot \mu^{-2} \mu \nabla_v e_Q - \mu^{-2} \lambda^{-1} \nabla_x e_Q \cdot \mu^2 \lambda^{-2} \nabla_v u \\ &= \lambda^{-3} \{u, e_Q\} = \lambda^{-3} (\mathcal{T} u)(x, v). \end{aligned}$$

This in turn leads to

$$\begin{aligned} (\mathcal{T}^2 \tilde{u})(\tilde{x}, \tilde{v}) &= \{\mathcal{T}\tilde{u}, e_{\tilde{Q}}\} = \nabla_{\tilde{x}}\, (\mathcal{T}\tilde{u}) \cdot \nabla_{\tilde{v}}\, e_{\tilde{Q}} - \nabla_{\tilde{x}}\, e_{\tilde{Q}} \cdot \nabla_{\tilde{v}}\, (\mathcal{T}\tilde{u}) \\ &= \lambda^{-4} \nabla_x (\mathcal{T} u) \cdot \mu^{-2} \mu \nabla_v e_Q - \mu^{-2} \lambda^{-1} \nabla_x e_Q \cdot \lambda^{-3} \mu \nabla_v (\mathcal{T} u) \\ &= \lambda^{-4} \mu^{-1} \{\mathcal{T} u, e_Q\} = \lambda^{-4} \mu^{-1} (\mathcal{T}^2 u)(x, v). \end{aligned} \tag{6.13}$$

Alternatively, if we put $\hat{u}(\tilde{x}, \tilde{v}) = u(\frac{\tilde{x}}{\lambda}, \mu\tilde{v})$, then $\tilde{u} = \mu\lambda^{-2}\hat{u}$, so (6.13) may be re-expressed as

$$(\mathcal{T}^2 \hat{u})(\tilde{x}, \tilde{v}) = \lambda^{-2} \mu^{-2} (\mathcal{T}^2 u)(x, v). \tag{6.14}$$

For the density induced by $\mathcal{T}\tilde{u}$, (6.12) yields

$$\rho_{\mathcal{T}\tilde{u}}(\tilde{x}) = \int (\mathcal{T}\tilde{u})(\tilde{x}, \tilde{v})\, d\tilde{v} = \lambda^{-3} \mu^{-3} \int (\mathcal{T} u)(x, v)\, dv = \lambda^{-3} \mu^{-3} \rho_{\mathcal{T} u}(x),$$

so that

$$U_{\mathcal{T}\tilde{u}}(\tilde{x}) = \lambda^{-1} \mu^{-3}\, U_{\mathcal{T} u}(x)$$

for the potential. In particular,

$$\nabla_{\tilde{x}} U_{\mathcal{T}\tilde{u}}(\tilde{x}) = \lambda^{-2} \mu^{-3}\, \nabla_x U_{\mathcal{T} u}(x),$$

and hence

$$\int |\nabla_{\tilde{x}} U_{T\tilde{u}}(\tilde{x})|^2 \, d\tilde{x} = \lambda^{-4}\mu^{-6}\lambda^3 \int |\nabla_x U_{Tu}(x)|^2 \, dx$$
$$= \lambda^{-1}\mu^{-6} \int |\nabla_x U_{Tu}(x)|^2 \, dx. \tag{6.15}$$

For

$$(Lu, u)_Q = \int\int \frac{dx\, dv}{|Q'(e_Q)|} |Tu|^2 - \frac{1}{4\pi} \int_{\mathbb{R}^3} |\nabla_x U_{Tu}|^2 \, dx$$

as given by (1.18), we then obtain from (6.5), (6.12) and (6.15):

$$\begin{aligned}(L\tilde{u}, \tilde{u})_{\tilde{Q}} &= \int\int \frac{d\tilde{x}\, d\tilde{v}}{|\tilde{Q}'(e_{\tilde{Q}})|} (T\tilde{u})^2 - \frac{1}{4\pi} \int |\nabla_{\tilde{x}} U_{T\tilde{u}}|^2 \, d\tilde{x} \\ &= \lambda^3\mu^{-3}\lambda^2\mu^{-3}\lambda^{-6} \int\int \frac{dx\, dv}{|Q'(e_Q)|} (Tu)^2 - \frac{1}{4\pi}\lambda^{-1}\mu^{-6} \int |\nabla_x U_{Tu}|^2 \, dx \\ &= \lambda^{-1}\mu^{-6}\,(Lu, u)_Q. \end{aligned} \tag{6.16}$$

In (1.20), the quantity

$$\lambda_* = \inf \{(Lu, u)_Q : u \in X^2_{\text{odd}}, \|u\|_Q = 1\}$$

is introduced. Therefore, owing to (6.16) and (6.11),

$$\begin{aligned}\tilde{\lambda}_* &= \inf \{(L\tilde{u}, \tilde{u})_{\tilde{Q}} : \tilde{u} \in X^2_{\text{odd}}, \|\tilde{u}\|_{\tilde{Q}} = 1\} \\ &= \lambda^{-1}\mu^{-6} \inf \{(Lu, u)_Q : u \in X^2_{\text{odd}}, \lambda\mu^{-4}\, \|u\|_Q = 1\} \\ &= \lambda^{-1}\mu^{-6}\lambda^{-1}\mu^4 \inf \{(L\hat{u}, \hat{u})_Q : \hat{u} \in X^2_{\text{odd}}, \|\hat{u}\|_Q = 1\} \\ &= \lambda^{-2}\mu^{-2}\lambda_*, \end{aligned} \tag{6.17}$$

by setting $u = \lambda^{-1/2}\mu^2\, \hat{u}$; it maybe checked that $u \in X^2_{\text{odd}}$ if and only if $\tilde{u} \in X^2_{\text{odd}}$ w.r. to the transformed variables.

Using (6.7), the function $A(r) = \frac{U_Q'(r)}{r}$ from (A.27) is found to scale as

$$\tilde{A}(\tilde{r}) = \frac{U'_{\tilde{Q}}(\tilde{r})}{\tilde{r}} = \lambda^{-1}\mu^{-2}\, \frac{U'_Q(r)}{\lambda r} = \lambda^{-2}\mu^{-2} A(r) \tag{6.18}$$

for $\tilde{r} \in [0, r_{\tilde{Q}}]$, with $r_{\tilde{Q}} = \lambda r_Q$ denoting the end of the support of $\rho_{\tilde{Q}}$, if r_Q denotes the end of the support of ρ_Q.

Similarly, denoting $B(r) = 4\pi\rho_Q(r) + A(r)$ as in Lemma A.7(d), owing to (6.18) and (6.6) one gets

$$\tilde{B}(\tilde{r}) = \lambda^{-2}\mu^{-2} B(r).$$

Now we turn to the operators $\mathcal{Q}_\nu$ from Chap. 4 and their first eigenvalues $\mu_1(\nu)$ for $\nu \in]-\infty, \delta_1^2[$; note the change in notation here for the parameter of the operators, since the letter λ is already occupied from $\tilde{x} = \lambda x, \tilde{r} = \lambda r$. Let

$$\tilde{\nu} = \frac{1}{\lambda^2\mu^2}\,\nu.$$

If $\nu \in]-\infty, \delta_1^2[$, then $\tilde{\nu} \in]-\infty, \tilde{\delta}_1^2[$ due to (6.9). For $\Psi = \Psi(r)$ let $\tilde{\Psi}(\tilde{r}) = \Psi(\frac{\tilde{r}}{\lambda})$. Since $\tilde{p}_r = \frac{\tilde{x}\cdot\tilde{v}}{|\tilde{x}|} = \mu^{-1}\frac{x\cdot v}{|x|} = \mu^{-1}p_r$, we obtain from (6.5):

$$\begin{aligned}\tilde{\psi}(\tilde{r}, \tilde{p}_r, \tilde{\ell}) &= |\tilde{Q}'(e_{\tilde{Q}})|\,\tilde{p}_r\,\tilde{\Psi}(\tilde{r}) = \mu^3\lambda^{-2}|Q'(e_Q)|\,\mu^{-1}p_r\,\Psi(r)\\ &= \mu^2\lambda^{-2}\,|Q'(e_Q)|\,p_r\,\Psi(r) = \mu^2\lambda^{-2}\,\psi(r, p_r, \ell).\end{aligned} \tag{6.19}$$

First we determine the scaling of $(-\mathcal{T}^2 - z)^{-1}\psi$. Defining

$$\tilde{z} = \frac{1}{\lambda^2\mu^2}\,z,$$

we assert that

$$((-\mathcal{T}^2 - \tilde{z})^{-1}\tilde{\psi})(\tilde{x}, \tilde{v}) = \mu^4\,((-\mathcal{T}^2 - z)^{-1}\psi)(x, v). \tag{6.20}$$

To see this, let $\tilde{g} = (-\mathcal{T}^2 - \tilde{z})^{-1}\tilde{\psi}$ and $g = (-\mathcal{T}^2 - z)^{-1}\psi$. Then, (6.20) is equivalent to $\tilde{g} = \mu^4 g$, but $\tilde{g}$ and g are not necessarily related by (6.10); in fact $\tilde{g} = \mu^4\hat{g}$ or $(-\mathcal{T}^2 - \tilde{z})\tilde{g} = \mu^4(-\mathcal{T}^2 - \tilde{z})\hat{g}$ is to be shown. For, owing to (6.14) and (6.19) we have

$$\begin{aligned}\mu^4(-\mathcal{T}^2 - \tilde{z})\hat{g} &= \mu^4(-\lambda^{-2}\mu^{-2}\mathcal{T}^2 g - \lambda^{-2}\mu^{-2}zg) = \mu^2\lambda^{-2}(-\mathcal{T}^2 - z)g = \mu^2\lambda^{-2}\,\psi\\ &= \tilde{\psi} = (-\mathcal{T}^2 - \tilde{z})\tilde{g},\end{aligned}$$

which completes the proof of (6.20). From (4.22) together with (6.20), we obtain

$$\begin{aligned}(\tilde{\mathcal{Q}}_{\tilde{z}}\tilde{\Psi})(\tilde{r}) &= 4\pi\int \tilde{p}_r\,((-\mathcal{T}^2 - \tilde{z})^{-1}\tilde{\psi})(\tilde{x}, \tilde{v})\,d\tilde{v}\\ &= 4\pi\mu^4\int \frac{\tilde{x}\cdot\tilde{v}}{|\tilde{x}|}\,((-\mathcal{T}^2 - z)^{-1}\psi)(\lambda^{-1}\tilde{x}, \mu\tilde{v})\,d\tilde{v}\\ &= 4\pi\int \frac{\lambda^{-1}\tilde{x}\cdot v}{|\lambda^{-1}\tilde{x}|}\,((-\mathcal{T}^2 - z)^{-1}\psi)(\lambda^{-1}\tilde{x}, v)\,dv\\ &= (\mathcal{Q}_z\Psi)(r).\end{aligned}$$

Thus, if we define

$$\tilde{\mu}_1(\tilde{\nu}) = \mu_1(\lambda^2\mu^2\tilde{\nu}), \quad \tilde{\nu} \in]-\infty, \tilde{\delta}_1^2[,$$

then $\tilde{\mu}_1(\tilde{\nu})$ is the first eigenvalue of $\widetilde{Q}_{\tilde{\nu}}$, and $\tilde{\Psi} = \tilde{\Psi}(\tilde{r})$ is an associated eigenfunction if and only if $\Psi = \Psi(r)$ is an eigenfunction of Q_ν for the eigenvalue $\mu_1(\nu)$. Due to (4.33) it follows that

$$\tilde{\mu}_* = \lim_{\tilde{\nu}\to\tilde{\delta}_1^2-} \tilde{\mu}_1(\tilde{\nu}) = \lim_{\nu\to\delta_1^2-} \mu_1(\nu) = \mu_*.$$

As already noted at the beginning of this chapter, it can be expected that quantities that are unaffected by the scaling do have a special relevance. Hence, μ_* is one such quantity. In addition, the condition $\lambda_* < \delta_1^2$ is invariant, as a consequence of (6.17) and (6.9). Further, we would like to mention

$$\frac{2\pi}{\sqrt{\lambda_*}}\sqrt{\rho_Q(0)},$$

cf. [59, Remark, p. 555], for which we deduce from (6.17) and (6.8):

$$\frac{2\pi}{\sqrt{\tilde{\lambda}_*}}\sqrt{\rho_{\tilde{Q}}(0)} = \frac{2\pi}{\lambda^{-1}\mu^{-1}\sqrt{\lambda_*}}\lambda^{-1}\mu^{-1}\sqrt{\rho_Q(0)} = \frac{2\pi}{\sqrt{\lambda_*}}\sqrt{\rho_Q(0)}.$$

This is called the Eddington-Ritter relation; also see [17, (27), p. 15] and [70, Section 4]. The relevance of the number $\frac{2\pi}{\sqrt{\lambda_*}}$ is that it is the 'linear period' of the system, in the sense that the linearized system about Q has a periodic solution of this period (if λ_* is an eigenvalue of L); recall Lemma 1.3.

Moreover, for any $r \in [0, r_Q]$ and $\tilde{r} = \lambda r$ one in fact has

$$\frac{\rho_{\tilde{Q}}(\tilde{r})}{\tilde{\lambda}_*} = \frac{\rho_Q(r)}{\lambda_*}, \quad \frac{\tilde{A}(\tilde{r})}{\tilde{\lambda}_*} = \frac{A(r)}{\lambda_*}, \quad \frac{\tilde{B}(\tilde{r})}{\tilde{\lambda}_*} = \frac{B(r)}{\lambda_*}.$$

Appendix A
Spherical Symmetry and Action-Angle Variables

A.1 General Theory

A function $g = g(x, v)$ is said to be spherically symmetric, if $g(Ax, Av) = g(x, v)$ for all $A \in \mathrm{SO}(3)$ and $x, v \in \mathbb{R}^3$. In this case $\rho_g(x) = \rho_g(r)$ and $U_g(x) = U_g(r)$ are radially symmetric; here $r = |x|$. More explicitly,

$$U_g(r) = -\frac{4\pi}{r}\int_0^r s^2\rho_g(s)\,ds - 4\pi\int_r^\infty s\rho_g(s)\,ds, \tag{A.1}$$

$$U_g'(r) = \frac{4\pi}{r^2}\int_0^r s^2\rho_g(s)\,ds = \frac{1}{r^2}\int_{|x|\le r}\rho_g(x)\,dx; \tag{A.2}$$

see [73]. Then, the mass in a ball of radius r is given by $m(r) = r^2 U_g'(r) = 4\pi\int_0^r s^2\rho_g(s)\,ds = \int_{|x|\le r}\rho_g(x)\,dx$.

Remark A.1 One can also write

$$U_g'(r) = \frac{4\pi}{3}\,r\rho_g(r) - \frac{4\pi}{3r^2}\int_0^r s^3\rho_g'(s)\,ds,$$

as follows from an integration by parts. ◇

Let g be spherically symmetric. Then, $g = g(x, v) = \tilde{g}(|x|, |v|, x\cdot v)$ does in fact depend only upon three variables. By the spherical symmetry, one can use a canonical change of variables

$$(x, v) \mapsto (p_r, L_3, \ell; r, \varphi, \chi) \tag{A.3}$$

on the support $K = \operatorname{supp} Q$ of a steady state solution Q as described in [9, Chap. 3.5.2] and [90, Sect. 5.3] to simplify matters considerably. Let

M. Kunze, *A Birman-Schwinger Principle in Galactic Dynamics*, Progress in Mathematical Physics 77, https://doi.org/10.1007/978-3-030-75186-9

$$r = |x|, \quad p_r = \frac{x \cdot v}{r} \quad \text{and} \quad L = x \wedge v.$$

In particular, then $L_3 = x_1 v_2 - x_2 v_1$. Thus, $\ell^2 = |L|^2 = |x|^2|v|^2 - (x \cdot v)^2 = r^2(|v|^2 - p_r^2)$ so that $|v|^2 = \frac{\ell^2}{r^2} + p_r^2$. For the particle energy $e_Q(x, v) = \frac{1}{2}|v|^2 + U_Q(r)$ this yields

$$e_Q = e_Q(r, p_r, \ell) = \frac{1}{2}\, p_r^2 + U_{\text{eff}}(r, \ell), \quad U_{\text{eff}}(r, \ell) = U_Q(r) + \frac{\ell^2}{2r^2} \tag{A.4}$$

being the effective potential, for which we have

$$U'_{\text{eff}}(r, \ell) = U'_Q(r) - \frac{\ell^2}{r^3}. \tag{A.5}$$

Since $\dot{r} = \frac{\partial e}{\partial p_r} = p_r$, the resulting equation of motion is

$$\ddot{r} = -U'_{\text{eff}}(r, \ell).$$

From (1.13), it also follows that

$$U''_{\text{eff}}(r, \ell) + \frac{2}{r}\, U'_{\text{eff}}(r, \ell) = 4\pi\rho_Q(r) + \frac{\ell^2}{r^4}. \tag{A.6}$$

Apart from r, the other "angular" variables φ and χ are determined by

$$\sin\varphi = \frac{L_1}{(\ell^2 - L_3^2)^{1/2}}, \quad \cos\varphi = \frac{L_2}{(\ell^2 - L_3^2)^{1/2}}, \tag{A.7}$$

$$\cos\chi = \frac{(e_3 \wedge L) \cdot x}{r(\ell^2 - L_3^2)^{1/2}}, \quad \sin\chi = \frac{\ell\, x_3}{r(\ell^2 - L_3^2)^{1/2}}. \tag{A.8}$$

The variable pairs $r \leftrightarrow p_r$, $\varphi \leftrightarrow L_3$, and $\chi \leftrightarrow \ell$ are conjugate, their Poisson brackets can be calculated explicitly; see [90, Sect. 5.3], also for an illustration of how the new coordinates can be read off.

Our goal is now to pass to action-angle variables; see [9, Chap. 3.5] and [90, p. 224]. The new variables $(p_r, L_3, \ell; r, \varphi, \chi)$ in (A.3) are not yet the desired action-angle variables, since $e = e(r, p_r, \ell)$ depends upon r, which plays the role of an angle. Therefore, a further canonical transformation

$$(r, p_r) \to (\theta, I) \quad \text{at a fixed } \ell \tag{A.9}$$

will be made. As ℓ is fixed, the potential to consider is $r \mapsto U_{\text{eff}}(r, \ell)$, and for this the change of variables (A.9) in a region where the orbits of $U_{\text{eff}}(\cdot, \ell)$ are periodic is standard; it is achieved by means of a generating function. Good general accounts of this procedure are [6, Chap. 50] and [21, Chap. 11.3]. Let $0 < r_-(e, \ell) < r_+(e, \ell)$

denote the zeros of $0 = 2(e - U_{\text{eff}}(r, \ell))$; see [7, Sect. 2], [30, Sect. 3.2], [50, Lemma 2.1] and [88, Theorem 2.4] for further discussions. The angle $\theta \in [0, \pi]$ corresponds to one half-turn of the periodic orbit γ in the potential $U_{\text{eff}}(\cdot, \ell)$, connecting the "pericenter" r_- to the "apocenter" r_+; here $\dot{r} = p_r > 0$ for $r \in]r_-, r_+[$ and $p_r(r_\pm) = 0$. Therefore if $\theta \in [\pi, 2\pi]$, then

$$r(\theta, I, \ell) = r(2\pi - \theta, I, \ell) \quad \text{and} \quad p_r(\theta, I, \ell) = -p_r(2\pi - \theta, I, \ell). \tag{A.10}$$

In other words, we need to determine the (inverse) transformation $(\theta, I) \mapsto (r, p_r)$ only for $\theta \in [0, \pi]$, where we have $p_r \geq 0$.

Let $E = E(I, \ell)$ be the solution to

$$I = \frac{1}{2\pi} \int_\gamma p_r \, dr = \frac{1}{\pi} \int_{r_-(E,\ell)}^{r_+(E,\ell)} \sqrt{2(E - U_{\text{eff}}(r, \ell))} \, dr, \tag{A.11}$$

where γ is as before. Then consider the generating function

$$S(r, I, \ell) = \int_{r_-(E(I,\ell),\ell)}^{r} \sqrt{2(E(I, \ell) - U_{\text{eff}}(r', \ell))} \, dr', \tag{A.12}$$

which we view as a generating function for (A.9); in the terminology used in physics books [21], it is a generating function of the second type, and it depends on one "old variable", r, and one "new variable", I. The rules for determining the full transformation from S are

$$\theta = \partial_I S, \quad p_r = \partial_r S. \tag{A.13}$$

Let us do a short and formal calculation to explain the use of (A.13). Firstly, $p_r = \partial_r S$ and (A.12) yield $e = \frac{p_r^2}{2} + U_{\text{eff}}(r, \ell) = E(I, \ell)$, so E will only depend upon action variables after the transformation (A.9), which leads to the overall transformation

$$(x, v) \mapsto (p_r, L_3, \ell; r, \varphi, \chi) \mapsto (I, L_3, \ell; \theta, \varphi, \chi), \tag{A.14}$$

cf. (A.3). Secondly, the transformation (A.9) is symplectic. To see this, differentiating $p_r = \partial_r S(r, I)$ w.r.t. to p_r implies that $1 = (\partial^2_{rI} S)(\partial_{p_r} I)$. Therefore, we deduce from $\theta = \partial_I S(r, I)$ that

$$\begin{aligned} d\theta \wedge dI &= \Big[(\partial^2_{rI} S)\, dr + (\partial^2_{II} S)\, dI\Big] \wedge dI = (\partial^2_{rI} S)\, dr \wedge dI \\ &= (\partial^2_{rI} S)\, dr \wedge \Big[(\partial_r I)\, dr + (\partial_{p_r} I)\, dp_r\Big] = (\partial^2_{rI} S)(\partial_{p_r} I)\, dr \wedge dp_r = dr \wedge dp_r, \end{aligned}$$

which means that (A.9) is indeed symplectic.

Let us now be somewhat more careful with the dependencies and the definition of the transformation (A.9). The equation

$$\theta = \partial_I S(r, I, \ell) \tag{A.15}$$

has a solution $r = r(\theta, I, \ell)$. In addition, put

$$p_r = p_r(\theta, I, \ell) = \partial_r S(r(\theta, I, \ell), I, \ell).$$

Thus, more explicitly

$$p_r(\theta, I, \ell) = \sqrt{2(E(I, \ell) - U_{\text{eff}}(r(\theta, I, \ell), \ell))}, \tag{A.16}$$

which yields

$$E(I, \ell) = \frac{1}{2} p_r(\theta, I, \ell)^2 + U_{\text{eff}}(r(\theta, I, \ell), \ell) = e(r(\theta, I, \ell), p_r(\theta, I, \ell), \ell). \tag{A.17}$$

Hence after applying the canonical transformation (A.14) the particle energy does only depend upon I and ℓ, both of which are actions. The associated frequencies are

$$\omega_1(I, \ell) = \frac{\partial E(I, \ell)}{\partial I}, \quad \omega_2(I, \ell) = \frac{\partial E(I, \ell)}{\partial L_3} = 0, \quad \omega_3(I, \ell) = \frac{\partial E(I, \ell)}{\partial \ell}, \tag{A.18}$$

and the period functions are

$$T_1(I, \ell) = \frac{2\pi}{\omega_1(I, \ell)}, \quad T_3(I, \ell) = \frac{2\pi}{\omega_3(I, \ell)}.$$

Also (A.15) yields

$$\theta = \partial_I S(r, I, \ell) = \omega_1(I, \ell) \int_{r_-(E(I, \ell), \ell)}^{r} \frac{dr'}{\sqrt{2(E(I, \ell) - U_{\text{eff}}(r', \ell))}}. \tag{A.19}$$

Since $\theta = 0$ at r_- and $\theta = \pi$ at r_+ (recall that $\dot{r} = p_r > 0$ along this part of the orbit), we obtain

$$\pi = \frac{2\pi}{T_1(I, \ell)} \int_{r_-(E(I, \ell), \ell)}^{r_+(E(I, \ell), \ell)} \frac{dr}{\sqrt{2(E(I, \ell) - U_{\text{eff}}(r, \ell))}},$$

or explicitly

$$T_1(I, \ell) = 2 \int_{r_-(E(I, \ell), \ell)}^{r_+(E(I, \ell), \ell)} \frac{dr}{\sqrt{2(E(I, \ell) - U_{\text{eff}}(r, \ell))}} \tag{A.20}$$

for the period function. In particular, $T_1(I, \ell) = T_1(E, \ell)$ by abuse of notation. Also (A.19) implies that

$$\frac{\partial\theta}{\partial r} = \frac{\omega_1}{p_r} \tag{A.21}$$

as long as $p_r > 0$, i.e., along the half-orbit.

More systematically, the relations $\theta = \partial_I S(r, I, \ell)$ for $r = r(\theta, I, \ell)$ and

$$p_r(\theta, I, \ell) = \partial_r S(r(\theta, I, \ell), I, \ell)$$

can be differentiated to obtain explicit formulas for the derivatives. This way it is found that

$$\begin{aligned}
1 &= (\partial^2_{rI} S)(\partial_\theta r),\\
0 &= (\partial^2_{rI} S)(\partial_I r) + \partial^2_{II} S,\\
\partial_\theta p_r &= (\partial^2_{rr} S)(\partial_\theta r),\\
\partial_I p_r &= (\partial^2_{rr} S)(\partial_I r) + \partial^2_{rI} S.
\end{aligned}$$

From (A.19), it follows that

$$\partial^2_{rI} S(r, I, \ell) = \frac{\omega_1(I, \ell)}{\sqrt{2(E(I, \ell) - U_{\text{eff}}(r, \ell))}} = \frac{\omega_1(I, \ell)}{p_r(\theta, I, \ell)},$$

which leads to

$$\frac{\partial r}{\partial\theta} = \frac{p_r}{\omega_1} \quad \text{and} \quad \frac{\partial r}{\partial I} = -(\partial^2_{II} S)\,\frac{p_r}{\omega_1}. \tag{A.22}$$

In addition, by (A.16),

$$\partial^2_{rr} S = -\frac{1}{p_r}\, U'_{\text{eff}}$$

and this gives

$$\frac{\partial p_r}{\partial\theta} = -\frac{1}{\omega_1}\, U'_{\text{eff}}.$$

Finally,

$$\frac{\partial p_r}{\partial I} = \frac{1}{\omega_1}\, U'_{\text{eff}}(\partial^2_{II} S) + \frac{\omega_1}{p_r}.$$

To summarize, spherically symmetric functions $g = g(x, v) = \tilde{g}(|x|, |v|, x \cdot v)$ may also be expressed as $g = \hat{g}(r, p_r, \ell) = g^*(\theta, I, \ell)$. Explicitly,

$$\tilde{g}(r, s, u) = \hat{g}\left(r, \frac{u}{r}, \sqrt{r^2 s^2 - u^2}\right), \quad \hat{g}(r, p_r, \ell) = \tilde{g}\left(r, \sqrt{p_r^2 + \frac{\ell^2}{r^2}}, r p_r\right). \tag{A.23}$$

Most of the time all versions of a function g will be denoted by the same symbol.

The Antonov stability estimate is valid for spherically symmetric functions $u = u(x, v)$ that are odd in v. Therefore, we need to have a closer look at this class of functions.

Remark A.2 (Parity)

(a) If $(x, v) \mapsto (p_r, L_3, \ell; r, \varphi, \chi)$ under the above transformation (A.3), then $(x, -v) \mapsto (-p_r, -L_3, \ell; r, \varphi + \pi, \pi - \chi)$. This follows from $p_r = \frac{x \cdot v}{r}$, $L = x \wedge v$, (A.7), and (A.8), or alternatively from the figure in [90, p. 223]. In addition, $e(x, -v) = e(x, v)$ and $\ell(x, -v) = \ell(x, v)$.

(b) From (A.10), we see that if

$$(x, v) \mapsto (p_r, L_3, \ell; r, \varphi, \chi) \mapsto (I, L_3, \ell; \theta, \varphi, \chi)$$

under the transformation (A.14), then

$$\begin{aligned}(x, -v) &\mapsto (-p_r, -L_3, \ell; r, \varphi + \pi, \pi - \chi)\\ &\mapsto (I, -L_3, \ell; 2\pi - \theta, \varphi + \pi, \pi - \chi).\end{aligned}$$

Thus if $g = g(x, v)$ is spherically symmetric, then

(a) g is even in v if and only if $\hat{g}(r, -p_r, \ell) = \hat{g}(r, p_r, \ell)$ if and only if $g^*(2\pi - \theta, I, \ell) = g^*(\theta, I, \ell)$,
(b) g is odd in v if and only if $\hat{g}(r, -p_r, \ell) = -\hat{g}(r, p_r, \ell)$ if and only if $g^*(2\pi - \theta, I, \ell) = -g^*(\theta, I, \ell)$,

as will be convenient to determine the parity in v of a function g. ◇

We are going to note some further useful relations, and we will be writing $\beta = \ell^2$, so that $d\beta = \ell\, d\ell$. Recall from above that the radii $r_\pm = r_\pm(e, \ell) = r_\pm(e, \beta)$ are the zeros of $0 = 2(e - U_{\text{eff}}(r, \ell))$ and satisfy $0 < r_-(e, \ell) < r_+(e, \ell)$ for the effective potential

$$U_{\text{eff}}(r, \ell) = U_{\text{eff}}(r, \beta) = U_Q(r) + \frac{\ell^2}{2r^2} = U_Q(r) + \frac{\beta}{2r^2}.$$

As in [50, Lemma 2.1], one knows that for every $\beta > 0$

$$\inf\{U_{\text{eff}}(r, \beta) : r \geq 0\}$$

is attained at some unique $r_0(\beta) \in]r_-(e, \beta), r_+(e, \beta)[$. Then $U_{\text{eff}}(\cdot, \beta)$ is decreasing in $[r_-(e, \beta), r_0(\beta)[$, increasing in $]r_0(\beta), r_+(e, \beta)]$, and we have $U'_{\text{eff}}(r_0(\beta), \beta) = 0$.

Remark A.3 We recall that

$$D = \{(e, \beta) : \beta \in [0, \beta_*], e \in [e_{\min}(\beta), e_0]\},$$

and

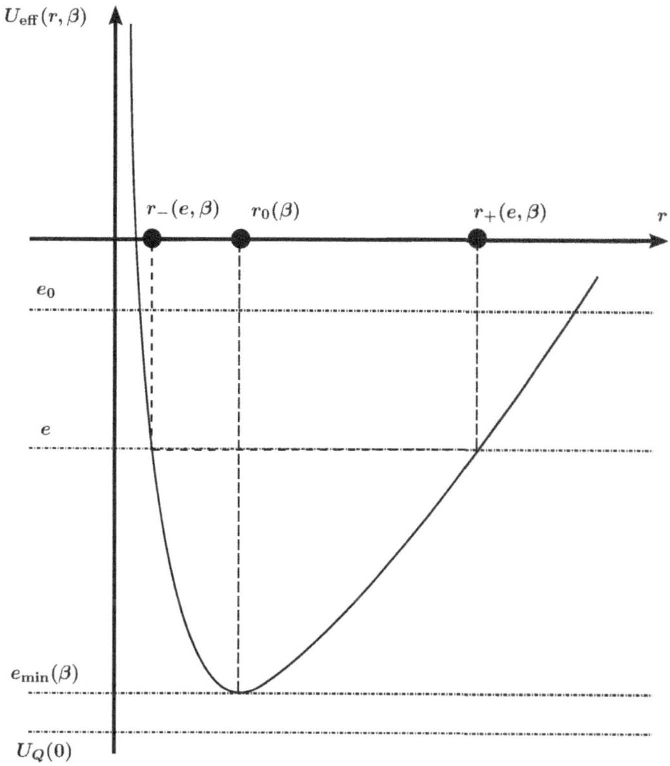

Fig. A.1 The effective potential $U_{\text{eff}}(r, \beta)$

$$\mathring{D} = \{(e, \beta) : \beta \in]0, \beta_*[, e \in]e_{\min}(\beta), e_0[\}$$

is its interior. Then

$$r_\pm \in C^2(\mathring{D}) \quad \text{and} \quad r_0 \in C^1(]0, \beta_*[).$$

This follows from the implicit function theorem and $U_Q \in C^2$, since $r_\pm$ solve $U_Q(r_\pm(e, \beta)) + \frac{\beta}{2r_\pm(e,\beta)^2} = e$, whereas r_0 is the solution to $r_0(\beta)^3 U_Q'(r_0(\beta)) = \beta$; see Lemma A.7(a) below for the latter.

Lemma A.4 *One has*

$$\frac{\partial r_\pm}{\partial \beta}(e, \beta) = -\frac{1}{2r_\pm(e, \beta)^2\, U_{\text{eff}}'(r_\pm(e, \beta), \beta)}, \tag{A.24}$$

and hence in particular

$$\frac{\partial r_-}{\partial \beta}(e, \beta) > 0 \quad \textit{and} \quad \frac{\partial r_+}{\partial \beta}(e, \beta) < 0.$$

Furthermore,

$$\frac{\partial r_\pm}{\partial e}(e,\beta) = \frac{1}{U'_{\text{eff}}(r_\pm(e,\beta),\beta)}, \tag{A.25}$$

so that also

$$\frac{\partial r_-}{\partial e}(e,\beta) < 0 \quad \textit{and} \quad \frac{\partial r_+}{\partial e}(e,\beta) > 0.$$

Proof With $U_{\text{eff}}(r,\beta) = U_Q(r) + \frac{\beta}{2r^2}$, we have $U_{\text{eff}}(r_\pm(e,\beta),\beta) = e$ so that $2r_\pm^2(e - U_Q(r_\pm)) = \beta$. Upon differentiation w.r.t. to β, we obtain

$$\begin{aligned}
1 &= -2r_\pm^2 U'_Q(r_\pm)\,\frac{\partial r_\pm}{\partial\beta} + 4r_\pm(e - U_Q(r_\pm))\,\frac{\partial r_\pm}{\partial\beta} \\
&= \Big(-2r_\pm^2 U'_Q(r_\pm) + \frac{2\beta}{r_\pm}\Big)\frac{\partial r_\pm}{\partial\beta} \\
&= -2r_\pm^2\Big(U'_Q(r_\pm) - \frac{\beta}{r_\pm^3}\Big)\frac{\partial r_\pm}{\partial\beta} \\
&= -2r_\pm^2\, U'_{\text{eff}}(r_\pm,\beta)\,\frac{\partial r_\pm}{\partial\beta},
\end{aligned}$$

which yields (A.24). Since $U'_{\text{eff}}(r_-,\beta) < 0$ and $U'_{\text{eff}}(r_+,\beta) > 0$, the claim concerning the signs follows. In order to derive (A.25), we differentiate $U_{\text{eff}}(r_\pm,\beta) = e$ w.r.t. to e and get $1 = U'_{\text{eff}}(r_\pm,\beta)\,\frac{\partial r_\pm}{\partial e}$. □

Now we determine the asymptotics of $\frac{\beta}{r_-^2 r_+^2}$ as $\beta \to 0$.

Lemma A.5 *We have*

$$\begin{aligned}
&\frac{\beta}{2r_-(e,\beta)^2 r_+(e,\beta)^2} - \frac{1}{2}\,U''_Q(0) \\
&= \frac{1}{r_+(e,\beta) + r_-(e,\beta)}\int_0^1 ds\,(sr_+(e,\beta) + (1-s)r_-(e,\beta)) \\
&\qquad \times \int_0^1 d\tau\,[U''_Q(\tau s r_+(e,\beta) + \tau(1-s)r_-(e,\beta)) - U''_Q(0)],
\end{aligned}$$

and in particular

$$\left|\frac{\beta}{2r_-(e,\beta)^2 r_+(e,\beta)^2} - \frac{1}{2}\,U''_Q(0)\right| \le \frac{1}{2}\sup_{r\in[0,r_+(e,\beta)]}|U''_Q(r) - U''_Q(0)|. \tag{A.26}$$

Proof Since $U_Q(r_\pm) + \frac{\beta}{2r_\pm^2} = e$, we can write, using that $U'_Q(0) = 0$,

$$\begin{aligned}\frac{\beta}{2r_-^2 r_+^2}(r_+^2 - r_-^2) &= e - \frac{\beta}{2r_+^2} - e + \frac{\beta}{2r_-^2}\\ &= U_Q(r_+) - U_Q(r_-)\\ &= (r_+ - r_-)\int_0^1 [U_Q'(sr_+ + (1-s)r_-) - U_Q'(0)]\,ds\\ &= (r_+ - r_-)\int_0^1 ds\,(sr_+ + (1-s)r_-)\\ &\qquad\times\int_0^1 d\tau\, U_Q''(\tau s r_+ + \tau(1-s)r_-).\end{aligned}$$

Therefore,

$$\begin{aligned}&\frac{\beta}{2r_-^2 r_+^2} - \frac{1}{2}U_Q''(0)\\ &= \frac{1}{r_+ + r_-}\int_0^1 ds\,(sr_+ + (1-s)r_-)\int_0^1 d\tau\, U_Q''(\tau s r_+ + \tau(1-s)r_-) - \frac{1}{2}U_Q''(0)\\ &= \frac{1}{r_+ + r_-}\int_0^1 ds\,(sr_+ + (1-s)r_-)\int_0^1 d\tau\,[U_Q''(\tau s r_+ + \tau(1-s)r_-) - U_Q''(0)],\end{aligned}$$

as claimed. Then (A.26) is a consequence of $0 < r_- \le r_+$. □

We need to introduce two more important functions.

Lemma A.6 *The following assertions hold.*

(a) The function

$$A(r) = \frac{U_Q'(r)}{r} \tag{A.27}$$

is C^1 and strictly decreasing in $r \in [0, r_Q]$. In addition, $A(0) = U_Q''(0) = \frac{4\pi}{3}\rho_Q(0)$ and $A(r_Q) = \frac{1}{r_Q^3}\|Q\|_{L^1(\mathbb{R}^6)}$. We also have

$$A'(r) = \frac{4\pi}{r^4}\int_0^r s^3\rho_Q'(s)\,ds \quad \text{and} \quad \lim_{r\to 0^+} A'(r) = \pi\rho_Q'(0). \tag{A.28}$$

(b) The function

$$B(r) = 4\pi\rho_Q(r) + A(r) = 4\pi\rho_Q(r) + \frac{U_Q'(r)}{r}$$

is C^1 and strictly decreasing in $r \in [0, r_Q]$. In addition, $B(0) = \frac{16\pi}{3}\rho_Q(0)$ as well as $B(r_Q) = A(r_Q) = \frac{1}{r_Q^3}\|Q\|_{L^1(\mathbb{R}^6)}$.

(c) *We have*

$$U''_{\text{eff}}(r,\beta)+\frac{3}{r}\,U'_{\text{eff}}(r,\beta)=B(r)\quad\textit{and}\quad U''_{\text{eff}}(r,\beta)=\frac{3\beta}{r^4}+B(r)-3A(r).$$

In particular,

$$\begin{cases} U''_{\text{eff}}(r,\beta)\ge B(r)\ge B(r_Q) & : \; r\in[r_-(e,\beta),r_0(\beta)] \\ U''_{\text{eff}}(r,\beta)\le B(r)\le B(0) & : \; r\in[r_0(\beta),r_+(e,\beta)] \end{cases} \tag{A.29}$$

and

$$|U''_{\text{eff}}(r,\beta)|\le\frac{3\beta}{r^4}+\frac{28\pi}{3}\,\rho_Q(0)\quad\textit{for}\quad r\in[r_-(e,\beta),r_+(e,\beta)]. \tag{A.30}$$

Proof (a) From the differential equation (1.13) for Q one has $U''_Q+\frac{2}{r}\,U'_Q=4\pi\rho_Q$. Therefore,

$$A'(r)=\frac{rU''_Q(r)-U'_Q(r)}{r^2}=\frac{4\pi r\rho_Q(r)-3U'_Q(r)}{r^2}. \tag{A.31}$$

Now by (A.2) and the hypotheses:

$$\rho'_Q(r)=\int_{\mathbb{R}^3}\frac{d}{dr}\,Q\,dv=\int_{\mathbb{R}^3}Q'(e_Q)\,\frac{d}{dr}\,e_Q\,dv=\int_{\mathbb{R}^3}Q'(e_Q)\,U'_Q\,dv\le 0, \tag{A.32}$$

so that ρ_Q is radially decreasing. It thus follows from Remark A.1 that

$$U'_Q(r)=\frac{4\pi}{3}\,r\rho_Q(r)-\frac{4\pi}{3r^2}\int_0^r s^3\rho'_Q(s)\,ds\ge\frac{4\pi}{3}\,r\rho_Q(r). \tag{A.33}$$

Hence $4\pi r\rho_Q(r)\le 3U'_Q(r)$ and $A'(r)\le 0$, in fact $A'(r)<0$ for $r\in]0,r_Q]$. From the left side of (A.33) we also obtain $\lim_{r\to 0^+}A(r)=\frac{4\pi}{3}\,\rho_Q(0)$, which shows that A is continuous on $[0,r_Q]$. Next we calculate

$$\begin{aligned} U'_Q(r_Q)&=\frac{4\pi}{r_Q^2}\int_0^{r_Q}s^2\rho_Q(s)\,ds=\frac{4\pi}{r_Q^2}\int_0^{\infty}s^2\rho_Q(s)\,ds=\frac{1}{r_Q^2}\int_{\mathbb{R}^3}\rho_Q(x)\,dx \\ &=\frac{1}{r_Q^2}\int_{\mathbb{R}^3}\int_{\mathbb{R}^3}Q(x,v)\,dx\,dv=\frac{1}{r_Q^2}\,\|Q\|_{L^1(\mathbb{R}^6)}, \end{aligned}$$

which is as desired. The expression for $A'(r)$ in (A.28) is a consequence of (A.31) and (A.33), whereas $\lim_{r\to 0^+}A'(r)=\pi\rho'_Q(0)$ is gotten from this relation by changing variables $s=r\tau$, $ds=r\,d\tau$. In particular, A is C^1 on $[0,r_Q]$.

(b) This follows from (a).

(c) Using (A.6) and (A.5), we deduce

$$
\begin{aligned}
U''_{\rm eff}(r,\beta)+\frac{3}{r}\,U'_{\rm eff}(r,\beta) &= 4\pi\rho_Q(r)+\frac{1}{r}\Big(\frac{\beta}{r^3}+U'_{\rm eff}(r,\beta)\Big)\\
&= 4\pi\rho_Q(r)+\frac{U'_Q(r)}{r}=B(r).
\end{aligned}
$$

Also

$$
\frac{3}{r}\,U'_{\rm eff}(r,\beta)=\frac{3}{r}\,\Big(U'_Q(r)-\frac{\beta}{r^3}\Big)=3A(r)-\frac{3\beta}{r^4},
$$

which yields the second claim. The estimates in (A.29) are due to $U'_{\rm eff}(r,\beta)\le 0$ for $r\in[r_-(e,\beta),r_0(\beta)]$, $U'_{\rm eff}(r,\beta)\ge 0$ for $r\in[r_0(\beta),r_+(e,\beta)]$ and the monotonicity of B. Lastly, (A.30) is obtained from $|U''_{\rm eff}(r,\beta)|\le\frac{3\beta}{r^4}+B(r)+3A(r)\le\frac{3\beta}{r^4}+B(0)+3A(0)=\frac{3\beta}{r^4}+\frac{28\pi}{3}\,\rho_Q(0)$. □

Lemma A.7 *We have*

(a) $r_0(\beta)^3U'_Q(r_0(\beta))=\beta$, *and in particular for every* $\beta_0\in]0,\beta_*[$ *there exists* $c_0>0$ *such that* $r_0(\beta)\ge c_0$ *for* $\beta\in[\beta_0,\beta_*]$;
(b) $r^2U'_{\rm eff}(r,\beta)=\int_{r_0(\beta)}^r(4\pi s^2\rho_Q(s)+\frac{\beta}{s^2})\,ds$;
(c) $\frac{de_{\rm min}}{d\beta}(\beta)=\frac{1}{2r_0(\beta)^2}$, *where* $e_{\rm min}(\beta)=U_{\rm eff}(r_0(\beta),\beta)$;
(d) $U''_{\rm eff}(r_0(\beta),\beta)=B(r_0(\beta))$;
(e) $\frac{dr_0}{d\beta}(\beta)=\frac{1}{r_0(\beta)^3\,B(r_0(\beta))}$;
(f) *as* $\beta\to 0^+$,

$$
\begin{aligned}
r_0(\beta)^4 &= \frac{1}{A(0)}\,\beta+O(\beta^{5/4}), \qquad (A.34)\\
e_{\rm min}(\beta) &= U_Q(0)+\sqrt{U''_Q(0)\sqrt{\beta}}+O(\beta^{3/4}),\\
\frac{de_{\rm min}}{d\beta}(\beta) &= \frac{\sqrt{A(0)}}{2\sqrt{\beta}}+O(\beta^{-1/4}).
\end{aligned}
$$

Proof (a) The relation $r_0(\beta)^3U'_Q(r_0(\beta))=\beta$ follows from

$$
U'_{\rm eff}(r_0(\beta),\beta)=0 \qquad (A.35)
$$

and (A.5). To establish the second claim, suppose on the contrary that there is $\beta_0\in]0,\beta_*[$ and a sequence $(\beta_j)\subset[\beta_0,\beta_*]$ so that $r_0(\beta_j)\le 1/j$. W.l.o.g. we may assume that $\beta_j\to\hat\beta\in[\beta_0,\beta_*]$. Then $\beta_j=r_0(\beta_j)^3U'_Q(r_0(\beta_j))\le j^{-3}U'_Q(r_0(\beta_j))$ and $0\le U'_Q(r)=\frac{4\pi}{r^2}\int_0^r s^2\rho_Q(s)\,ds\le\frac{4\pi r}{3}\rho_Q(0)\le\frac{4\pi r_Q}{3}\rho_Q(0)$ yield a contradiction as $j\to\infty$.

(b) Owing to (A.6) we have that

$$
(r^2U'_{\rm eff})'=4\pi r^2\rho_Q+\frac{\beta}{r^2}, \qquad (A.36)
$$

so it remains to integrate this equation, using (A.35).

(c) We obtain

$$\frac{d}{d\beta}\left[U_{\text{eff}}(r_0(\beta),\beta)\right] = U'_{\text{eff}}(r_0(\beta),\beta)\,\frac{dr_0}{d\beta}(\beta) + \frac{\partial U_{\text{eff}}}{\partial\beta}(r_0(\beta),\beta) = \frac{\partial U_{\text{eff}}}{\partial\beta}(r_0(\beta),\beta).$$

Since $\frac{\partial U_{\text{eff}}}{\partial\beta}(r,\beta) = \frac{1}{2r^2}$, the claim follows.

(d) From (A.6), (A.35) and (a), one finds that

$$\begin{aligned} U''_{\text{eff}}(r_0(\beta),\beta) &= 4\pi\rho_Q(r_0(\beta)) - \frac{2}{r_0(\beta)}\,U'_{\text{eff}}(r_0(\beta),\beta) + \frac{\beta}{r_0(\beta)^4} \\ &= 4\pi\rho_Q(r_0(\beta)) + \frac{\beta}{r_0(\beta)^4} \\ &= 4\pi\rho_Q(r_0(\beta)) + \frac{U'_Q(r_0(\beta))}{r_0(\beta)} \\ &= B(r_0(\beta)). \end{aligned}$$

(e) Differentiating (a), we get

$$1 = r_0^3\,U''_Q(r_0)\,\frac{dr_0}{d\beta} + 3r_0^2\,\frac{dr_0}{d\beta}\,U'_Q(r_0) = (r_0^3\,U''_Q(r_0) + 3r_0^2\,U'_Q(r_0))\,\frac{dr_0}{d\beta}.$$

But

$$\begin{aligned} r_0^3\,U''_Q(r_0) + 3r_0^2\,U'_Q(r_0) &= r_0^3\left(4\pi\rho_Q(r_0) - \frac{2}{r_0}\,U'_Q(r_0)\right) + 3r_0^2\,U'_Q(r_0) \\ &= 4\pi r_0^3\,\rho_Q(r_0) + r_0^2\,U'_Q(r_0) \\ &= (4\pi\,\rho_Q(r_0) + A(r_0))\,r_0^3. \end{aligned}$$

(f) From (a) and Lemma A.6(a), we derive that

$$r_0(\beta)^4 A(r_0(\beta)) = r_0(\beta)^3 U'_Q(r_0(\beta)) = \beta.$$

As A is bounded from below by $A(r_Q) > 0$, we get $r_0(\beta) \le (\beta/A(r_Q))^{1/4}$, and in particular $\lim_{\beta\to 0^+} r_0(\beta) = 0$. In addition,

$$\left| r_0(\beta)^4 - \frac{1}{A(0)}\,\beta \right| = \left| \frac{1}{A(r_0(\beta))} - \frac{1}{A(0)} \right| \beta \le \frac{1}{A(r_Q)^2}\,|A(r_0(\beta)) - A(0)|\,\beta.$$

Now (A.28) from Lemma A.6(a) together with (Q4) implies that

$$|A'(r)| \le \frac{4\pi}{r^4}\int_0^r s^3|\rho'_Q(s)|\,ds \le \pi\|\rho'_Q\|_\infty,$$

and this yields

$$\left| r_0(\beta)^4 - \frac{1}{A(0)}\,\beta \right| \le \frac{\pi \|\rho_Q'\|_\infty}{A(r_Q)^2}\, r_0(\beta)\,\beta \le \frac{\pi \|\rho_Q'\|_\infty}{A(r_Q)^{9/4}}\, \beta^{5/4}.$$

In other words,

$$r_0(\beta)^4 = \frac{1}{A(0)}\,\beta + O(\beta^{5/4}), \quad \beta \to 0^+.$$

Therefore, since $U_Q'(0) = 0$ by (A.2) and the boundedness of ρ_Q on $[0, r_Q]$,

$$\begin{aligned} U_{\rm eff}(r_0(\beta), \beta) &= U_Q(r_0(\beta)) + \frac{\beta}{2r_0(\beta)^2} \\ &= U_Q(0) + \frac{1}{2}\, U_Q''(0)\, r_0(\beta)^2 + O(r_0(\beta)^3) + \frac{\beta}{2\sqrt{\frac{1}{A(0)}\,\beta + O(\beta^{5/4})}}. \end{aligned}$$

Next we expand

$$r_0(\beta)^2 = \sqrt{\frac{1}{A(0)}\,\beta + O(\beta^{5/4})} = \sqrt{\frac{\beta}{A(0)}}\left(1 + O(\beta^{1/4})\right)$$

and recall $A(0) = U_Q''(0)$ from Lemma A.6(a) to obtain the second claim. The third relation follows from (c) and (A.34). □

Lemma A.8 *We have*

$$e - e_{\min}(\beta) = (r_+(e, \beta) - r_0(\beta))^2 \int_0^1 \tau\, U_{\rm eff}''(\tau r_0(\beta) + (1-\tau) r_+(e, \beta), \beta)\, d\tau,$$

$$e - e_{\min}(\beta) = (r_0(\beta) - r_-(e, \beta))^2 \int_0^1 \tau\, U_{\rm eff}''(\tau r_0(\beta) + (1-\tau) r_-(e, \beta), \beta)\, d\tau.$$

Proof By means of (A.35) we calculate

$$\begin{aligned} e - e_{\min}(\beta) &= U_{\rm eff}(r_+, \beta) - U_{\rm eff}(r_0, \beta) \\ &= \int_{r_0}^{r_+} U_{\rm eff}'(s, \beta)\, ds \\ &= \int_{r_0}^{r_+} [U_{\rm eff}'(s, \beta) - U_{\rm eff}'(r_0, \beta)]\, ds \\ &= \int_{r_0}^{r_+} ds \int_{r_0}^{s} d\tau\, U_{\rm eff}''(\tau, \beta) \\ &= \int_{r_0}^{r_+} (r_+ - \tau)\, U_{\rm eff}''(\tau, \beta)\, d\tau \end{aligned}$$

$$= \int_0^{r_+ - r_0} s\, U''_{\text{eff}}(r_+ - s, \beta)\, ds$$
$$= (r_+ - r_0)^2 \int_0^1 \tau\, U''_{\text{eff}}(\tau r_0 + (1-\tau) r_+, \beta)\, d\tau,$$

as claimed. The second relation is established in the same way. □

Lemma A.9 *For* $s \in [r_-(e,\beta) - r_0(\beta), r_+(e,\beta) - r_0(\beta)]$, *we have*

$$U_{\text{eff}}(r_0(\beta) + s, \beta) - e_{\min}(\beta) = s^2 \int_0^1 (1-\rho)\, U''_{\text{eff}}(r_0(\beta) + \rho s, \beta)\, d\rho,$$
$$U'_{\text{eff}}(r_0(\beta) + s, \beta) = s \int_0^1 U''_{\text{eff}}(r_0(\beta) + \rho s, \beta)\, d\rho.$$

In particular,

$$U_{\text{eff}}(r_0(\beta) + s, \beta) - e_{\min}(\beta) \geq \frac{1}{2} s^2 B(r_Q) \tag{A.37}$$

for $s \in [r_-(e,\beta) - r_0(\beta), 0]$ *and also*

$$|U'_{\text{eff}}(r_0(\beta) + s, \beta)| \leq |s| \left(3\beta \int_0^1 \frac{d\rho}{(r_0(\beta) + \rho s)^4} + \frac{28\pi}{3} \rho_Q(0) \right) \tag{A.38}$$

for $s \in [r_-(e,\beta) - r_0(\beta), r_+(e,\beta) - r_0(\beta)]$.

Proof We write $r_0 = r_0(\beta)$ for short and introduce $V(s,\beta) = U_{\text{eff}}(r_0 + s, \beta) - e_{\min}(\beta)$. Then $V(0,\beta) = 0$ by the definition of $e_{\min}(\beta)$, cf. Lemma A.7(c). In addition, $\partial_s V(s,\beta) = U'_{\text{eff}}(r_0 + s, \beta)$ yields $\partial_s V(0,\beta) = 0$. Hence, by Taylor expansion,

$$V(s,\beta) = \int_0^s (s-\tau)\, \partial_s^2 V(\tau,\beta)\, d\tau$$
$$= \int_0^s \sigma\, \partial_s^2 V(s-\sigma,\beta)\, d\sigma = s^2 \int_0^1 \rho\, \partial_s^2 V((1-\rho)s,\beta)\, d\rho,$$

which gives the first relation. The second relation is shown analogously. Concerning (A.37), this follows from (A.29) in Lemma A.6, and similarly, (A.38) is obtained from (A.30). □

The last result is taken from [50, Lemma 2.1], but nevertheless a proof is included to make the presentation self-contained. Note that the bounds obtained in Lemma A.10 blow up as $\ell \to 0$.

Lemma A.10 *The following assertions hold, where we write* $r_\pm = r_\pm(e,\ell)$*:*

(a) $e - U_{\mathrm{eff}}(r,\ell) \geq \frac{\ell^2}{2r^2 r_- r_+}(r-r_-)(r_+-r)$ *for* $r \in [r_-, r_+]$.
(b) $T_1(e,\ell) \leq \pi \frac{\sqrt{r_- r_+}}{\ell}(r_- + r_+)$.
(c) $U_Q(r) \geq \max\{U_Q(0), -\frac{1}{r}\|Q\|_{L^1(\mathbb{R}^6)}\}$ *for* $r \in [0,\infty[$.
(d) Let $e < 0$. *Then*

$$\frac{\ell^2}{2\|Q\|_{L^1(\mathbb{R}^6)}} \leq r_- < r_+ \leq \frac{\|Q\|_{L^1(\mathbb{R}^6)}}{-e}.$$

In particular,

$$T_1(e,\ell) \leq 2\pi \frac{\|Q\|^2_{L^1(\mathbb{R}^6)}}{e^2}\frac{1}{\ell} \quad \textit{and} \quad \omega_1(e,\ell) \geq \frac{e^2}{\|Q\|^2_{L^1(\mathbb{R}^6)}}\ell.$$

Proof (a) Let

$$w(r) = e - U_{\mathrm{eff}}(r,\ell) - \frac{\ell^2}{2r^2 r_- r_+}(r-r_-)(r_+-r).$$

Then

$$\begin{aligned} w' &= -U'_{\mathrm{eff}} + \frac{\ell^2}{2r_- r_+}\frac{r(r_+ + r_-) - 2r_- r_+}{r^3}, \\ w'' &= -U''_{\mathrm{eff}} + \frac{\ell^2}{r_- r_+}\frac{-r(r_+ + r_-) + 3r_- r_+}{r^4}, \\ (rw)'' &= rw'' + 2w' = -r\,U''_{\mathrm{eff}} - 2\,U'_{\mathrm{eff}} + \frac{\ell^2}{r^3} = \frac{1}{r}\left(-(r^2 U'_{\mathrm{eff}})' + \frac{\ell^2}{r^2}\right). \end{aligned}$$

From (A.36) we recall that $(r^2 U'_{\mathrm{eff}})' = 4\pi r^2 \rho_Q + \frac{\ell^2}{r^2}$. As a consequence, $(rw(r))'' = -4\pi r \rho_Q(r) \leq 0$. Since $rw(r)$ vanishes both at $r = r_-$ and $r = r_+$, it follows that $w(r) \geq 0$ for $r \in [r_-, r_+]$.

(b) By (A.20) and (a),

$$\begin{aligned} T_1(e,\ell) &= 2\int_{r_-(e,\ell)}^{r_+(e,\ell)} \frac{dr}{\sqrt{2(e - U_{\mathrm{eff}}(r,\ell))}} \\ &\leq 2\frac{\sqrt{r_- r_+}}{\ell}\int_{r_-}^{r_+} \frac{r\,dr}{\sqrt{(r-r_-)(r_+-r)}} = \pi\frac{\sqrt{r_- r_+}}{\ell}(r_- + r_+). \end{aligned}$$

(c) Relation (A.2) implies that $r^2\,U'_Q(r) = 4\pi\int_0^r s^2 \rho_Q(s)\,ds \geq 0$, so U_Q is radially increasing. Since ρ_Q has compact support by hypothesis, it moreover follows from (A.1) that

$$\lim_{r\to\infty}[rU_Q(r)] = \lim_{r\to\infty}\left[-4\pi\int_0^r s^2\rho_Q(s)\,ds - 4\pi r\int_r^\infty s\rho_Q(s)\,ds\right] = -4\pi\int_0^\infty s^2\rho_Q(s)\,ds$$
$$= -\int_{\mathbb{R}^3}\rho_Q(x)\,dx = -\|Q\|_{L^1(\mathbb{R}^6)}.$$

Also

$$[rU_Q(r)]' = rU_Q'(r) + U_Q(r) = -4\pi\int_r^\infty s\rho_Q(s)\,ds \le 0,$$

so that $rU_Q(r) \ge -\|Q\|_{L^1(\mathbb{R}^6)}$, and hence (c) is obtained.

(d) Observe that

$$]r_-, r_+[= \{r : e - U_{\text{eff}}(r,\ell) > 0\} = \left\{r : e - U_Q(r) - \frac{\ell^2}{2r^2} > 0\right\}.$$

Thus if $r \in]r_-, r_+[$, then (c) yields

$$0 < e - U_Q(r) - \frac{\ell^2}{2r^2} \le e + \frac{1}{r}\|Q\|_{L^1(\mathbb{R}^6)} - \frac{\ell^2}{2r^2},$$

which using $e < 0$ can be rewritten as

$$\frac{\|Q\|^2_{L^1(\mathbb{R}^6)}}{-2e} - \ell^2 \ge \left(\sqrt{-2e}\,r - \frac{\|Q\|_{L^1(\mathbb{R}^6)}}{\sqrt{-2e}}\right)^2, \quad r \in]r_-, r_+[. \tag{A.39}$$

Thus, by (c),

$$\begin{aligned} U_Q(r) + \frac{\ell^2}{2r^2} &\ge \inf_{s\in[0,\infty[} U_{\text{eff}}(s) = U_Q(r_0) + \frac{\ell^2}{2r_0^2} \\ &\ge \max\left\{U_Q(0), -\frac{1}{r_0}\|Q\|_{L^1(\mathbb{R}^6)} + \frac{\ell^2}{2r_0^2}\right\} \\ &\ge \max\left\{U_Q(0), -\frac{\|Q\|^2_{L^1(\mathbb{R}^6)}}{2\ell^2}\right\}, \quad r \in [0,\infty[. \end{aligned}$$

In particular, if $r \in]r_-, r_+[$, then $U_Q(r) + \frac{\ell^2}{2r^2} < e$ implies that $e \ge -\frac{\|Q\|^2_{L^1(\mathbb{R}^6)}}{2\ell^2}$, or equivalently, $\frac{\|Q\|^2_{L^1(\mathbb{R}^6)}}{-2e} - \ell^2 \ge 0$. Going back to (A.39) and solving for r, we have shown that $r \in]r_-, r_+[$ yields

$$\frac{\|Q\|_{L^1(\mathbb{R}^6)} - \sqrt{\|Q\|^2_{L^1(\mathbb{R}^6)} + 2\ell^2 e}}{\sqrt{-2e}} \le \sqrt{-2e}\,r \le \frac{\|Q\|_{L^1(\mathbb{R}^6)} + \sqrt{\|Q\|^2_{L^1(\mathbb{R}^6)} + 2\ell^2 e}}{\sqrt{-2e}}.$$

Hence, owing to $e < 0$,

$$
\begin{aligned}
\frac{\ell^2}{2\|Q\|_{L^1(\mathbb{R}^6)}} \leq \frac{\ell^2}{\|Q\|_{L^1(\mathbb{R}^6)} + \sqrt{\|Q\|^2_{L^1(\mathbb{R}^6)} + 2\ell^2 e}} &\leq r \\
\leq \frac{\|Q\|_{L^1(\mathbb{R}^6)} + \sqrt{\|Q\|^2_{L^1(\mathbb{R}^6)} + 2\ell^2 e}}{-2e} &\leq \frac{\|Q\|_{L^1(\mathbb{R}^6)}}{-e},
\end{aligned}
$$

which proves the claims, noting that $T_1(e, \ell) \leq 2\pi \frac{r_+^2}{\ell}$ by (b) and $\omega_1 = \frac{2\pi}{T_1}$. □

A.2 Some Transformation Rules

In this section, we consider the transformation of integrals and Poisson brackets. We continue to write spherically symmetric functions $g = g(x, v) = \tilde{g}(|x|, |v|, x \cdot v)$ as $g = \hat{g}(r, p_r, \ell) = g^*(\theta, I, \ell)$.

If $G(x) = \int_{\mathbb{R}^3} \tilde{g}(|x|, |v|, x \cdot v)\, dv$ and $A \in \mathrm{SO}(3)$, then $G(Ax) = G(x)$, i.e.,

$$
\begin{aligned}
G(x) = G(0, 0, r) =: \tilde{G}(r) &= \int_{\mathbb{R}^3} \tilde{g}(r, |v|, r v_3)\, dv = \int_0^\infty ds\, s^2 \int_{|\omega|=1} dS(\omega)\, \tilde{g}(r, s, r s \omega_3) \\
&= 2\pi \int_0^\infty ds\, s^2 \int_0^\pi d\theta\, \sin\theta\, \tilde{g}(r, s, rs\cos\theta) = 2\pi \int_0^\infty ds\, s^2 \int_{-1}^1 dt\, \tilde{g}(r, s, rst).
\end{aligned}
$$

In particular,

$$
\begin{aligned}
\int_{\mathbb{R}^3}\int_{\mathbb{R}^3} g(x, v)\, dx\, dv &= \int_{\mathbb{R}^3} G(x)\, dx = 4\pi \int_0^\infty r^2 \tilde{G}(r)\, dr \\
&= 8\pi^2 \int_0^\infty \int_0^\infty dr\, ds\, r^2 s^2 \int_{-1}^1 dt\, \tilde{g}(r, s, rst).
\end{aligned}
$$

Furthermore, by (A.23),

$$
\begin{aligned}
G(x) = \tilde{G}(r) &= 2\pi \int_0^\infty ds\, s^2 \int_{-1}^1 dt\, \tilde{g}(r, s, rst) \\
&= 2\pi \int_0^\infty ds\, s^2 \int_{-1}^1 dt\, \hat{g}(r, st, rs\sqrt{1 - t^2}) \\
&= 2\pi \int_0^\infty ds\, s^2 \int_0^\pi d\theta\, \sin\theta\, \hat{g}(r, s\cos\theta, rs\sin\theta) \\
&= 2\pi \int_{\mathbb{R}} dp_r \int_0^\infty dR\, R\, \hat{g}(r, p_r, rR) = \frac{2\pi}{r^2} \int_{\mathbb{R}} dp_r \int_0^\infty d\ell\, \ell\, \hat{g}(r, p_r, \ell).
\end{aligned}
\tag{A.40}
$$

This in turn implies

$$\int_{\mathbb{R}^3}\int_{\mathbb{R}^3} g(x,v)\,dx\,dv = 4\pi\int_0^\infty r^2\tilde{G}(r)\,dr = 8\pi^2\int_0^\infty dr\int_{\mathbb{R}} dp_r\int_0^\infty d\ell\,\ell\,\hat{g}(r,p_r,\ell).$$

For the transformation to $g^* = g^*(\theta, I, \ell)$ one can use the fact that $(r, p_r) \to (\theta, I)$ is canonical at fixed ℓ. Hence also

$$\int_{\mathbb{R}^3}\int_{\mathbb{R}^3} g(x,v)\,dx\,dv = 8\pi^2\int_0^{2\pi} d\theta\int_0^\infty dI\int_0^\infty d\ell\,\ell\,g^*(\theta,I,\ell).$$

To summarize,

$$dx = 4\pi r^2\,dr,\quad dv = \frac{2\pi}{r^2}\,dp_r\,d\ell\,\ell,\quad dx\,dv = 8\pi^2\,dr\,dp_r\,d\ell\,\ell = 8\pi^2\,d\theta\,dI\,d\ell\,\ell. \tag{A.41}$$

The Poisson bracket $\{g,h\}_{xv} = \nabla_x g\cdot\nabla_v h - \nabla_x h\cdot\nabla_v g$ of two such spherically symmetric functions is also simpler in the other coordinates. For this we write $\Phi = (r,\varphi,\chi)$, $\mathfrak{A} = (p_r, L_3, \ell)$, $\Theta = (\theta,\varphi,\chi)$, and $\mathfrak{J} = (I, L_3, \ell)$; see (A.14). Then

$$\{g,h\}_{xv} = \{\hat{g},\hat{h}\}_{\Phi\mathfrak{A}} = \{g^*,h^*\}_{\Theta\mathfrak{J}},$$

since the coordinate changes are canonical. But the functions do depend only upon (r, p_r, ℓ) and (θ, I, ℓ), respectively. Hence

$$\begin{aligned}\{\hat{g},\hat{h}\}_{\Phi\mathfrak{A}} &= (\partial_r\hat{g})(\partial_{p_r}\hat{h}) - (\partial_r\hat{h})(\partial_{p_r}\hat{g}),\\ \{g^*,h^*\}_{\Theta\mathfrak{J}} &= (\partial_\theta g^*)(\partial_I h^*) - (\partial_\theta h^*)(\partial_I g^*).\end{aligned} \tag{A.42}$$

The equality of these Poisson brackets could also be verified by a direct calculation. Identifying all versions of a function g, thus for instance,

$$\{r,g\} = \nabla_x r\cdot\nabla_v g = \frac{x}{r}\cdot\nabla_v g = \partial_{p_r} g,$$

and hence in particular $\{r, e_Q\} = \partial_{p_r} e_Q = \partial_{p_r} E = p_r$ by (A.17), as expected. Next recall from (1.11) that

$$\mathcal{T}g = \{g, e_Q\} = v\cdot\nabla_x g - \nabla_v g\cdot\nabla_x U_Q.$$

Then if $g = g(r)$ is a function of r alone, we have

$$\mathcal{T}(g(r)) = p_r g'(r).$$

Furthermore, $e_Q = e = E(I,\ell)$, see (A.17). Hence by (A.42) and (A.18),

$$\mathcal{T}g = \{g, e_Q\} = (\partial_\theta g)(\partial_I E) - (\partial_\theta E)(\partial_I g) = \omega_1 \partial_\theta g \tag{A.43}$$

is appealingly simple in the coordinates (θ, I, ℓ). Since ω_1 is independent of θ, see (A.18), it also follows that

$$\mathcal{T}^2 g = \omega_1 \partial_\theta(\omega_1 \partial_\theta g) = \omega_1^2\, \partial_\theta^2 g. \tag{A.44}$$

A.3 Variational Equation for the Effective Potential

We recall that

$$D = \{(e, \beta) : \beta \in [0, \beta_*], e \in [e_{\min}(\beta), e_0]\},$$

and

$$\mathring{D} = \{(e, \beta) : \beta \in]0, \beta_*[, e \in]e_{\min}(\beta), e_0[\}$$

is its interior. For $(e, \beta) \in \mathring{D}$ we consider in some more detail the linearization (variational equation) associated with

$$\ddot{r} = -U'_{\text{eff}}(r, \beta), \quad r(0) = r_-(e, \beta), \quad \dot{r}(0) = 0,$$

where $r(t) = r(t, e, \beta)$.

Lemma A.11 *(a) The function*

$$\mathbb{R} \times \mathring{D} \ni (t, e, \beta) \mapsto r(t, e, \beta) \in]0, \infty[$$

is C^2 in t and C^1 in (e, β). In addition,

$$\mathbb{R} \times \mathring{D} \ni (t, e, \beta) \mapsto \dot{r}(t, e, \beta) \in]0, \infty[$$

is in $C^1(\mathbb{R} \times \mathring{D})$.

(b) Denote

$$\begin{aligned} z = z(t, e, \beta) &= \frac{\partial r}{\partial t}(t, e, \beta), \\ y = y(t, e, \beta) &= \frac{\partial r}{\partial e}(t, e, \beta), \\ w = w(t, e, \beta) &= \frac{\partial r}{\partial \beta}(t, e, \beta). \end{aligned}$$

Then z is T_1-periodic in t, whereas only $y(T_1) = y(0)$ and $w(T_1) = w(0)$ holds. These functions are the solutions to

$$\ddot{z}(t) + U''_{\rm eff}(r(t), \beta)z(t) = 0, \quad z(0) = 0, \quad \dot{z}(0) = -U'_{\rm eff}(r_-(e, \beta), \beta) = -\frac{1}{\frac{\partial r_-}{\partial e}(e, \beta)},$$

$$\ddot{y}(t) + U''_{\rm eff}(r(t), \beta)y(t) = 0, \quad y(0) = \frac{\partial r_-}{\partial e}(e, \beta), \quad \dot{y}(0) = 0,$$

$$\ddot{w}(t) + U''_{\rm eff}(r(t), \beta)w(t) = \frac{1}{r(t)^3}, \quad w(0) = \frac{\partial r_-}{\partial \beta}(e, \beta), \quad \dot{w}(0) = 0.$$

Proof (a) By definition, r solves the initial value problem

$$\ddot{r}(t, e, \beta) = -U'_Q(r(t, e, \beta)) + \frac{\beta}{r(t, e, \beta)^3}, \quad r(0, e, \beta) = r_-(e, \beta), \quad \dot{r}(0, e, \beta) = 0.$$

Since U_Q is C^2-regular and so is r_-, the latter by Remark A.3, ODE theory implies that r is C^2-regular in t and C^1-regular in (e, β). Similarly, $\dot{r}$ is C^1 in (t, e, β).

(b) Also these statements follow from general ODE theory. □

Lemma A.12 *(a)* $\{z, y\}$ *is a fundamental system for* $\ddot{u} + U''_{\rm eff}(r(t), \beta)u = 0$ *with Wronskian determinant* 1.

(b) One has

$$\frac{\partial T_1}{\partial e}(e, \beta) = \frac{1}{U'_{\rm eff}(r_-(e, \beta), \beta)}\, \dot{y}(T_1(e, \beta), e, \beta) = \frac{\partial r_-}{\partial e}(e, \beta)\, \dot{y}(T_1(e, \beta), e, \beta),$$

$$\frac{\partial T_1}{\partial \beta}(e, \beta) = \frac{1}{U'_{\rm eff}(r_-(e, \beta), \beta)}\, \dot{w}(T_1(e, \beta), e, \beta) = \frac{\partial r_-}{\partial e}(e, \beta)\, \dot{w}(T_1(e, \beta), e, \beta).$$

(c) One has

$$\frac{\partial T_1}{\partial \beta}(e, \beta) = -\frac{1}{2}\frac{\partial}{\partial e}\int_0^{T_1(e,\beta)} \frac{ds}{r(s, e, \beta)^2}.$$

Proof (a) Both z and y are solutions, hence

$$z(t)\dot{y}(t) - y(t)\dot{z}(t) = z(0)\dot{y}(0) - y(0)\dot{z}(0) = \frac{\partial r_-}{\partial e}(e, \beta)\, \frac{1}{\frac{\partial r_-}{\partial e}(e, \beta)} = 1$$

is constant and the non-vanishing Wronskian determinant.

(b) From the T_1-periodicity of r, we deduce that

$$\dot{r}(T_1(e, \beta), e, \beta) = 0. \tag{A.45}$$

Differentiating w.r. to e, it follows that

$$
\begin{aligned}
0 &= \ddot{r}(T_1(e,\beta), e, \beta)\,\frac{\partial T_1}{\partial e}(e,\beta) + \frac{\partial^2 r}{\partial e\,\partial t}(T_1(e,\beta), e, \beta)\\
&= -\,U_{\rm eff}'(r(T_1(e,\beta), e, \beta), \beta)\,\frac{\partial T_1}{\partial e}(e,\beta) + \dot{y}(T_1(e,\beta), e, \beta)\\
&= -\,U_{\rm eff}'(r_-(e,\beta), \beta)\,\frac{\partial T_1}{\partial e}(e,\beta) + \dot{y}(T_1(e,\beta), e, \beta).
\end{aligned}
$$

Similarly, upon differentiating (A.45) w.r. to β, we obtain

$$
\begin{aligned}
0 &= \ddot{r}(T_1(e,\beta), e, \beta)\,\frac{\partial T_1}{\partial \beta}(e,\beta) + \frac{\partial^2 r}{\partial \beta\,\partial t}(T_1(e,\beta), e, \beta)\\
&= -\,U_{\rm eff}'(r(T_1(e,\beta), e, \beta), \beta)\,\frac{\partial T_1}{\partial \beta}(e,\beta) + \dot{w}(T_1(e,\beta), e, \beta)\\
&= -\,U_{\rm eff}'(r_-(e,\beta), \beta)\,\frac{\partial T_1}{\partial \beta}(e,\beta) + \dot{w}(T_1(e,\beta), e, \beta).
\end{aligned}
$$

For the second relations one just has to use (A.25).

(c) Here, we calculate

$$
\begin{aligned}
\frac{d}{dt}\,(\dot{w}(t)y(t) - w(t)\dot{y}(t)) &= \ddot{w}(t)y(t) - w(t)\ddot{y}(t)\\
&= \Big[\frac{1}{r(t)^3} - U_{\rm eff}''(r(t), \beta)w(t)\Big]y(t)\\
&\quad\; - w(t)\Big[-U_{\rm eff}''(r(t), \beta)y(t)\Big]\\
&= \frac{1}{r(t)^3}\,y(t).
\end{aligned}
$$

Since $\dot{w}(0)y(0) - w(0)\dot{y}(0) = 0$, this yields $\dot{w}(t)y(t) = w(t)\dot{y}(t) + \int_0^t \frac{y(s)}{r(s)^3}\,ds$. At $t = T_1 = T_1(e,\beta)$ we deduce that

$$
\begin{aligned}
\dot{w}(T_1)y(0) &= \dot{w}(T_1)y(T_1)\\
&= w(T_1)\dot{y}(T_1) + \int_0^{T_1} \frac{y(s)}{r(s)^3}\,ds\\
&= w(0)\dot{y}(T_1) + \int_0^{T_1} \frac{y(s)}{r(s)^3}\,ds.
\end{aligned}
$$

Therefore,

$$
\dot{w}(T_1)\,\frac{\partial r_-}{\partial e}(e,\beta) = \frac{\partial r_-}{\partial \beta}(e,\beta)\,\dot{y}(T_1) + \int_0^{T_1} \frac{y(s)}{r(s)^3}\,ds,
$$

and equivalently, using (A.25) as well as (A.24),

$$\dot{w}(T_1)\,\frac{1}{U'_{\text{eff}}(r_-,\beta)} = -\frac{1}{2r_-^2\,U'_{\text{eff}}(r_-,\beta)}\,\dot{y}(T_1) + \int_0^{T_1}\frac{y(s)}{r(s)^3}\,ds,$$

which can be restated as

$$\dot{w}(T_1) = -\frac{\dot{y}(T_1)}{2r_-^2} + U'_{\text{eff}}(r_-,\beta)\int_0^{T_1}\frac{y(s)}{r(s)^3}\,ds. \tag{A.46}$$

Next observe that

$$\frac{\partial}{\partial e}\int_0^{T_1}\frac{ds}{r(s)^2} = \frac{1}{r(T_1)^2}\frac{\partial T_1}{\partial e} - 2\int_0^{T_1}\frac{ds}{r(s)^3}\,y(s),$$

and consequently $r(T_1) = r(0) = r_-$ and (b) implies that

$$\int_0^{T_1}\frac{y(s)}{r(s)^3}\,ds = \frac{1}{2r_-^2}\,\frac{1}{U'_{\text{eff}}(r_-,\beta)}\,\dot{y}(T_1) - \frac{1}{2}\frac{\partial}{\partial e}\int_0^{T_1}\frac{ds}{r(s)^2}.$$

Going back to (A.46), we see that

$$\dot{w}(T_1) = -\frac{1}{2}\,U'_{\text{eff}}(r_-,\beta)\,\frac{\partial}{\partial e}\int_0^{T_1}\frac{ds}{r(s)^2}$$

holds. Using once again (b), we have established the claim. □

Appendix B
Function Spaces and Operators

B.1 Fourier Expansion

Spherically symmetric functions $g(x, v) = g(r, p_r, \ell) = g(\theta, I, \ell)$ of (θ, I, ℓ) that are defined on K, the support of Q, can be expanded into a Fourier series

$$g(\theta, I, \ell) = \sum_{k\in\mathbb{Z}} g_k(I, \ell)\, e^{ik\theta},$$

since

$$K = \{(\theta, E, \ell) : \theta \in [0, 2\pi], \ell \in [0, l_*], E \in [e_{\min}(\ell), e_0]\}$$

in the variables (θ, E, ℓ) by (1.24), and θ is 2π-periodic; recall that $I = I(E, \ell)$ is the inverse function to $E = E(I, \ell)$ at fixed ℓ. Thus, K can be equally expressed in the variables (θ, I, ℓ) and we will mostly be using those which are more convenient. The Fourier coefficients are

$$g_k(I, \ell) = \frac{1}{2\pi} \int_0^{2\pi} g(\theta, I, \ell)\, e^{-ik\theta}\, d\theta. \tag{B.1}$$

This motivates the following.

Definition B.1 *(X^α-spaces)* For $\alpha \geq 0$ denote

$$X^\alpha = \Big\{g = \sum_{k\in\mathbb{Z}} g_k(I, \ell)\, e^{ik\theta} : \|g\|_{X^\alpha}^2 = 16\pi^3 \sum_{k\in\mathbb{Z}} (1 + k^2)^\alpha\, \|g_k\|^2_{L^2_{\frac{1}{|Q'|}}(D)} < \infty\Big\},$$

where

$$D = \{(E, \ell) : \ell \in [0, \ell_*], E \in [e_{\min}(\ell), e_0]\}$$

is from (1.23) and expressed in (I, ℓ), and moreover

M. Kunze, *A Birman-Schwinger Principle in Galactic Dynamics*, Progress in Mathematical Physics 77, https://doi.org/10.1007/978-3-030-75186-9

$$(\phi, \psi)_{L^2_{\frac{1}{|Q'|}}(D)} = \iint_D dI\, d\ell\, \ell \, \frac{1}{|Q'(e)|}\, \overline{\phi(I,\ell)}\, \psi(I,\ell)$$

for suitable functions ϕ, ψ on D; note $e = e(I, \ell)$. The associated scalar product on the Hilbert space X^α is given by

$$(g, h)_{X^\alpha} = 16\pi^3 \sum_{k\in\mathbb{Z}} (1 + k^2)^\alpha\, (g_k, h_k)_{L^2_{\frac{1}{|Q'|}}(D)} \tag{B.2}$$

for $g = \sum_{k\in\mathbb{Z}} g_k\, e^{ik\theta}$ and $h = \sum_{k\in\mathbb{Z}} h_k\, e^{ik\theta}$. We also let

$$X_0^\alpha = \{g \in X^\alpha : g_0 = 0\} \tag{B.3}$$

as well as

$$X^{00} = \{g \in X^0 : g_k \neq 0 \text{ for only finitely many } k\}.$$

Remark B.2 (a) Note that for $\alpha = 0$ the scalar product (B.2) agrees with the scalar product in $L^2_{\mathrm{sph},\frac{1}{|Q'|}}(K)$ of spherically symmetric functions given by

$$(g, h)_Q := (g, h)_{L^2_{\mathrm{sph},\frac{1}{|Q'|}}(K)} = \iint_K \frac{1}{|Q'(e_Q)|}\, \overline{g(x, v)}\, h(x, v)\, dx\, dv,$$

recall (1.12); the associated norm is denoted by $\|g\|_Q = (g, g)_Q$. In fact, extending all functions to be zero outside of K (to simplify the notation), this follows, using (A.41), from

$$\begin{aligned}
(g, h)_Q &= \int_{\mathbb{R}^3}\int_{\mathbb{R}^3} \frac{1}{|Q'(e_Q)|}\, \overline{g(x, v)}\, h(x, v)\, dx\, dv \\
&= 8\pi^2 \int_0^{2\pi} d\theta \int_0^\infty dI \int_0^\infty d\ell\, \ell\, \frac{1}{|Q'(e_Q)|}\, \overline{g(\theta, I, \ell)}\, h(\theta, I, \ell) \\
&= 8\pi^2 \sum_{k,m\in\mathbb{Z}} \int_0^\infty dI \int_0^\infty d\ell\, \ell\, \frac{1}{|Q'(e_Q)|}\, \overline{g_k(I, \ell)}\, h_m(I, \ell) \int_0^{2\pi} d\theta\, e^{i(m-k)\theta} \\
&= 16\pi^3 \sum_{k\in\mathbb{Z}} \int_0^\infty dI \int_0^\infty d\ell\, \ell\, \frac{1}{|Q'(e_Q)|}\, \overline{g_k(I, \ell)}\, h_k(I, \ell) \\
&= 16\pi^3 \sum_{k\in\mathbb{Z}} (g_k, h_k)_{L^2_{\frac{1}{|Q'|}}(D)} = (g, h)_{X^0};
\end{aligned} \tag{B.4}$$

observe that $e = E(I, \ell)$ is independent of θ. In particular, we see that $X^0 = L^2_{\mathrm{sph},\frac{1}{|Q'|}}(K)$.

(b) If $\beta \geq \alpha \geq 0$, then $X^\beta \subset X^\alpha$ and $\|g\|_{X^\alpha} \leq \|g\|_{X^\beta}$ for $g \in X^\beta$.

(c) $X^{00} \subset X^\alpha$ is dense for all $\alpha \geq 0$. In particular, $X^\beta \subset X^\alpha$ is dense for $\beta \geq \alpha \geq 0$. To verify this, let $g \in X^\alpha$ and introduce

$$g^{(N)} = \sum_{|k|\leq N} g_k\, e^{ik\theta} \in X^{00} \tag{B.5}$$

for $N \in \mathbb{N}$. Then

$$\|g^{(N)} - g\|_{X^\alpha}^2 = 16\pi^3 \sum_{|k|\geq N+1} (1+k^2)^\alpha\, \|g_k\|^2_{L^2_{\frac{1}{|Q'|}}(D)} \to 0, \quad N \to \infty,$$

so that $g^{(N)} \to g$ in X^α as $N \to \infty$.
(d) A set $B \subset X^0$ is relatively compact if and only if

(i) B is bounded, and
(ii) for every $\varepsilon > 0$ there is $N = N(\varepsilon) \in \mathbb{N}$ such that

$$\sup_{g\in B} \Bigg(\sum_{|k|\geq N} \|g_k\|^2_{L^2_{\frac{1}{|Q'|}}(D)} \Bigg)^{1/2} \leq \varepsilon\,, \quad \text{and}$$

(iii) for every $k \in \mathbb{Z}$ the set $\{g_k : g \in B\} \subset L^2_{\frac{1}{|Q'|}}(D)$ is relatively compact, where $g = \sum_{k\in\mathbb{Z}} g_k\, e^{ik\theta}$.

Due to $X^0 \cong l^2_{\mathbb{Z}}(L^2_{\frac{1}{|Q'|}}(D))$, this follows from [52, Thm. 5.1] for $p = 2$. ◇

Lemma B.3 (Parity) *Write the spherically symmetric function $g = g(x, v)$ as $g = g(\theta, I, \ell) = \sum_{k\in\mathbb{Z}} g_k(I, \ell)\, e^{ik\theta}$. Then*

(a) g is even in v if and only if $g_{-k} = g_k$ for all $k \in \mathbb{Z}$. If g is real-valued, then $g_k(I, \ell) \in \mathbb{R}$.
(b) g is odd in v if and only if $g_{-k} = -g_k$ for all $k \in \mathbb{Z}$, and in particular $g_0 = 0$. If g is real-valued, then $g_k(I, \ell) \in i\mathbb{R}$.

Proof We use Remark A.2. For instance, g is even in v if and only if $g(\theta, I, \ell) = g(2\pi - \theta, I, \ell)$, which is equivalent to

$$\sum_{k\in\mathbb{Z}} g_k(I, \ell)\, e^{ik\theta} = \sum_{k\in\mathbb{Z}} g_k(I, \ell)\, e^{ik(2\pi-\theta)} = \sum_{k\in\mathbb{Z}} g_k(I, \ell)\, e^{-ik\theta} = \sum_{j\in\mathbb{Z}} g_{-j}(I, \ell)\, e^{ij\theta},$$

and comparing coefficients, we arrive at the first part of (a). Furthermore, if g is real-valued, then by (B.1):

$$\overline{g_k(I, \ell)} = \frac{1}{2\pi} \int_0^{2\pi} g(\theta, I, \ell)\, e^{ik\theta}\, d\theta = g_{-k}(I, \ell) = g_k(I, \ell),$$

which means that g_k is real-valued. The proof of (b) is analogous. □

Thus if we need to restrict ourselves to odd functions later, then we have to pass to a subspace $X^\alpha_{\rm odd}$ of X^α, as introduced in

Definition B.4 *($X^\alpha_{\rm odd}$-spaces and $X^\alpha_{\rm even}$-spaces)* For $\alpha \geq 0$ denote

$$\begin{aligned} X^\alpha_{\rm odd} &= \{g \in X^\alpha : g_{-k} = -g_k \ for \ k \in \mathbb{Z}\}, \\ X^\alpha_{\rm even} &= \{g \in X^\alpha : g_{-k} = g_k \ for \ k \in \mathbb{Z}\}. \end{aligned} \tag{B.6}$$

We derive some further useful relations.

Lemma B.5 *For appropriate functions $\Psi = \Psi(r)$ and $\varphi = \varphi(e)$ consider*

$$\psi(r, p_r, \ell) = \varphi(e_Q)\, p_r \Psi(r). \tag{B.7}$$

Then

$$\psi_k(I, \ell) = -\frac{i}{\pi}\,\varphi(e)\,\omega_1(e, \ell) \int_{r_-(e,\ell)}^{r_+(e,\ell)} \Psi(r) \sin(k\theta(r, e, \ell))\, dr$$

for the Fourier coefficients of ψ, and in particular $\psi_0 = 0$.

Proof Since $e_Q = E(I, \ell)$ is independent of θ by (A.11), using (A.10) the Fourier coefficients are calculated to be

$$\begin{aligned} \psi_k(I, \ell) &= \frac{1}{2\pi} \int_0^{2\pi} \psi\, e^{-ik\theta}\, d\theta \\ &= \frac{1}{2\pi}\,\varphi(E) \Bigg(\int_0^\pi p_r(\theta, I) \Psi(r(\theta, I))\, e^{-ik\theta}\, d\theta \\ &\qquad - \int_\pi^{2\pi} p_r(2\pi - \theta, I) \Psi(r(2\pi - \theta, I))\, e^{-ik\theta}\, d\theta \Bigg) \\ &= \frac{1}{2\pi}\,\varphi(E) \int_0^\pi p_r \Psi(r)\, (e^{-ik\theta} - e^{ik\theta})\, d\theta \\ &= -\frac{i}{\pi}\,\varphi(E) \int_0^\pi p_r \Psi(r) \sin(k\theta)\, d\theta \\ &= -\frac{i}{\pi}\,\varphi(E)\,\omega_1 \int_{r_-}^{r_+} \Psi(r) \sin(k\theta)\, dr, \end{aligned}$$

where we applied the transformation $[0, \pi] \ni \theta \to r \in [r_-, r_+]$, $dr = \frac{p_r}{\omega_1}\, d\theta$, cf. (A.22), in the last step. The dependencies are $e = E = E(I, \ell)$, $\omega_1(e, \ell) = \omega_1(I, \ell)$ and $\theta(r, e, \ell) = \theta(r, I, \ell)$. □

Remark B.6 (Projection of $\phi = \phi(r)$) Observe that the zero'th Fourier coefficient of a function $\phi = \phi(r)$ depending only upon r is

$$\begin{aligned}\phi_0(I,\ell) &= \frac{1}{2\pi}\int_0^{2\pi}\phi(r(\theta,I,\ell))\,d\theta = \frac{1}{\pi}\int_0^{\pi}\phi(r(\theta,I,\ell))\,d\theta \\ &= \frac{\omega_1(e,\ell)}{\pi}\int_{r_-(e,\ell)}^{r_+(e,\ell)}\frac{\phi(r)}{p_r}\,dr = \frac{2}{T_1(e,\ell)}\int_{r_-(e,\ell)}^{r_+(e,\ell)}\frac{\phi(r)}{p_r}\,dr,\end{aligned}$$

where $p_r = \sqrt{2(e-U_Q(r)) - \frac{\ell^2}{r^2}}$ and $e = E = E(I,\ell)$ This agrees with the relation from [29, equ. (29)] for the projection onto $\ker\mathcal{T}$ of such a radial function. ◇

Next we will re-express $U_g'(r)$ for $g = \sum_{k\in\mathbb{Z}} g_k\, e^{ik\theta}$ in terms of the g_k.

Lemma B.7 *For $g = \sum_{k\in\mathbb{Z}} g_k\, e^{ik\theta}$ we have*

$$\begin{aligned}U_g'(r) &= \frac{16\pi^3}{r^2}\iint_D d\ell\,\ell\,de\,\mathbf{1}_{\{(e,\ell):r_+(e,\ell)\le r\}}\frac{1}{\omega_1(e,\ell)}\,g_0(I,\ell) \\ &\quad + \frac{16\pi^2}{r^2}\sum_{k\in\mathbb{Z}}\iint_D d\ell\,\ell\,de\,\mathbf{1}_{\{(e,\ell):r_-(e,\ell)\le r\le r_+(e,\ell)\}} \\ &\qquad\qquad \times\frac{1}{\omega_1(e,\ell)}\,\frac{\sin(k\theta(r,e,\ell))}{k}\,g_k(I,\ell),\end{aligned}$$

where $I = I(e,\ell)$. In particular, if $g = \sum_{k\neq 0} g_k\, e^{ik\theta}$, then

$$U_g'(r) = \frac{16\pi^2}{r^2}\sum_{k\neq 0}\iint_D d\ell\,\ell\,de\,\mathbf{1}_{\{(e,\ell):r_-(e,\ell)\le r\le r_+(e,\ell)\}}\frac{1}{\omega_1(e,\ell)}\,\frac{\sin(k\theta(r,e,\ell))}{k}\,g_k(I,\ell). \tag{B.8}$$

Proof By linearity it suffices to consider the special case that $g(r,p_r,\ell) = g(\theta,I,\ell) = h(I,\ell)\,e^{ik\theta}$, where $k\in\mathbb{Z}$ and h is defined for $(I,\ell)\in D$, cf. Definition B.1. For the density, using (A.41) we have

$$\rho_g(r) = \int_{\mathbb{R}^3} g\,dv = \frac{2\pi}{r^2}\int_{\mathbb{R}} dp_r\int_0^\infty d\ell\,\ell\,g(r,p_r,\ell). \tag{B.9}$$

To analyze the domain of integration in p_r and ℓ at fixed r, we note that

$$e_0 \ge \frac{1}{2}\,p_r^2 + U_Q(r) + \frac{\ell^2}{2r^2} \ge U_Q(0) + \frac{\ell^2}{2r^2}$$

holds. Therefore, $2r^2(e_0 - U_Q(0)) \ge \ell^2$ and we get the restriction

$$\ell \le \hat{l}(r) = \sqrt{2r^2(e_0 - U_Q(0))}. \tag{B.10}$$

If ℓ is fixed, then $U_{\text{eff}}(r_\pm, \ell) = e$ yields $p_r = \pm\sqrt{2(e - U_Q(r)) - \frac{\ell^2}{r^2}} = 0$ at $r_\pm$. Thus, we must have $-\hat{p} \le p_r \le \hat{p}$ for

$$\hat{p}(r, \ell) = \sqrt{2(e_0 - U_Q(r)) - \frac{\ell^2}{r^2}}. \tag{B.11}$$

Hence, we obtain from (B.9) that

$$\rho_{he^{ik\theta}}(r) = \frac{2\pi}{r^2} \int_0^{\hat{l}(r)} d\ell\, \ell \int_{-\hat{p}}^{\hat{p}} dp_r\, h(I, \ell)\, e^{ik\theta}, \tag{B.12}$$

where $\theta = \theta(r, p_r, \ell)$ and $I = I(r, p_r, \ell)$. Now we are going to apply the transformation $p_r \mapsto e(r, p_r, \ell) = \frac{1}{2}p_r^2 + U_Q(r) + \frac{\ell^2}{2r^2}$, which is quadratic. Observe that

$$e(r, \pm\hat{p}, \ell) = e_0 \quad \text{and} \quad e(r, 0, \ell) = U_Q(r) + \frac{\ell^2}{2r^2} < e_0.$$

If $p_r \in [-\hat{p}, 0[$, then $\dot{r} = p_r < 0$ and $e(r, p_r, \ell) \in]\frac{\ell^2}{2r^2} + U_Q(r), e_0]$ with the inverse transformation given by $e \mapsto p_r = p_r(r, e, l) = -\sqrt{2(e - U_Q(r)) - \frac{\ell^2}{r^2}}$. Similarly, if $p_r \in]0, \hat{p}]$, then $\dot{r} = p_r > 0$ and $e(r, p_r, \ell) \in]\frac{\ell^2}{2r^2} + U_Q(r), e_0]$, whereas the inverse transformation is $e \mapsto p_r = p_r(r, e, l) = \sqrt{2(e - U_Q(r)) - \frac{\ell^2}{r^2}}$. Noting that $\frac{\partial e}{\partial p_r} = p_r$, we deduce from (B.12),

$$\begin{aligned}
\rho_{he^{ik\theta}}(r) &= \frac{2\pi}{r^2} \int_0^{\hat{l}(r)} d\ell\, \ell \left(\int_{-\hat{p}}^{0} dp_r\, h(I, \ell)\, e^{ik\theta} + \int_0^{\hat{p}} dp_r\, h(I, \ell)\, e^{ik\theta} \right) \\
&= \frac{2\pi}{r^2} \int_0^{\hat{l}(r)} d\ell\, \ell \left(\int_{e_0}^{\frac{\ell^2}{2r^2} + U_Q(r)} \frac{de}{-\sqrt{2(e - U_Q(r)) - \frac{\ell^2}{r^2}}}\, h(I, \ell)\, e^{ik\theta(r, p_r^-, \ell)} \right. \\
&\qquad \left. + \int_{\frac{\ell^2}{2r^2} + U_Q(r)}^{e_0} \frac{de}{\sqrt{2(e - U_Q(r)) - \frac{\ell^2}{r^2}}}\, h(I, \ell)\, e^{ik\theta(r, p_r^+, \ell)} \right) \\
&= \frac{2\pi}{r^2} \int_0^{\hat{l}(r)} d\ell\, \ell \int_{\frac{\ell^2}{2r^2} + U_Q(r)}^{e_0} \frac{de}{\sqrt{2(e - U_Q(r)) - \frac{\ell^2}{r^2}}} \\
&\qquad \times h(I, \ell)\, (e^{ik\theta(r, p_r^-, \ell)} + e^{ik\theta(r, p_r^+, \ell)})
\end{aligned} \tag{B.13}$$

for $p_r^\pm = p_r^\pm(r, e, \ell) = \pm\sqrt{2(e - U_Q(r)) - \frac{\ell^2}{r^2}}$. Using (A.10), we obtain $\theta(r, p_r^-, \ell) = 2\pi - \theta(r, p_r^+, \ell)$. Thus, (B.13) simplifies to

$$\rho_{he^{ik\theta}}(r) = \frac{4\pi}{r^2} \int_0^{\hat{l}(r)} d\ell\, \ell \int_{\frac{\ell^2}{2r^2} + U_Q(r)}^{e_0} \frac{de}{\sqrt{2(e - U_Q(r)) - \frac{\ell^2}{r^2}}}\, h(I, \ell)\, \cos(k\theta),$$

where $I = I(e, \ell)$ and $\theta = \theta(r, p_r^+, \ell)$. Then, (A.2) implies that

$$
\begin{aligned}
&U'_{he^{ik\theta}}(R)\\
&= \frac{4\pi}{R^2}\int_0^R r^2 \rho_{he^{ik\theta}}(r)\,dr\\
&= \frac{16\pi^2}{R^2}\int_0^R dr \int_0^{\hat{l}(r)} d\ell\,\ell \int_{\frac{\ell^2}{2r^2}+U_Q(r)}^{e_0} \frac{de}{\sqrt{2(e-U_Q(r))-\frac{\ell^2}{r^2}}}\, h(I,\ell)\cos(k\theta)\\
&= \frac{16\pi^2}{R^2}\int_0^R dr \int_0^{l_*} d\ell\,\ell \int_{e_{\min}(\ell)}^{e_0} \frac{de}{\sqrt{2(e-U_Q(r))-\frac{\ell^2}{r^2}}}\\
&\qquad\qquad \times \mathbf{1}_{\{0\le\ell\le\hat{l}(r),\ e_0\le\frac{\ell^2}{2r^2}+U_Q(r)\le e\}}\, h(I,\ell)\cos(k\theta)
\end{aligned}
\tag{B.14}
$$

for $I = I(e, \ell)$ and $\theta = \theta(r, p_r^+, \ell)$. We claim that

$$
\mathbf{1}_{\{0\le\ell\le\hat{l}(r),\ e_0\le\frac{\ell^2}{2r^2}+U_Q(r)\le e\}} = \mathbf{1}_{[r_-,r_+]}(r), \tag{B.15}
$$

where $r_\pm = r_\pm(e, \ell)$ as before. Recall that $r_\pm$ are the solutions to $2(e - U_Q(r)) - \frac{\ell^2}{r^2} = 0$, whereas $2(e - U_Q(r)) - \frac{\ell^2}{r^2} > 0$ in $]r_-, r_+[$ and $2(e - U_Q(r)) - \frac{\ell^2}{r^2} < 0$ outside $[r_-, r_+]$. Hence if $\mathbf{1}_{\{\dots\}} = 1$, then $r \in [r_-, r_+]$. Conversely, if $r \in [r_-, r_+]$, then $e \ge \frac{\ell^2}{2r^2} + U_Q(r)$ and consequently $e_0 \ge e \ge \frac{\ell^2}{2r^2} + U_Q(0)$, so that $2r^2(e_0 - U_Q(0)) \ge \ell^2$, which means that $\ell \le \hat{l}(r)$. This completes the argument for (B.15). Going back to (B.14) and recalling (1.23), it yields

$$
\begin{aligned}
U'_{he^{ik\theta}}(R) &= \frac{16\pi^2}{R^2}\int_0^R dr \iint_D d\ell\,\ell\,de \frac{1}{\sqrt{2(e-U_Q(r))-\frac{\ell^2}{r^2}}}\,\mathbf{1}_{[r_-,r_+]}(r)\,h(I,\ell)\cos(k\theta)\\
&= \frac{16\pi^2}{R^2}\iint_D d\ell\,\ell\,de\,h(I,\ell)\int_{r_-}^{r_+} dr\,\mathbf{1}_{[0,R]}(r)\frac{\cos(k\theta)}{\sqrt{2(e-U_Q(r))-\frac{\ell^2}{r^2}}};
\end{aligned}
$$

note that $I = I(e, \ell)$ does not depend upon r. To calculate the integral

$$
\mathcal{I} = \int_{r_-}^{r_+} dr\,\mathbf{1}_{[0,R]}(r)\frac{\cos(k\theta)}{\sqrt{2(e-U_Q(r))-\frac{\ell^2}{r^2}}}
$$

we use the transformation $[0, \pi] \ni \theta \mapsto r(\theta, I, \ell) = r(\theta, e, \ell) \in [r_-, r_+]$, which has $dr = \frac{p_r^+}{\omega_1}\,d\theta$ by (A.22). Therefore,

$$
\begin{aligned}
\mathcal{I} &= \mathbf{1}_{\{(e,\ell):\, r_+(e,\ell)\le R\}} \int_{r_-}^{r_+} dr\, \frac{\cos(k\theta)}{\sqrt{2(e-U_Q(r))-\frac{\ell^2}{r^2}}} \\
&\quad + \mathbf{1}_{\{(e,\ell):\, r_-(e,\ell)\le R\le r_+(e,\ell)\}} \int_{r_-}^{R} dr\, \frac{\cos(k\theta)}{\sqrt{2(e-U_Q(r))-\frac{\ell^2}{r^2}}} \\
&= \mathbf{1}_{\{(e,\ell):\, r_+(e,\ell)\le R\}} \frac{1}{\omega_1}\int_0^{\pi} d\theta\, \cos(k\theta) \\
&\quad + \mathbf{1}_{\{(e,\ell):\, r_-(e,\ell)\le R\le r_+(e,\ell)\}} \frac{1}{\omega_1}\int_0^{\theta(R,e,\ell)} d\theta\, \cos(k\theta) \\
&= \mathbf{1}_{\{(e,\ell):\, r_+(e,\ell)\le R\}} \frac{\pi}{\omega_1}\,\delta_{k0} \\
&\quad + \mathbf{1}_{\{(e,\ell):\, r_-(e,\ell)\le R\le r_+(e,\ell)\}} \frac{1}{\omega_1}\,\frac{\sin(k\theta(R,e,\ell))}{k}\,;
\end{aligned}
$$

here $\omega_1=\omega_1(I,\ell)=\omega_1(e,\ell)$. Thus, we arrive at

$$
\begin{aligned}
U'_{he^{ik\theta}}(r) &= \frac{16\pi^3}{r^2}\,\delta_{k0}\iint_D d\ell\,\ell\, de\, h(I,\ell)\,\mathbf{1}_{\{(e,\ell):\, r_+(e,\ell)\le r\}}\,\frac{1}{\omega_1(e,\ell)} \\
&\quad + \frac{16\pi^2}{r^2}\iint_D d\ell\,\ell\, de\, h(I,\ell)\,\mathbf{1}_{\{(e,\ell):\, r_-(e,\ell)\le r\le r_+(e,\ell)\}} \\
&\qquad\qquad \times \frac{1}{\omega_1(e,\ell)}\,\frac{\sin(k\theta(r,e,\ell))}{k},
\end{aligned}
$$

which yields the asserted formula for $U'_g(r)$. □

B.2 Operators

The next definition is consistent with the fact that $\mathcal{T}g=\{g,e_Q\}=\omega_1\partial_\theta g$ and $\mathcal{T}^2 g=\omega_1^2\,\partial_\theta^2 g$, cf. (A.43) and (A.44).

Lemma B.8 *The following assertions hold.*

(a) Let

$$
(\mathcal{T}g)_k(I,\ell)=ik\,\omega_1(I,\ell)g_k(I,\ell)\quad \text{for}\quad k\in\mathbb{Z}. \tag{B.16}
$$

If $\alpha\ge 0$, then $\mathcal{T}: X^{\alpha+1}\to X^\alpha$ is well-defined and

$$
\|\mathcal{T}g\|_{X^\alpha}\le \Delta_1\,\|g\|_{X^{\alpha+1}}\quad \text{for}\quad g\in X^{\alpha+1},
$$

where Δ_1 is from Theorem 3.5. In addition, $\mathcal{T}(X^{\alpha+1})\subset X_0^\alpha$.

(b) Let $\mathcal{D}(\mathcal{T}^2) = X^2$ and define

$$\mathcal{T}^2 : X^2 \to X^0, \quad (\mathcal{T}^2 g)_k(I, \ell) = -k^2 \omega_1^2(I, \ell)\, g_k(I, \ell) \quad \text{for} \quad k \in \mathbb{Z}. \tag{B.17}$$

Then $\mathcal{T}^2$ is a self-adjoint operator on X^0 and

$$\|\mathcal{T}^2 g\|_{X^\alpha} \leq \Delta_1^2 \, \|g\|_{X^{\alpha+2}} \quad \text{for} \quad g \in X^{\alpha+2}. \tag{B.18}$$

(c) We have

$$(-\mathcal{T}^2 g, g)_{X^0} = 16\pi^3 \sum_{k \neq 0} k^2 \|\omega_1 g_k\|^2_{L^2_{\frac{1}{|Q'|}}(D)} = \|\mathcal{T} g\|^2_{X^0} \quad \text{for} \quad g \in X^2.$$

Proof (a) By the definition of $\mathcal{T}$ and of the norms,

$$\begin{aligned} \|\mathcal{T} g\|^2_{X^\alpha} &= 16\pi^3 \sum_{k \in \mathbb{Z}} (1 + k^2)^\alpha \, \|(\mathcal{T} g)_k\|^2_{L^2_{\frac{1}{|Q'|}}(D)} \\ &= 16\pi^3 \sum_{k \in \mathbb{Z}} (1 + k^2)^\alpha k^2 \, \|\omega_1 g_k\|^2_{L^2_{\frac{1}{|Q'|}}(D)} \\ &\leq 16\pi^3 \Delta_1^2 \sum_{k \in \mathbb{Z}} (1 + k^2)^{\alpha+1} \, \|g_k\|^2_{L^2_{\frac{1}{|Q'|}}(D)} = \Delta_1^2 \, \|g\|^2_{X^{\alpha+1}}. \end{aligned} \tag{B.19}$$

The fact that $\mathcal{T}(X^{\alpha+1}) \subset X_0^\alpha$ follows from $(\mathcal{T} g)_0 = 0$. (b) First observe that $\mathcal{T}^2$ is densely defined in X^0 by Remark B.2(c). To show that $\mathcal{T}^2$ is symmetric, we remark that for $g, h \in X^2$,

$$\begin{aligned} (\mathcal{T}^2 g, h)_{X^0} &= 16\pi^3 \sum_{k \in \mathbb{Z}} ((\mathcal{T}^2 g)_k, h_k)_{L^2_{\frac{1}{|Q'|}}(D)} \\ &= -16\pi^3 \sum_{k \in \mathbb{Z}} \iint_D dI\, d\ell\, \ell \, \frac{1}{|Q'(e)|} \, k^2 \omega_1^2(I, \ell) \, \overline{g_k(I, \ell)} \, h_k(I, \ell) \end{aligned} \tag{B.20}$$

$$= 16\pi^3 \sum_{k \in \mathbb{Z}} (g_k, (\mathcal{T}^2 h)_k)_{L^2_{\frac{1}{|Q'|}}(D)} = (g, \mathcal{T}^2 h)_{X^0}. \tag{B.21}$$

Next we verify that $\operatorname{ran}(\mathcal{T}^2 \pm i) = X^0$. For, let $h \in X^0$ be given and define $g^\pm$ by

$$g_k^\pm(I, \ell) = -\frac{h_k(I, \ell)}{k^2 \omega_1^2(I, \ell) \mp i} \quad \text{for} \quad k \in \mathbb{Z}. \tag{B.22}$$

Then $g^\pm \in X^2$. To see this, we are going to use estimate (4.3) from Remark 4.2(a). It implies (note $a = 0$ and $b = 1$) that

$$
\begin{aligned}
\|g^{\pm}\|_{X^2}^2 &= 16\pi^3 \sum_{k\in\mathbb{Z}} (1+k^2)^2 \, \|g_k^{\pm}\|^2_{L^2_{\frac{1}{|Q'|}}(D)} \\
&= 16\pi^3 \sum_{k\in\mathbb{Z}} (1+k^2)^2 \iint_D dI\, d\ell\, \ell \, \frac{1}{|Q'(e)|} \, \frac{|h_k(I,\ell)|^2}{|k^2\omega_1^2(I,\ell) \mp i|^2} \\
&\le 16\pi^3 \iint_D dI\, d\ell\, \ell \, \frac{1}{|Q'(e)|} \, |h_0(I,\ell)|^2 \\
&\quad + 16\pi^3 \sum_{k\neq 0} (1+k^2)^2 \iint_D dI\, d\ell\, \ell \, \frac{1}{|Q'(e)|} \, \frac{|h_k(I,\ell)|^2}{\frac{1}{4}k^4\delta_1^4 + 1} \\
&\le 16\pi^3 \iint_D dI\, d\ell\, \ell \, \frac{1}{|Q'(e)|} \, |h_0(I,\ell)|^2 \\
&\quad + 32\pi^3 (1 + 4\delta_1^{-4}) \sum_{k\neq 0} \iint_D dI\, d\ell\, \ell \, \frac{1}{|Q'(e)|} \, |h_k(I,\ell)|^2 \\
&\le 2(1+4\delta_1^{-4}) \, \|h\|^2_{X^0}.
\end{aligned}
$$

This proves that $g^{\pm} \in X^2$ is well-defined through (B.22), and then $(\mathcal{T}^2 \pm i)g^{\pm} = h$ is obtained from the definition of $\mathcal{T}^2$. Thus, $\mathcal{T}^2$ is self-adjoint. The bound (B.18) is derived analogously to (B.19). (c) Here, we have

$$
\begin{aligned}
(-\mathcal{T}^2 g, g)_{X^0} &= 16\pi^3 \sum_{k\in\mathbb{Z}} ((-\mathcal{T}^2 g)_k, g_k)_{L^2_{\frac{1}{|Q'|}}(D)} = 16\pi^3 \sum_{k\in\mathbb{Z}} (k^2\omega_1^2 \, g_k, g_k)_{L^2_{\frac{1}{|Q'|}}(D)} \\
&= 16\pi^3 \sum_{k\in\mathbb{Z}} \|ik\,\omega_1 g_k\|^2_{L^2_{\frac{1}{|Q'|}}(D)} = 16\pi^3 \sum_{k\in\mathbb{Z}} \|(\mathcal{T}g)_k\|^2_{L^2_{\frac{1}{|Q'|}}(D)} = \|\mathcal{T}g\|^2_{X^0},
\end{aligned}
$$

as was to be shown. □

Lemma B.9 *The following assertions hold.*

(a) Let $g = g(I,\ell) \in L^2_{\frac{1}{|Q'|}}(D)$. Then $\mathcal{T}g = 0$. In particular, $\mathcal{T}\,|Q'(e_Q)| = 0$.

(b) Let L^2_r be defined as in Chapter 4. For $\Psi \in L^2_r$ let $F(r) = F(0) + \int_0^r \Psi(s)\,ds$ for $r \in [0, r_Q]$ and denote by F_0 the zero'th Fourier coefficient of F. Then $|Q'(e_Q)|(F - F_0) \in X^1_{\text{even}}$ and

$$
\mathcal{T}(|Q'(e_Q)|(F - F_0)) = |Q'(e_Q)| \, p_r \, \Psi. \tag{B.23}
$$

Proof (a) If $g = g(I,\ell)$, then $g_0 = g$ and $g_k = 0$ for $k \neq 0$ by definition of the Fourier coefficients, cf. (B.1). In particular, $g \in X^1$ and hence (B.16) from Lemma B.8 yields $\mathcal{T}g = 0$. The second part follows from $e_Q = e_Q(I,\ell)$ and

$$
\||Q'(e_Q)|\|^2_{L^2_{\frac{1}{|Q'|}}(D)} = \iint_D dI\, d\ell\, \ell \, |Q'(e)| < \infty,
$$

the latter due to $e \in [U_Q(0), e_0]$ and (Q3). (b) First note that

$$\int_0^{2\pi} |Q'(e_Q)|\, F_0(I,\ell)\, e^{-ik\theta}\, d\theta = 2\pi\, |Q'(e_Q)|\, F_0(I,\ell)\delta_{k0}$$

for $k \in \mathbb{Z}$ as well as

$$\int_0^{2\pi} |Q'(e_Q)|\, F(\theta, I,\ell)\, d\theta = 2\pi\, |Q'(e_Q)|\, F_0(I,\ell).$$

If $k \neq 0$, then through integration by parts

$$\begin{aligned}
&\int_0^{2\pi} |Q'(e_Q)|\, F(\theta, I,\ell)\, e^{-ik\theta}\, d\theta \\
&= |Q'(e_Q)| \int_0^{2\pi} F(r(\theta, I,\ell))\, e^{-ik\theta}\, d\theta \\
&= -\frac{1}{ik}\, |Q'(e_Q)| \left[F(r(\theta, I, l))e^{-ik\theta}\Big|_0^{2\pi} - \int_0^{2\pi} e^{-ik\theta}\, F'(r)\, \frac{dr}{d\theta}\, d\theta \right] \\
&= \frac{1}{ik}\, \frac{|Q'(e_Q)|}{\omega_1(I,\ell)} \int_0^{2\pi} p_r\, \Psi(r)\, e^{-ik\theta}\, d\theta,
\end{aligned}$$

where we used $r(2\pi, I,\ell) = r(0, I,\ell)$ and (A.22). Thus if we let $\psi(r, p_r, \ell) = |Q'(e_Q)|\, p_r\, \Psi(r)$, then

$$\int_0^{2\pi} |Q'(e_Q)|\, F(\theta, I,\ell)\, e^{-ik\theta}\, d\theta = \frac{2\pi}{ik}\, \frac{1}{\omega_1(I,\ell)}\, \psi_k(I,\ell)$$

for $k \neq 0$. To summarize, we have shown that

$$\begin{aligned}
&[|Q'(e_Q)|(F - F_0)]_0(I,\ell) = 0, \\
&[|Q'(e_Q)|(F - F_0)]_k(I,\ell) = \frac{1}{ik}\, \frac{1}{\omega_1(I,\ell)}\, \psi_k(I,\ell) \text{ for } k \neq 0.
\end{aligned} \tag{B.24}$$

Since ψ is odd, we have $\psi_{-k} = -\psi_k$ for $k \in \mathbb{Z}$ by Lemma B.3(b); hence Lemma B.3(a) implies that $|Q'(e_Q)|(F - F_0)$ is even. Moreover, recalling from Lemma B.5 that $\psi_0 = 0$ and using (4.20) from Lemma 4.4, we get

$$\begin{aligned}
\| |Q'(e_Q)|(F - F_0)\|_{X^1}^2 &= 16\pi^3 \sum_{k\in\mathbb{Z}} (1+k^2)\, \|[Q'(e_Q)|(F - F_0)]_k\|_{L^2_{\frac{1}{|Q'|}}(D)}^2 \\
&= 16\pi^3 \sum_{k\neq 0} \frac{1+k^2}{k^2} \left\| \frac{1}{\omega_1}\, \psi_k \right\|_{L^2_{\frac{1}{|Q'|}}(D)}^2 \\
&\leq 32\pi^3 \delta_1^{-2} \sum_{k\neq 0} \|\psi_k\|_{L^2_{\frac{1}{|Q'|}}(D)}^2 \\
&= 32\pi^3 \delta_1^{-2} \|\psi\|_{X^0}^2 \leq 32\pi^3 \delta_1^{-2} \rho_Q(0)\, \|\Psi\|_{L^2_r}^2 < \infty.
\end{aligned}$$

Therefore, it follows that $|Q'(e_Q)|(F - F_0) \in X^1_{\text{even}}$. To establish (B.23), it suffices to observe that

$$[\mathcal{T}(|Q'(e_Q)|(F - F_0))]_k = ik\,\omega_1\,[|Q'(e_Q)|(F - F_0)]_k = \psi_k$$

for $k \in \mathbb{Z}$ by (B.16) and (B.24). □

From (B.3) in Definition B.1 recall that, for $\alpha \geq 0$,

$$X^\alpha_0 = \{g \in X^\alpha : g_0 = 0\}$$

is the space of functions with vanishing zero'th Fourier coefficient. It is a Hilbert space under the scalar product defined in (B.2). If we restrict $\mathcal{T}^2$ to such spaces, then we get the following.

Corollary B.10 *Let $\mathcal{D}(\mathcal{T}^2) = X^2_0$ and define $\mathcal{T}^2 : X^2_0 \to X^0_0$ as before. Then $\mathcal{T}^2$ is a self-adjoint operator on X^0_0 such that $-\mathcal{T}^2 \geq \delta_1^2$ as operators. In particular, we have $\sigma(-\mathcal{T}^2) \subset [\delta_1^2, \infty[$ and $\Omega = \mathbb{C} \setminus [\delta_1^2, \infty[\subset \rho(-\mathcal{T}^2)$. For $z \in \Omega$, the resolvent $R_{-\mathcal{T}^2}(z) = (-\mathcal{T}^2 - z)^{-1}$ is given by*

$$(-\mathcal{T}^2 - z)^{-1} : X^0_0 \to X^2_0, \quad ((-\mathcal{T}^2 - z)^{-1}h)_k(I, \ell) = \frac{h_k(I, \ell)}{k^2\omega_1^2(I, \ell) - z} \quad \text{for } k \neq 0 \tag{B.25}$$

and $((-\mathcal{T}^2 - z)^{-1}h)_0 = 0$. We also have

$$((-\mathcal{T}^2 - z)^{-1}h, h)_{X^0} = 16\pi^3 \sum_{k\neq 0} \iint_D dI\, d\ell\, \ell\, \frac{1}{|Q'(e)|}\, \frac{|h_k(I, \ell)|^2}{k^2\omega_1^2(I, \ell) - \bar{z}} \tag{B.26}$$

for $h \in X^0_0$.

Proof Clearly $X^{00}_0 \subset X^\alpha_0$ is dense for $\alpha \geq 0$, where $X^{00}_0 = X^{00} \cap \{g : g_0 = 0\}$; thus in particular, $X^2_0 \subset X^0_0$ is dense. The symmetry of $\mathcal{T}^2$ is shown as in (B.21). Also $\operatorname{ran}(\mathcal{T}^2 \pm i) = X^0_0$ holds, since if $h \in X^0_0$ is given, then $g^\pm$ defined via (B.22) yields functions $g^\pm \in X^2_0$. Thus, $\mathcal{T}^2 : X^2_0 \to X^0_0$ is self-adjoint. To establish that $-\mathcal{T}^2 \geq \delta_1^2$, let $g \in X^2_0$. Then, by (B.20),

$$\begin{aligned}
(-\mathcal{T}^2 g, g)_{X^0} &= 16\pi^3 \sum_{k\neq 0} \iint_D dI\, d\ell\, \ell\, \frac{1}{|Q'(e)|}\, k^2\omega_1^2(I, \ell)\, |g_k(I, \ell)|^2 \\
&\geq 16\pi^3\delta_1^2 \sum_{k\neq 0} \iint_D dI\, d\ell\, \ell\, \frac{1}{|Q'(e)|}\, |g_k(I, \ell)|^2 \\
&= \delta_1^2\, \|g\|^2_{X^0}.
\end{aligned}$$

From $-\mathcal{T}^2 \geq \delta_1^2$ it follows that $\sigma(-\mathcal{T}^2) \subset [\delta_1^2, \infty[$, see [37, Prop. 5.12], which is equivalent to $\rho(-\mathcal{T}^2) \supset \mathbb{C} \setminus [\delta_1^2, \infty[$. For the last assertions, they are derived in a similar fashion as the estimates in Lemma B.8(b), but for completeness we include the details. Fix $z \in \Omega$, let $h \in X_0^0$ be given and define $g_k(I, \ell) = \frac{h_k(I,\ell)}{k^2\omega_1^2(I,\ell)-z}$ for $k \in \mathbb{Z}$. Note that $g_0 = 0$, and we are going to establish that $g \in X_0^2$; then $(-\mathcal{T}^2 - z)g = h$ will be a direct consequence of the definition of $\mathcal{T}^2$. According to Remark 4.2(a), there is $\alpha_0 > 0$ such that $|k^2\omega_1^2(I, \ell) - z| \geq \alpha_0$ for $|k| \geq 1$ and $(I, \ell) \in D$. In addition, by (4.3) and (4.4) there is $k_0 \in \mathbb{N}$ so that $|k^2\omega_1^2(I, \ell) - z| \geq \frac{1}{2}k^2\delta_1^2$ for $|k| \geq k_0$ and $(I, \ell) \in D$. Hence, we can bound

$$\begin{aligned}
\|g\|_{X_0^2}^2 &= 16\pi^3 \sum_{k\neq 0}(1+k^2)^2 \iint_D dI\,d\ell\,\ell\,\frac{1}{|Q'(e)|}\,\frac{|h_k(I,\ell)|^2}{|k^2\omega_1^2(I,\ell)-z|^2}\\
&\leq 16\pi^3\alpha_0^{-2}(1+k_0^2)^2 \sum_{1\leq|k|\leq k_0-1} \iint_D dI\,d\ell\,\ell\,\frac{1}{|Q'(e)|}\,|h_k(I,\ell)|^2\\
&\quad + 16\pi^3 \sum_{|k|\geq k_0}(1+k^2)^2 \iint_D dI\,d\ell\,\ell\,\frac{1}{|Q'(e)|}\,\frac{|h_k(I,\ell)|^2}{\frac{1}{4}k^4\delta_1^4}\\
&\leq 16\pi^3\alpha_0^{-2}(1+k_0^2)^2 \sum_{1\leq|k|\leq k_0-1} \iint_D dI\,d\ell\,\ell\,\frac{1}{|Q'(e)|}\,|h_k(I,\ell)|^2\\
&\quad + 256\pi^3\,\delta_1^{-4} \sum_{|k|\geq k_0} \iint_D dI\,d\ell\,\ell\,\frac{1}{|Q'(e)|}\,|h_k(I,\ell)|^2\\
&\leq (\alpha_0^{-2}(1+k_0^2)^2 + 16\,\delta_1^{-4})\|h\|_{X_0^0}^2;
\end{aligned}$$

observe that both α_0 and k_0 will depend upon z. For (B.26), we calculate from (B.2)

$$\begin{aligned}
((-\mathcal{T}^2 - z)^{-1}h, h)_{X^0} &= 16\pi^3 \sum_{k\in\mathbb{Z}} ([(-\mathcal{T}^2 - z)^{-1}h]_k, h_k)_{L^2_{\frac{1}{|Q'|}}(D)}\\
&= 16\pi^3 \sum_{k\neq 0} \iint_D dI\,d\ell\,\ell\,\frac{1}{|Q'(e)|}\,\frac{|h_k(I,\ell)|^2}{k^2\omega_1^2(I,\ell)-\bar{z}},
\end{aligned}$$

which completes the proof. □

Remark B.11 If we consider $\mathcal{T}^2 : X^2 \to X^0$, then the resolvent $(-\mathcal{T}^2 - z)^{-1} : X^0 \to X^2$ is not defined at $z = 0$, since the associated multiplier $\frac{1}{k^2\omega_1^2(I,\ell)-z}$ blows up for $k = 0$. This problem does not occur for $\mathcal{T}^2 : X_0^2 \to X_0^0$, as we omit the coefficient $k = 0$ for $(-\mathcal{T}^2 - z)^{-1} : X_0^0 \to X_0^2$. From the context it will always be clear which operator and which resolvent we are dealing with. ◇

For the next result, we remind that $\delta_1 = \inf_{\mathring{D}} \omega_1$ and $\Delta_1 = \sup_{\mathring{D}} \omega_1$. We also recall that the discrete spectrum of a self-adjoint operator A in a Hilbert space, called $\sigma_d(A)$, consists of all eigenvalues of A of finite multiplicity that are isolated points of the spectrum $\sigma(A)$. Its complement $\sigma_{\text{ess}}(A) = \sigma(A) \setminus \sigma_d(A)$ is the essential spectrum.

Lemma B.12 *We have*

$$\sigma_{\text{ess}}(-\mathcal{T}^2) = \bigcup_{k=1}^{\infty} k^2[\delta_1^2, \Delta_1^2] \tag{B.27}$$

for the essential spectrum, and in particular

$$\delta_1^2 = \min \sigma_{\text{ess}}(-\mathcal{T}^2). \tag{B.28}$$

If in addition (ω_1-1) is satisfied, then $\delta_1^2 \notin \sigma_p(-\mathcal{T}^2)$, the point spectrum of $-\mathcal{T}^2$.

Proof To establish (B.27), we first show that $k_0^2\,]\delta_1^2, \Delta_1^2[\subset \sigma_{\text{ess}}(-\mathcal{T}^2)$ for all $k_0 \in \mathbb{N}$. For, let $\mu_0^2 \in]\delta_1^2, \Delta_1^2[$. Using [37, Thm. 7.2], it suffices to construct a Weyl sequence $(g^{(j)})$ for $\lambda_0 = k_0^2\mu_0^2$, i.e., a sequence $(g^{(j)}) \subset X_0^2$ such that $\|g^{(j)}\|_{X^0} = 1$ for $j \in \mathbb{N}$, $g^{(j)} \rightharpoonup 0$ and $(-\mathcal{T}^2 - \lambda_0)g^{(j)} \to 0$ in X^0 as $j \to \infty$.

Since ω_1 is continuous by Theorem 3.6 and $\mathring{D}$ is connected, $\omega_1(\mathring{D})$ is an interval that satisfies $]\delta_1, \Delta_1[\subset \omega_1(\mathring{D}) \subset [\delta_1, \Delta_1]$. In particular, $\mu_0 = \omega_1(\hat{I}, \hat{\ell})$ for some $(\hat{I}, \hat{l}) \in \mathring{D}$. Let $\hat{e} = \hat{e}(\hat{I}, \hat{l}) \in]U_Q(0), e_0[$ denote the associated energy and select $\varepsilon > 0$ according to (Q2), i.e., with the property that

$$m = \inf\{|Q'(e)| : e \in [\hat{e} - \varepsilon, \hat{e} + \varepsilon]\} > 0. \tag{B.29}$$

From the continuity of Q' in $]-\infty, e_0[$ we furthermore have

$$M = \sup\{|Q'(e)| : e \in [\hat{e} - \varepsilon, \hat{e} + \varepsilon]\} < \infty. \tag{B.30}$$

Defining $\hat{\beta} = \hat{l}^2$ and assuming $\varepsilon > 0$ to be sufficiently small, we may suppose that the square $S = \{(e, \beta) : e \in [\hat{e} - \varepsilon, \hat{e} + \varepsilon], \beta \in [\hat{\beta} - \varepsilon, \hat{\beta} + \varepsilon]\} \subset \mathring{D}$. Let $\eta_1 = \eta_1(x) \in C_0^\infty(\mathbb{R}^2)$ be a function such that $\eta_1 \geq 0$, η_1 has its support in $B_1(0)$, and $\int_{\mathbb{R}^2} \eta_1(x)\, dx = 1$. We will use the standard mollifiers $\eta_j(x) = j^2\eta_1(jx)$ for $j \in \mathbb{N}$, they have their supports in $B_{1/j}(0)$ and satisfy $\int_{\mathbb{R}^2} \eta_j(x)\, dx = 1$. Then the functions $\chi_j(e, \beta) = \eta_j(e - \hat{e}, \beta - \hat{\beta})^{1/2}$ have their supports in $B_{1/j}(\hat{e}, \hat{\beta}) \subset S$ for j large enough, w.l.o.g. for $j \geq 1$. For $\tilde{g}_j(e, \beta) = |Q'(e)|\chi_j(e, \beta)$, we deduce from $\beta = \ell^2$ and (A.18) that

$$\begin{aligned}
\|\tilde{g}_j\|^2_{L^2_{\frac{1}{|Q'|}}(D)} &= \iint_D dI\, d\ell\, \ell\, \frac{1}{|Q'(e)|}\, |\tilde{g}_j(I, \ell)|^2 \\
&= \frac{1}{2} \iint_D de\, d\beta\, \frac{1}{\omega_1(e, \beta)\, |Q'(e)|}\, |\tilde{g}_j(e, \beta)|^2 \\
&= \frac{1}{2} \iint_D de\, d\beta\, \frac{|Q'(e)|}{\omega_1(e, \beta)}\, \eta_j(e - \hat{e}, \beta - \hat{\beta}).
\end{aligned}$$

Since

$$\iint\limits_{D} de\,d\beta\,\eta_j(e-\hat{e},\beta-\hat{\beta}) = \iint\limits_{S} de\,d\beta\,\eta_j(e-\hat{e},\beta-\hat{\beta}) = \iint\limits_{[-\varepsilon,\varepsilon]^2} \eta_j(x)\,dx = 1,$$

it follows from (B.29) and (B.30) that

$$\frac{m}{2\Delta_1} \le \|\tilde{g}_j\|^2_{L^2_{\frac{1}{|Q'|}}(D)} \le \frac{M}{2\delta_1}. \tag{B.31}$$

In addition, if $\phi \in C_0^\infty(\mathring{D})$, then

$$\begin{aligned}
|(\tilde{g}_j,\phi)_{L^2_{\frac{1}{|Q'|}}(D)}| &= \frac{1}{2}\left|\iint\limits_{D} de\,d\beta\,\frac{1}{\omega_1(e,\beta)\,|Q'(e)|}\,\overline{\tilde{g}_j(e,\beta)}\,\phi(e,\ell)\right| \\
&= \frac{j}{2}\left|\iint\limits_{D} de\,d\beta\,\frac{1}{\omega_1(e,\beta)}\,\eta_1(j(e-\hat{e},\beta-\hat{\beta}))^{1/2}\,\phi(e,\ell)\right| \\
&\le \frac{j}{2\delta_1}\,\|\phi\|_{L^\infty}\iint\limits_{S} de\,d\beta\,\eta_1(j(e-\hat{e},\beta-\hat{\beta}))^{1/2} \\
&\le \frac{j}{2\delta_1}\,\|\phi\|_{L^\infty}\iint\limits_{[-\varepsilon,\varepsilon]^2} dy\,\eta_1(jy)^{1/2} \\
&\le \frac{1}{2\delta_1 j}\,\|\phi\|_{L^\infty}\iint\limits_{B_1(0)} dx\,\eta_1(x)^{1/2} \\
&\le Cj^{-1}\|\phi\|_{L^\infty}.
\end{aligned} \tag{B.32}$$

Due to (B.29), (B.30) and (B.31), and taking into account the support properties of $\tilde{g}_j$, we conclude from (B.32) that also

$$\lim_{j\to\infty}(\tilde{g}_j,\phi)_{L^2_{\frac{1}{|Q'|}}(D)} = 0, \quad \phi \in L^2_{\frac{1}{|Q'|}}(D). \tag{B.33}$$

Next, we also note that

$$\begin{aligned}
&\iint\limits_{D} de\,d\beta\,|Q'(e)|\,(\omega_1(e,\beta)^2-\mu_0^2)^2\,\chi_j(e,\beta)^2 \\
&\le C\iint\limits_{S} de\,d\beta\,(\omega_1(e,\beta)-\mu_0)^2\,\eta_j(e-\hat{e},\beta-\hat{\beta}) \\
&= C\iint\limits_{[-\varepsilon,\varepsilon]^2} dy\,(\omega_1(\hat{e}+y_1,\hat{\beta}+y_2)-\omega_1(\hat{e},\hat{\beta}))^2\,\eta_j(y) \\
&= C\iint\limits_{B_1(0)} dx\,(\omega_1(\hat{e}+j^{-1}x_1,\hat{\beta}+j^{-1}x_2)-\omega_1(\hat{e},\hat{\beta}))^2\,\eta_1(x) \to 0, \quad j\to\infty.
\end{aligned} \tag{B.34}$$

For the Weyl sequence, consider $\tilde{g}^{(j)}(\theta, I, \ell) = \tilde{g}_j(e, \beta)\, e^{ik_0\theta} - \tilde{g}_j(e, \beta)\, e^{-ik_0\theta}$. Then $\tilde{g}^{(j)}_{k_0} = \tilde{g}_j$, $\tilde{g}^{(j)}_{-k_0} = -\tilde{g}_j$ and $\tilde{g}^{(j)}_k = 0$ for $k \neq \pm k_0$; in particular, $\tilde{g}^{(j)}$ is odd. Furthermore,

$$\|\tilde{g}^{(j)}\|^2_{X^0} = 16\pi^3 \sum_{k\in\mathbb{Z}} \|\tilde{g}^{(j)}_k\|^2_{L^2_{\frac{1}{|Q'|}}(D)} = 32\pi^3 \|\tilde{g}_j\|^2_{L^2_{\frac{1}{|Q'|}}(D)}$$

in conjunction with (B.31) shows that

$$16\pi^3 m \Delta_1^{-1} \leq \|\tilde{g}^{(j)}\|^2_{X^0} \leq 16\pi^3 M \delta_1^{-1} \tag{B.35}$$

for all j. Next let $h \in X^0$. Then $h_{\pm k_0} \in L^2_{\frac{1}{|Q'|}}(D)$ and

$$\begin{aligned}(\tilde{g}^{(j)}, h)_{X^0} &= 16\pi^3 \sum_{k\neq 0} (\tilde{g}^{(j)}_k, h_k)_{L^2_{\frac{1}{|Q'|}}(D)} \\ &= 16\pi^3 (\tilde{g}_j, h_{k_0})_{L^2_{\frac{1}{|Q'|}}(D)} - 16\pi^3 (\tilde{g}_j, h_{-k_0})_{L^2_{\frac{1}{|Q'|}}(D)} \to 0\end{aligned}$$

as $j \to \infty$ due to (B.33). Therefore, $\tilde{g}^{(j)} \rightharpoonup 0$ in X^0 for $j \to \infty$. In addition, using (B.34),

$$\begin{aligned}\|(-\mathcal{T}^2 - \lambda_0)\tilde{g}^{(j)}\|^2_{X^0} &= 8\pi^3 \sum_{k\in\mathbb{Z}} \iint_D de\, d\beta\, \frac{1}{\omega_1(e, \beta)\, |Q'(e_Q)|} \\ &\qquad \times |((-\mathcal{T}^2 - \lambda_0)\tilde{g}^{(j)})_k(I, \ell)|^2 \\ &= 16\pi^3 \iint_D de\, d\beta\, \frac{(k_0^2\omega_1(e, \beta)^2 - \lambda_0)^2}{\omega_1(e, \beta)\, |Q'(e_Q)|}\, |\tilde{g}^{(j)}_{k_0}(I, \ell)|^2 \\ &= 16\pi^3 k_0^4 \iint_D de\, d\beta\, |Q'(e)|\, \frac{(\omega_1(e, \beta)^2 - \mu_0^2)^2}{\omega_1(e, \beta)}\, \chi_j(e, \beta)^2 \\ &\leq C \iint_D de\, d\beta\, |Q'(e)|\, (\omega_1(e, \beta)^2 - \mu_0^2)^2\, \chi_j(e, \beta)^2 \to 0\end{aligned}$$

as $j \to \infty$.

Thus, by (B.35), we can normalize the sequence $(\tilde{g}^{(j)})$ to obtain the desired Weyl sequence $(g^{(j)})$.

Hence, we have verified that $k_0^2\,]\delta_1^2, \Delta_1^2[\subset \sigma_{\text{ess}}(-\mathcal{T}^2)$ for all $k_0 \in \mathbb{N}$. Since $\sigma_{\text{ess}}(-\mathcal{T}^2)$ is closed [37, Problem 10.5], it follows that even $k_0^2\, [\delta_1^2, \Delta_1^2] \subset \sigma_{\text{ess}}(-\mathcal{T}^2)$ holds for $k_0 \in \mathbb{N}$, which proves '$\supset$' in (B.27).

For the converse '$\subset$', take $\lambda \in \sigma_{\text{ess}}(-\mathcal{T}^2)$ and let $(g^{(j)}) \subset X^2_0$ be an associated Weyl sequence, i.e., we have $\|g^{(j)}\|_{X^0} = 1$ for $j \in \mathbb{N}$, $g^{(j)} \rightharpoonup 0$ and $h^{(j)} = (-\mathcal{T}^2 - \lambda)g^{(j)} \to 0$ in X^0 as $j \to \infty$. First note that $\sigma_{\text{ess}}(-\mathcal{T}^2) \subset \sigma(-\mathcal{T}^2) \subset [\delta_1^2, \infty[$ by

Corollary B.10 yields $\lambda \geq \delta_1^2$. For $k \in \mathbb{N}$ define $\varepsilon_k = \operatorname{dist}(\frac{\lambda}{k^2}, [\delta_1^2, \Delta_1^2])$. If we assume that $\varepsilon_k > 0$ for all $k \in \mathbb{N}$, then $\frac{\lambda}{k^2} \to 0$ as $k \to \infty$ implies that $\hat{\varepsilon} = \inf\{\varepsilon_k : k \in \mathbb{N}\} > 0$. Since $\omega_1^2(I, \ell) \in [\delta_1^2, \Delta_1^2]$,

$$|k^2\omega_1^2(I, \ell) - \lambda| = k^2\Big|\omega_1^2(I, \ell) - \frac{\lambda}{k^2}\Big| \geq k^2\hat{\varepsilon}, \quad k \in \mathbb{N}.$$

If we write $g^{(j)} = \sum_{k\neq 0} g_k^{(j)}(I, \ell)\, e^{ik\theta}$ and $h^{(j)} = \sum_{k\neq 0} h_k^{(j)}(I, \ell)\, e^{ik\theta}$, then

$$g_k^{(j)}(I, \ell) = \frac{h_k^{(j)}(I, \ell)}{k^2\omega_1^2(I, \ell) - \lambda}$$

due to (B.25). It follows that

$$\begin{aligned}
1 &= \|g^{(j)}\|_{X^0}^2 \\
&= 16\pi^3 \sum_{k\neq 0} \iint_D dI\, d\ell\, \ell\, \frac{1}{|Q'(e)|}\, |g_k^{(j)}(I, \ell)|^2 \\
&= 16\pi^3 \sum_{k\neq 0} \iint_D dI\, d\ell\, \ell\, \frac{1}{|Q'(e)|}\, \frac{|h_k^{(j)}(I, \ell)|^2}{|k^2\omega_1^2(I, \ell) - \lambda|^2} \\
&\leq \frac{16\pi^3}{\hat{\varepsilon}^2} \sum_{k\neq 0} \frac{1}{k^4} \iint_D dI\, d\ell\, \ell\, \frac{1}{|Q'(e)|}\, |h_k^{(j)}(I, \ell)|^2 \\
&\leq \frac{16\pi^3}{\hat{\varepsilon}^2}\, \|h^{(j)}\|_{X^0}^2 \to 0, \quad j \to \infty,
\end{aligned}$$

which is impossible. As a consequence, we must have $\varepsilon_{k_0} = 0$ for some $k_0 \in \mathbb{N}$. But then $\frac{\lambda}{k_0^2} \in [\delta_1^2, \Delta_1^2]$, which means that $\lambda \in k_0^2[\delta_1^2, \Delta_1^2]$.

Concerning (B.28), as $\sigma_{\rm ess}(-\mathcal{T}^2) \subset [\delta_1^2, \infty[$, we have $\inf \sigma_{\rm ess}(-\mathcal{T}^2) \geq \delta_1^2$. For the converse, $[\delta_1^2, \Delta_1^2] \subset \sigma_{\rm ess}(-\mathcal{T}^2)$ due to (B.27), and therefore $\inf \sigma_{\rm ess}(-\mathcal{T}^2) \leq \delta_1^2$. The infimum is a minimum, owing to $\delta_1^2 \in \sigma_{\rm ess}(-\mathcal{T}^2)$.

For the last claim, assume on the contrary that $g \in X_0^2$ is such that $g \neq 0$ and $-\mathcal{T}^2 g = \delta_1^2 g$. For the components this means that $k^2\omega_1^2\, g_k = \delta_1^2\, g_k$ in $L^2_{\frac{1}{|Q'|}}(D)$ for all $k \neq 0$. Since $\omega_1^2(I, \ell) \geq \delta_1^2$, it follows that $g_k = 0$ in $L^2_{\frac{1}{|Q'|}}(D)$ for $|k| \geq 2$. According to (ω_1-1) the set $\{(I, \ell) \in D : \omega_1(I, \ell) = \delta_1\}$ has Lebesgue measure zero. Therefore, $g_{\pm 1}(I, \ell) = 0$ for a.e. $(I, \ell) \in D$, which means that also $g_{\pm 1} = 0$ in $L^2_{\frac{1}{|Q'|}}(D)$, and hence $g = 0$. □

Lemma B.13 *Define*

$$(\mathcal{T}^{-1}h)_k(I, \ell) = \frac{1}{ik\, \omega_1(I, \ell)}\, h_k(I, \ell) \textit{ for } k \neq 0 \quad \textit{and} \quad (\mathcal{T}^{-1}h)_0(I, \ell) = 0. \tag{B.36}$$

(a) If $\alpha \geq 0$, then $\mathcal{T}^{-1} : X_0^\alpha \to X_0^{\alpha+1}$ is well-defined and

$$\|\mathcal{T}^{-1}h\|_{X^{\alpha+1}} \leq \sqrt{2}\,\delta_1^{-1}\|h\|_{X^\alpha} \quad \text{for} \quad h \in X_0^\alpha.$$

(b) $\mathcal{T}^{-1}\mathcal{T}h = g - g_0$ for $g \in X^{\alpha+1}$, and in particular $\mathcal{T}^{-1}\mathcal{T}^2 g = \mathcal{T}g$ for $g \in X^{\alpha+2}$.
(c) If $\mathcal{T}g = h$ for $g \in X^{\alpha+1}$, then $\mathcal{T}^{-1}h = g - g_0$.
(d) $\mathcal{T}\mathcal{T}^{-1}h = h$ for $h \in X_0^\alpha$.
(e) $\mathcal{T}(-\mathcal{T}^2)^{-1}h = -\mathcal{T}^{-1}h$ for $h \in X_0^0$.

Proof (a) To begin with, $(\mathcal{T}^{-1}h)_0 = 0$ holds by definition. In addition,

$$\begin{aligned}\|\mathcal{T}^{-1}h\|_{X^{\alpha+1}}^2 &= 16\pi^3 \sum_{k\neq 0} (1+k^2)^{\alpha+1} \,\|(\mathcal{T}^{-1}h)_k\|_{L^2_{\frac{1}{|Q'|}}(D)}^2 \\ &= 16\pi^3 \sum_{k\neq 0} \frac{(1+k^2)^{\alpha+1}}{k^2} \left\|\frac{1}{\omega_1}h_k\right\|_{L^2_{\frac{1}{|Q'|}}(D)}^2 \\ &\leq 32\pi^3\delta_1^{-2} \sum_{k\neq 0} (1+k^2)^\alpha \,\|h_k\|_{L^2_{\frac{1}{|Q'|}}(D)}^2 = 2\delta_1^{-2}\,\|h\|_{X^\alpha}^2,\end{aligned}$$

and hence in particular $\mathcal{T}^{-1}h \in X_0^{\alpha+1}$.
(b) Note that $\mathcal{T}g \in X_0^\alpha$ by Lemma B.8(a). Hence, we have $\mathcal{T}^{-1}\mathcal{T}g = \sum_{k\neq 0} \frac{1}{ik\omega_1}(\mathcal{T}g)_k\, e^{ik\theta} = \sum_{k\neq 0} g_k\, e^{ik\theta} = g - g_0$. For the second statement one uses $(\mathcal{T}g)_0 = 0$.
(c) This follows from (b).
(d) Here, we calculate $\mathcal{T}\mathcal{T}^{-1}h = \sum_{k\in\mathbb{Z}} ik\omega_1\,(\mathcal{T}^{-1}h)_k\, e^{ik\theta} = \sum_{k\neq 0} ik\omega_1\, (\mathcal{T}^{-1}h)_k\, e^{ik\theta} = \sum_{k\neq 0} h_k\, e^{ik\theta} = h$, due to $h_0 = 0$.
(e) According to their definition in (B.25), the coefficients of the resolvent $(-\mathcal{T}^2)^{-1}h$ at $z = 0$ are given by $((-\mathcal{T}^2)^{-1}h)_k = \frac{h_k}{k^2\omega_1^2}$ for $k \neq 0$ and $((-\mathcal{T}^2)^{-1}h)_0 = 0$. Therefore,

$$\begin{aligned}\mathcal{T}(-\mathcal{T}^2)^{-1}h &= \sum_{k\neq 0} ik\omega_1\,((-\mathcal{T}^2)^{-1}h)_k\, e^{ik\theta} = \sum_{k\neq 0} ik\,\frac{h_k}{k^2\omega_1}\, e^{ik\theta} \\ &= -\sum_{k\neq 0}\frac{1}{ik\omega_1}\, h_k\, e^{ik\theta} = -\,\mathcal{T}^{-1}h,\end{aligned}$$

and this completes the proof. □

For the next result recall the operator $\mathcal{K}$ from (1.15), given by

$$\mathcal{K}g = -Q'(e_Q)\, p_r\, U_g'(r) = |Q'(e_Q)|\, p_r\, U_g'(r), \tag{B.37}$$

where we have used assumption (Q3) for the last step.

Corollary B.14 *For $z \in \Omega$ and $\psi(\theta, I, \ell) = \sum_{k\neq 0} \psi_k(I, \ell)e^{ik\theta} \in X_0^0$ we have*

$$\mathcal{KT}(-\mathcal{T}^2 - z)^{-1}\psi \\ = |Q'(e_Q)|\, p_r \frac{16\pi^2 i}{r^2} \sum_{k\neq 0} \iint_D d\ell\, \ell\, de\, \mathbf{1}_{[r_-(e,\ell),\, r_+(e,\ell)]}(r) \frac{\sin(k\theta(r, e, \ell))}{k^2\omega_1^2(e, \ell) - z} \psi_k(I, \ell),$$

where $I = I(e, \ell)$.

Proof If we let $g = (-\mathcal{T}^2 - z)^{-1}\psi \in X_0^2$, then

$$g_k(I, \ell) = \frac{\psi_k(I, \ell)}{k^2\omega_1^2(I, \ell) - z} \quad \text{for} \quad k \in \mathbb{Z}$$

by (B.25); in particular, $g_0 = 0$. Next, if $h = \mathcal{T} g$, then

$$h_k(I, \ell) = ik\,\omega_1(I, \ell)\, g_k(I, \ell) = \frac{ik\,\omega_1(I, \ell)}{k^2\omega_1^2(I, \ell) - z} \psi_k(I, \ell)$$

for the coefficients of $h \in X_0^1$. Therefore, $\mathcal{KT}(-\mathcal{T}^2 - z)^{-1}\psi = \mathcal{K}h = |Q'(e_Q)|$ $p_r\, U_h'(r)$ and (B.8) from Lemma B.7 yield the claim. □

Lemma B.15 *Define $\mathcal{KT} : X^0 \to X^0$ by*

$$(\mathcal{KT} g)_k(I, \ell) = 16\pi\, |Q'(e)|\, \omega_1(e, \ell) \sum_{m\neq 0} \iint_D d\tilde{\ell}\, \tilde{\ell}\, d\tilde{e}\, g_m(\tilde{I}, \tilde{\ell}) \\ \times \int_0^\infty \frac{dr}{r^2} \mathbf{1}_{[r_-(e,\ell),\, r_+(e,\ell)]\cap[r_-(\tilde{e},\tilde{\ell}),\, r_+(\tilde{e},\tilde{\ell})]}(r) \\ \times \sin(k\theta(r, e, \ell)) \sin(m\theta(r, \tilde{e}, \tilde{\ell})) \tag{B.38}$$

for $k \in \mathbb{Z}$. Then

(a) $\mathcal{KT}$ agrees with what is obtained from (B.37). In particular

$$\mathcal{KT} g = 4\pi\, |Q'(e_Q)|\, p_r \int_{\mathbb{R}^3} p_r\, g\, dv \tag{B.39}$$

and $\mathcal{KT}$ is a linear bounded operator on X^0.

(b) $\mathcal{KT}$ is symmetric and

$$(\mathcal{KT} g, g)_{X^0} = \frac{1}{4\pi} \int_{\mathbb{R}^3} |U_{\mathcal{T}g}'(r)|^2\, dx \geq 0 \quad for \quad g \in X^0. \tag{B.40}$$

(c) $\mathcal{KT} : X_0^0 \to X_0^0$ is well-defined, linear, bounded, symmetric and positive.

(d) If $g_j \rightharpoonup 0$ weakly in X^0 as $j \to \infty$, then $\mathcal{KT} g_j \rightharpoonup 0$ weakly in X^0 as $j \to \infty$.

Proof (a) For $\mathcal{KT}g = |Q'(e_Q)|\, p_r\, U'_{\mathcal{T}g}(r)$ we deduce from Lemma B.5 that

$$(\mathcal{KT}g)_k(I,\ell) = -\frac{i}{\pi}\,|Q'(e)|\,\omega_1(e,\ell)\int_{r_-(e,\ell)}^{r_+(e,\ell)} U'_{\mathcal{T}g}(r)\sin(k\theta(r,e,\ell))\,dr$$

where $e = e(I,\ell)$. Noting that $(\mathcal{T}g)_m(I,\ell) = im\,\omega_1(I,\ell)g_m(I,\ell)$, and in particular $(\mathcal{T}g)_0 = 0$, the claim thus follows from (B.8) in Lemma B.7. Regarding (B.39), Lemma 2.4 says that $U'_{\mathcal{T}g}(r) = 4\pi\int_{\mathbb{R}^3} p_r\, g\, dv$. Hence, (B.37) shows that (B.39) holds. For the boundedness, we write out (B.39) more explicitly:

$$(\mathcal{KT}g)(x,v) = 4\pi\,|Q'(e)|\,p_r\int_{\mathbb{R}^3}\tilde{p}_r\, g(x,\tilde{v})\,d\tilde{v} \tag{B.41}$$

for $p_r = x\cdot v/|x|$, $\tilde{p}_r = x\cdot\tilde{v}/|x|$, $e = \frac{1}{2}|v|^2 + U_Q(r)$. If we also let $\tilde{e} = \frac{1}{2}|\tilde{v}|^2 + U_Q(r)$, then we obtain from Hölder's inequality and Lemma 2.5

$$\begin{aligned}
&|(\mathcal{KT}g)(x,v)|\\
&\le 4\pi\,|Q'(e)|\,|p_r|\left(\int_{\mathbb{R}^3}|Q'(\tilde{e})|\,\tilde{p}_r^2\,d\tilde{v}\right)^{1/2}\left(\int_{\mathbb{R}^3}\frac{1}{|Q'(\tilde{e})|}\,|g(x,\tilde{v})|^2\,d\tilde{v}\right)^{1/2}\\
&\le 4\pi\,|Q'(e)|\,|p_r|\,\rho_Q(r)^{1/2}\left(\int_{\mathbb{R}^3}\frac{1}{|Q'(\tilde{e})|}\,|g(x,\tilde{v})|^2\,d\tilde{v}\right)^{1/2}.
\end{aligned}$$

Observe that $Q' \in L^\infty_{\text{loc}}(\mathbb{R})$ by (Q3) and $e \in [U_Q(0), e_0]$. Furthermore $|p_r| \le \max_K |v|$, the maximal $|v|$ for some $(x,v)\in K$. Also $\rho_Q(r) \le \rho_Q(0)$ due to (A.32). Therefore, (B.4) leads to

$$\begin{aligned}
\|\mathcal{KT}g\|^2_{X^0} &= \|\mathcal{KT}g\|^2_{L^2_{\text{sph},\frac{1}{|Q'|}}(K)}\\
&= \iint_K \frac{1}{|Q'(e)|}\,|(\mathcal{KT}g)(x,v)|^2\,dx\,dv\\
&\le \frac{64\pi^2}{3}\rho_Q(0)\,(\max_K |v|)^5\,\sup\{|Q'(e)| : e\in[U_Q(0),e_0]\}\\
&\qquad\times\int_{\mathbb{R}^3}\int_{\mathbb{R}^3}\frac{dx\,d\tilde{v}}{|Q'(\tilde{e})|}\,|g(x,\tilde{v})|^2\\
&=: C^2_{\mathcal{KT}}\,\|g\|^2_{L^2_{\text{sph},\frac{1}{|Q'|}}(K)} = C^2_{\mathcal{KT}}\,\|g\|^2_{X^0}.
\end{aligned} \tag{B.42}$$

(b) Here we calculate, for $g, h \in X^0$ and using (B.4) as well as (B.41),

$$\begin{aligned}
(\mathcal{KT}g, h)_{X^0} &= (\mathcal{KT}g, h)_Q\\
&= \int_{\mathbb{R}^3}\int_{\mathbb{R}^3}\frac{1}{|Q'(e)|}\,\overline{(\mathcal{KT}g)(x,v)}\,h(x,v)\,dx\,dv
\end{aligned}$$

$$
\begin{aligned}
&= 4\pi \int_{\mathbb{R}^3} dx \int_{\mathbb{R}^3} dv\, p_r \int_{\mathbb{R}^3} d\tilde{v}\, \tilde{p}_r\, \overline{g(x,\tilde{v})}\, h(x,v) \qquad \text{(B.43)}\\
&= 4\pi \int_{\mathbb{R}^3} dx \int_{\mathbb{R}^3} dv\, p_r \int_{\mathbb{R}^3} d\tilde{v}\, \tilde{p}_r\, \overline{g(x,v)}\, h(x,\tilde{v})\\
&= \int_{\mathbb{R}^3}\int_{\mathbb{R}^3} \frac{1}{|Q'(e)|}\, \overline{g(x,v)}\, (\mathcal{KT}h)(x,v)\, dx\, dv\\
&= (g, \mathcal{KT}h)_Q = (g, \mathcal{KT}h)_{X^0},
\end{aligned}
$$

which shows the symmetry. If we specify to $h = g$, then (B.43) and Lemma 2.4 yield

$$
\begin{aligned}
(\mathcal{KT}g, g)_{X^0} &= 4\pi \int_{\mathbb{R}^3} dx \left(\int_{\mathbb{R}^3} dv\, p_r\, g(x,v)\right)\left(\int_{\mathbb{R}^3} d\tilde{v}\, \tilde{p}_r\, \overline{g(x,\tilde{v})}\right)\\
&= 4\pi \int_{\mathbb{R}^3} dx \left|\int_{\mathbb{R}^3} dv\, p_r\, g(x,v)\right|^2 = \frac{1}{4\pi}\int_{\mathbb{R}^3} |U'_{\mathcal{T}g}(r)|^2\, dx,
\end{aligned}
$$

so that $\mathcal{KT}$ is positive.

(c) Note that $(\mathcal{KT}g)_0 = 0$ by (B.38), even if $g \in X^0$, and not $g \in X^0_0$. Thus, in particular $\mathcal{KT}(X^0_0) \subset X^0_0$, and the remaining assertions follow from (a) and (b).

(d) Let $h \in X^0$. Then $\mathcal{KT}h \in X^0$ and by (b):

$$
(\mathcal{KT}g_j, h)_{X^0} = (g_j, \mathcal{KT}h)_{X^0} \to 0, \quad j \to \infty,
$$

which means that $\mathcal{KT}g_j \rightharpoonup 0$ weakly in X^0 as $j \to \infty$.

□

Corollary B.16 *If $g \in X^0$, then*

$$
\|U'_{\mathcal{T}g}\|_{L^2_r} \le 4\pi \rho_Q(0)^{1/2}\, \|g\|_{X^0}.
$$

Proof The argument is very similar to the one for (B.42) above. First recall from Lemma 2.4 that $U'_{\mathcal{T}g} = 4\pi \int_{\mathbb{R}^3} p_r g\, dv$, whence this is defined for $g \in X^0$, and not only for $g \in X^1$. Here we have, using Lemma 2.5 and the monotonicity of ρ_Q,

$$
\begin{aligned}
|U'_{\mathcal{T}g}(x)| &\le 4\pi \int_{\mathbb{R}^3} |p_r|\, |g(x,v)|\, dv\\
&\le 4\pi \left(\int_{\mathbb{R}^3} |Q'(e)|\, p_r^2\, dv\right)^{1/2} \left(\int_{\mathbb{R}^3} \frac{1}{|Q'(e)|}\, |g(x,v)|^2\, dv\right)^{1/2}\\
&\le 4\pi \rho_Q(r)^{1/2} \left(\int_{\mathbb{R}^3} \frac{1}{|Q'(e)|}\, |g(x,v)|^2\, dv\right)^{1/2}\\
&\le 4\pi \rho_Q(0)^{1/2} \left(\int_{\mathbb{R}^3} \frac{1}{|Q'(e)|}\, |g(x,v)|^2\, dv\right)^{1/2}.
\end{aligned}
$$

Accordingly, we obtain

$$\|U'_{\mathcal{T}g}\|_{L^2_r}^2 \le 16\pi^2 \rho_Q(0) \int_{\mathbb{R}^3}\int_{\mathbb{R}^3} \frac{1}{|Q'(e)|}\, |g(x,v)|^2\, dx\, dv = 16\pi^2 \rho_Q(0)\, \|g\|_{X^0}^2,$$

where we (as always) extend g by zero outside K. □

Now we turn to $L = -\mathcal{T}^2 - \mathcal{K}\mathcal{T}$ from (1.16).

Lemma B.17 *The linear operator $L = -\mathcal{T}^2 - \mathcal{K}\mathcal{T}$ with domain $\mathcal{D}(L) = \mathcal{D}(-\mathcal{T}^2) = X_0^2$ is self-adjoint on X_0^0.*

Proof According to Corollary B.10, $-\mathcal{T}^2 : X_0^2 \to X_0^0$ is self-adjoint. In addition, $\mathcal{K}\mathcal{T} : X_0^0 \to X_0^0$ is symmetric and bounded by Lemma B.15(c), and in particular closed. Moreover, $-\mathcal{K}\mathcal{T}$ is surely $-\mathcal{T}^2$-bounded with relative bound zero. Hence L is self-adjoint, trivially by the Kato-Rellich-Theorem; see [37, Theorem 13.5]. □

Since the Antonov stability estimate concerns spherically symmetric functions $u = u(x, v)$ that are odd in v, we have to restrict X_0^0 further, in accordance with Definition B.4; recall from (B.6) that

$$X^\alpha_{\rm odd} = \{g \in X^\alpha : g_{-k} = -g_k \;\; for\;\; k \in \mathbb{Z}\}$$

are the odd functions, represented as a Fourier series.

Remark B.18 (a) The operators $\mathcal{T}$ and $\mathcal{T}^{-1}$ do reverse the parity, whereas $\mathcal{T}^2$ preserves the parity; this is a direct consequence of Lemma B.3 and (B.16), (B.36), (B.17). The resolvent $(-\mathcal{T}^2 - z)^{-1} : X^0_{\rm odd} \to X^2_{\rm odd}$ of $-\mathcal{T}^2$ on $X^2_{\rm odd}$ is still given by (B.25) for $z \in \Omega$.

(b) $\mathcal{K}\mathcal{T}g$ is always odd; this is due to Lemma B.3 and (B.38).

(c) The statements regarding the spectrum of $-\mathcal{T}^2$ from Lemma B.12 are not affected, if $-\mathcal{T}^2$ is considered on $X^2_{\rm odd}$ instead of X_0^2; for (B.27), note that the Weyl sequence as constructed in the proof of Lemma B.12 consists of odd functions.

Corollary B.19 *The linear operator $L = -\mathcal{T}^2 - \mathcal{K}\mathcal{T}$ with domain $\mathcal{D}(L) = X^2_{\rm odd}$ is self-adjoint on $X^0_{\rm odd}$. In addition,*

$$(Lu, u)_{X^0} = \|\mathcal{T}u\|_{X^0}^2 - \frac{1}{4\pi}\int_{\mathbb{R}^3} |U'_{\mathcal{T}u}(r)|^2\, dx \quad for \;\; u \in X^2_{\rm odd}. \tag{B.44}$$

For the essential spectrum, we have $\sigma_{\rm ess}(L) = \sigma_{\rm ess}(-\mathcal{T}^2)$. In particular,

$$\sigma_{\rm ess}(L) = \bigcup_{k=1}^{\infty} k^2[\delta_1^2, \Delta_1^2] \quad and \quad \delta_1^2 = \min \sigma_{\rm ess}(L). \tag{B.45}$$

If ω_1 is not constant, then there exists $\lambda_c > \delta_1^2$ such that $[\lambda_c, \infty[\subset \sigma_{\rm ess}(L)$.

Proof The self-adjointness follows from Lemma B.17 and Remark B.18(a) and (b). For (B.44), we can apply Lemma B.8(c) and Lemma B.15(b).

To establish that $\sigma_{\text{ess}}(L) = \sigma_{\text{ess}}(-\mathcal{T}^2)$ we are going to use Weyl's Theorem; see [37, Theorem 14.6]. For this we need to prove that $\mathcal{KT}$ is relatively L-compact. Since $\mathcal{D}(\mathcal{KT}) = X^0 \supset X^2_{\text{odd}} = \mathcal{D}(L)$ for the domains, this will follow once we can show that $\mathcal{KT}(L+i)^{-1} : X^0_{\text{odd}} \to X^0_{\text{odd}}$ is a compact operator. Due to Corollary C.6 we know that $\mathcal{K} : X^0 \to X^0$ is compact; recall that $L^2_{\text{sph}, \frac{1}{|Q'|}}(K) = X^0$. Hence, it suffices to prove that $\mathcal{T}(L+i)^{-1} : X^0_{\text{odd}} \to X^0$ is a bounded operator. By the second resolvent identity, [37, Prop. 1.9],

$$(L+i)^{-1} = (-\mathcal{T}^2+i)^{-1} + (-\mathcal{T}^2+i)^{-1}\mathcal{KT}(L+i)^{-1},$$

so that

$$\mathcal{T}(L+i)^{-1} = \mathcal{T}(-\mathcal{T}^2+i)^{-1} + \mathcal{T}(-\mathcal{T}^2+i)^{-1}\mathcal{KT}(L+i)^{-1}. \tag{B.46}$$

If $h \in X^0_{\text{odd}}$, then (B.16) and (B.25) yield

$$\|\mathcal{T}(-\mathcal{T}^2+i)^{-1}h\|^2_{X^0} = 16\pi^3 \sum_{k\neq 0} \iint_D dI\, d\ell\, \ell \,\frac{1}{|Q'(e)|}\, \frac{k^2\omega_1(I,\ell)^2 |h_k(I,\ell)|^2}{|k^2\omega_1^2(I,\ell)+i|^2}.$$

In particular, for $k \neq 0$ we have

$$\frac{k^2\omega_1(I,\ell)^2 |h_k(I,\ell)|^2}{|k^2\omega_1^2(I,\ell)+i|^2} = \frac{k^2\omega_1(I,\ell)^2 |h_k(I,\ell)|^2}{k^4\omega_1^4(I,\ell)+1} \le \delta_1^{-4}\Delta_1^2\, \frac{|h_k(I,\ell)|^2}{k^2} \le C|h_k(I,\ell)|^2.$$

It follows that

$$\|\mathcal{T}(-\mathcal{T}^2+i)^{-1}h\|^2_{X^0} \le C \sum_{k\neq 0} \iint_D dI\, d\ell\, \ell\, \frac{1}{|Q'(e)|}\, |h_k(I,\ell)|^2 \le C\|h\|^2_{X^0},$$

which shows that $\mathcal{T}(-\mathcal{T}^2+i)^{-1} : X^0_{\text{odd}} \to X^0$ is bounded. Also $\mathcal{KT} : X^0 \to X^0_{\text{odd}}$ is bounded, by Lemma B.15(a). Lastly, L is self-adjoint, so that $\sigma(L) \subset \mathbb{R}$. Hence, $\text{dist}(-i, \sigma(L)) \ge 1$ implies that for the resolvent $\|(L+i)^{-1}\|_{L(X^0)} = \text{dist}(-i, \sigma(L))^{-1} \le 1$. Altogether, $\mathcal{T}(L+i)^{-1} : X^0_{\text{odd}} \to X^0$ from (B.46) is seen to be bounded.

Regarding (B.45), the assertions follow from (B.27) and (B.28) in Lemma B.12, together with Remark B.18(c).

For the last claim, since ω_1 is not constant, there is $k_0 \in \mathbb{N}$ such that $\delta_1^2 < \frac{k^2}{(k+1)^2}\Delta_1^2$ for $k \ge k_0$. By induction w.r. to $k \ge k_0$ we are going to establish that $[k_0^2\delta_1^2, k^2\Delta_1^2] \subset \sigma_{\text{ess}}(L)$ for all $k \ge k_0$. For $k = k_0$ we have $[k_0^2\delta_1^2, k_0^2\Delta_1^2] \subset \sigma_{\text{ess}}(L)$ due to (B.45).

Suppose now that $[k_0^2\delta_1^2, k^2\Delta_1^2] \subset \sigma_{\text{ess}}(L)$ is verified for some $k \geq k_0$. Then $(k+1)^2\delta_1^2 < k^2\Delta_1^2$ in conjunction with (B.45) yields

$$[k^2\Delta_1^2, (k+1)^2\Delta_1^2] \subset](k+1)^2\delta_1^2, (k+1)^2\Delta_1^2] \subset \sigma_{\text{ess}}(L),$$

which leads to

$$[k_0^2\delta_1^2, (k+1)^2\Delta_1^2] = [k_0^2\delta_1^2, k^2\Delta_1^2] \cup [k^2\Delta_1^2, (k+1)^2\Delta_1^2] \subset \sigma_{\text{ess}}(L)$$

and completes the proof. □

Appendix C
An Evolution Equation

C.1 Summary of the Argument

The aim of this section is to show that $\lambda_* < \delta_1^2$ implies that λ_* is an eigenvalue of L (Theorem C.8). To outline the argument, we remark that λ_* can be expressed as

$$\lambda_* = \inf \{\Phi(u) : u \in X^1_{\text{odd}}, \|u\|_{X^0} = 1\}$$

for X^1_{odd} as defined in Appendix II, Sect. B.1, and

$$\Phi(u) = \|\mathcal{T} u\|^2_{X^0} - (\mathcal{K}\mathcal{T} u, u)_{X^0}.$$

For a given time interval $J = [0, a]$ or $J = [0, \infty[$ and a given continuous function $h : J \to X^1_{\text{odd}}$, we consider the family of operators

$$\mathcal{W}(t, s) : g \mapsto \mathcal{W}(t, s)g, \quad (\mathcal{W}(t, s)g)_k = \mathcal{W}_k(t, s)g_k \ (k \in \mathbb{Z}), \tag{C.1}$$

$$\mathcal{W}_k(t, s)(I, \ell) = \exp\Big(- \int_s^t [k^2\omega_1^2(I, \ell) - \Phi(h(\tau))]\, d\tau\Big), \tag{C.2}$$

for $t, s \in J$, $t \geq s$; to emphasize the dependence on h, we will at times also write $\mathcal{W}(t, s; h)$. Note the evolution system property

$$\mathcal{W}(t, s) \circ \mathcal{W}(s, \tau) = \mathcal{W}(t, \tau), \quad t, s, \tau \in J, \ t \geq s \geq \tau. \tag{C.3}$$

We will consider the evolution equation

$$g(t) = \mathcal{W}(t, 0)\psi + \int_0^t \mathcal{W}(t, s)\, \mathcal{K}\mathcal{T} g(s)\, ds \tag{C.4}$$

M. Kunze, *A Birman-Schwinger Principle in Galactic Dynamics*, Progress in Mathematical Physics 77, https://doi.org/10.1007/978-3-030-75186-9

for $t \geq 0$ and initial data ψ, where $\mathcal{W}(t, s) = \mathcal{W}(t, s; g)$. In Theorem C.4, we are going to establish that if $\psi \in X^2_{\text{odd}}$ is such that $\|\psi\|_{X^0} = 1$ and $\Phi(\psi) \leq \lambda_* + \varepsilon_*$ (for $\varepsilon_* > 0$ small enough), then there exists a global continuous solution $g : [0, \infty[\to X^1_{\text{odd}}$ of (C.4) that satisfies $\|g(t)\|_{X^0} = 1$ for $t \in [0, \infty[$. This result does not rely on $\lambda_* < \delta_1^2$, the condition $\lambda_* \leq \delta_1^2$ is enough. The point about (C.4) is the following. Differentiating (C.2) for $h = g$ w.r. to t, we get

$$\partial_t \mathcal{W}_k(t, s)(I, \ell) = -[k^2 \omega_1^2(I, \ell) - \Phi(g(t))]\, \mathcal{W}_k(t, s)(I, \ell)$$

and hence, at least formally,

$$\begin{aligned}\partial_t(\mathcal{W}(t, s)g) &\cong (\partial_t \mathcal{W}_k(t, s) g_k) = (-[k^2\omega_1^2 - \Phi(g(t))]\, \mathcal{W}_k(t, s) g_k) \\ &\cong \mathcal{T}^2 \mathcal{W}(t, s) g + \Phi(g(t))\, \mathcal{W}(t, s) g.\end{aligned}$$

Applying this relation to (C.4), it follows that

$$\begin{aligned} g'(t) &= \mathcal{T}^2 \mathcal{W}(t, s)\psi + \Phi(g(t))\psi \\ &\quad + \int_0^t [\mathcal{T}^2 \mathcal{W}(t, s) \mathcal{K}\mathcal{T} g(s) + \Phi(g(t))\, \mathcal{W}(t, s)\mathcal{K}\mathcal{T} g(s)]\, ds + \mathcal{K}\mathcal{T} g(t) \\ &= \mathcal{T}^2 g(t) + \Phi(g(t))\, g(t) + \mathcal{K}\mathcal{T} g(t) \\ &= -Lg(t) + \Phi(g(t))\, g(t). \end{aligned} \tag{C.5}$$

This implies that the $\|\cdot\|_{X^0}$-norm is preserved along the solution flow. Since $\Phi(u) = (Lu, u)_Q = (Lu, u)_{X^0}$ for $u \in X^2_{\text{odd}}$ and as the solution $g(t)$ is regular enough, we also deduce from (C.5) that

$$\begin{aligned}\frac{d}{dt}\Phi(g(t)) &= \frac{d}{dt}(Lg(t), g(t))_{X^0} = 2\,(Lg(t), g'(t))_{X^0} \\ &= 2\,(Lg(t), -Lg(t) + \Phi(g(t))\, g(t))_{X^0} = -2\,(\|Lg(t)\|^2_{X^0} - \Phi(g(t))^2).\end{aligned}$$

Now

$$\begin{aligned}\|g'(t)\|^2_{X^0} &= \| - Lg(t) + \Phi(g(t))g(t)\|^2_{X^0} \\ &= \|Lg(t)\|^2_{X^0} - 2\Phi(g(t))\,(Lg(t), g(t))_{X^0} + \Phi(g(t))^2\, \|g(t)\|^2_{X^0} \\ &= \|Lg(t)\|^2_{X^0} - \Phi(g(t))^2,\end{aligned}$$

which in turn yields

$$\frac{d}{dt}\Phi(g(t)) = -2\,\|g'(t)\|^2_{X^0} \leq 0.$$

Therefore, we see that Φ is a Lyapunov function for the evolution. Since $\|g(t)\|_{X^0} = 1$, we also have $\Phi(g(t)) = (Lg(t), g(t))_{X^0} \geq \lambda_*$, and it is a natural question to ask, if we can construct a minimizer of Φ in the following way. Consider a sequence of initial data $(\psi_j) \subset X^2_{\rm odd}$ such that $\Phi(\psi_j) \le \lambda_* + 1/j$ and let g_j denote the corresponding solution to (C.4) so that $g_j(0) = \psi_j$. Then $\lambda_* \le \Phi(g_j(t)) \le \Phi(\psi_j) \le \lambda_* + 1/j$ for all $t \in [0, \infty[$ and $j \in \mathbb{N}$. Hence, the key point is to find a sequence of times (t_j) with the properties that $t_j \to \infty$ and $\{g_j(t_j) : j \in \mathbb{N}\} \subset X^0$ is relatively compact. We will show that this goal can be accomplished, if the condition $\lambda_* < \delta_1^2$ is imposed; the limiting function φ_* will then be the desired eigenfunction of L for the eigenvalue λ_*.

C.2 Set Up

The best constant in the Antonov stability estimate is

$$\lambda_* = \inf\{(Lu, u)_{X^0} : u \in X^2_{\rm odd}, \|u\|_{X^0} = 1\} > 0,$$

cf. (1.20). We also introduce

$$\Phi : X^1_{\rm odd} \to \mathbb{R},$$
$$\Phi(u) = \|\mathcal{T}u\|^2_{X^0} - (\mathcal{K}\mathcal{T}u, u)_{X^0} = \|\mathcal{T}u\|^2_{X^0} - \frac{1}{4\pi}\int_{\mathbb{R}^3} |U'_{\mathcal{T}u}(r)|^2\, dx, \quad \text{(C.6)}$$

recall Lemma B.15(b). Then

$$\lambda_* = \inf\{\Phi(u) : u \in X^1_{\rm odd}, \|u\|_{X^0} = 1\}$$

by Lemma C.10 below. Let the operators $\mathcal{W}(t, s)$ be defined as in (C.1) and (C.2).

Remark C.1 (Parity) Since k enters as k^2 into the definition, one has $\mathcal{W}_{-k}(t, s) = \mathcal{W}_k(t, s)$. Therefore, it follows from Lemma B.3 that $\mathcal{W}(t, s)$ preserves the parity in v: if g is even in v, then $\mathcal{W}(t, s)g$ is even in v, and if g is odd in v, then $\mathcal{W}(t, s)g$ is odd in v. ◇

We will study the evolution equation

$$g(t) = \mathcal{W}(t, 0)\psi + \int_0^t \mathcal{W}(t, s)\mathcal{K}\mathcal{T}g(s)\, ds \quad \text{(C.7)}$$

for $t \ge 0$ and initial data ψ, defining $\mathcal{W}(t, s) = \mathcal{W}(t, s; g)$.

C.3 Existence of solutions

First we will study local existence and uniqueness for (C.7). Henceforth, we will always assume that the parameter $\varepsilon_* > 0$ satisfies $\varepsilon_* \le \min\{\frac{1}{4}, \frac{\delta_1^2}{2}\}$.

Lemma C.2 *Let $\psi \in X^2_{\rm odd}$ be such that $\Phi(\psi) \le \lambda_* + \varepsilon_*$. Then (C.7) has a unique solution $g \in C(J, X^1_{\rm odd})$ on some time interval $J = [0, a]$ with $a > 0$ such that $\Phi(g(t)) \le \lambda_* + 2\varepsilon_*$ for $t \in J$ and $\|g\|_{\infty,1} \le 10\,\|\psi\|_{X^1}$, where $\|g\|_{\infty,1} = \max\{\|g(t)\|_{X^1} : t \in J\}$.*

Proof Define

$$\mathcal{G} = \{g \in C(J, X^1_{\rm odd}) : g(0) = \psi,\ \ \Phi(g(t)) \le \lambda_* + 2\varepsilon_* \text{ for } t \in J,\\ \|g\|_{\infty,1} \le 10\,\|\psi\|_{X^1}\},$$

where $J = [0, a]$ with

$$a = \min\Bigg\{1, \frac{8\varepsilon_*}{125\,C^2_{\mathcal{KT}}}, \frac{\varepsilon_*^3}{2000\,C_\Phi^2\,\|\psi\|^2_{X^1}\,\|\psi\|^2_{X^2}\,(\Delta_1^2 + 1 + 20\,C_{\mathcal{KT}})^2}, \frac{1}{800\,C_\Phi\,\|\psi\|^2_{X^1}},\\ \frac{\varepsilon_*}{4\cdot 10^8 (C_\Phi\,\|\psi\|^2_{X^1} + C_\Phi C_{\mathcal{KT}}\,\|\psi\|^2_{X^1} + C_{\mathcal{KT}})^2}\Bigg\} > 0,$$

for $C_{\mathcal{KT}} > 0$ from (B.42), C_Φ from Lemma C.9 and Δ_1 from Theorem 3.5. By Lemma C.9, $\mathcal{G}$ is a closed subset of the Banach space $C(J, X^1_{\rm odd})$, which is equipped with the norm $\|\cdot\|_{\infty,1}$. If we set $g(t) = \psi$ for $t \in J$, then $g \in \mathcal{G}$, which shows that $\mathcal{G} \ne \emptyset$. Next let $F : \mathcal{G} \to \mathcal{G}$ be given by

$$(F(g))(t) = \mathcal{W}(t,0)\psi + \int_0^t \mathcal{W}(t,s)\,\mathcal{KT}g(s)\,ds, \quad g \in \mathcal{G}, \quad t \in J,$$

where $\mathcal{W}(t,s) = \mathcal{W}(t,s;g)$. First we are going to verify that F is well-defined, i.e., $F(\mathcal{G}) \subset \mathcal{G}$. Fix $g \in \mathcal{G}$ and write $h(t) = (F(g))(t)$. Then $h(0) = \psi$. The operators $\mathcal{W}(t,s)$ do preserve the parity in v, whereas $\mathcal{KT}$ is always odd; see Remarks C.1 and B.18(c). Thus, if $\psi \in X^2_{\rm odd}$ is odd in v and $g(s) \in X^1_{\rm odd}$ is odd in v, also $h = F(g)$ will be odd in v. In addition, by (C.68), (C.75) for $\alpha = 1$ and (B.42):

$$\begin{aligned}
\|h(t)\|_{X^1} &\le \|\mathcal{W}(t,0)\psi\|_{X^1} + \left\|\int_0^t \mathcal{W}(t,s)\,\mathcal{KT}g(s)\,ds\right\|_{X^1}\\
&\le e^{2\varepsilon_* t}\,\|\psi\|_{X^1} + \sqrt{\frac{2}{\varepsilon_*}}\,(1+e^{4\varepsilon_* t})^{1/2}\left(\int_0^t \|\mathcal{KT}g(s)\|^2_{X^0}\,ds\right)^{1/2}\\
&\le e^{2\varepsilon_* t}\,\|\psi\|_{X^1} + \sqrt{\frac{2}{\varepsilon_*}}\,(1+e^{4\varepsilon_* t})^{1/2} C_{\mathcal{KT}}\left(\int_0^t \|g(s)\|^2_{X^0}\,ds\right)^{1/2}\\
&\le e^{2\varepsilon_* t}\,\|\psi\|_{X^1} + \sqrt{\frac{2}{\varepsilon_*}}\,(1+e^{4\varepsilon_* t})^{1/2} C_{\mathcal{KT}}\,\sqrt{t}\,\|g\|_{\infty,1} \qquad \text{(C.8)}\\
&\le e^{2\varepsilon_* a}\,\|\psi\|_{X^1} + \sqrt{\frac{2}{\varepsilon_*}}\,(1+e^{4\varepsilon_* a})^{1/2} C_{\mathcal{KT}}\,\sqrt{a}\,\|g\|_{\infty,1}
\end{aligned}$$

for $t \in [0, a]$. Since $\|g\|_{\infty,1} \leq 10\,\|\psi\|_{X^1}$, $a \leq 1$, $\varepsilon_* \leq 1/4$ and $a \leq \frac{8\varepsilon_*}{125\, C_{\mathcal{KT}}^2}$, it follows that also $\|h\|_{\infty,1} \leq 10\,\|\psi\|_{X^1}$ is verified. To prove that h is continuous at $t_0 = 0$, note that by (C.84) and (C.8):

$$\begin{aligned}
&\|h(t) - \psi\|_{X^1} \\
&\quad \leq \|[\mathcal{W}(t,0) - \mathcal{W}(0,0)]\psi\|_{X^1} + \left\| \int_0^t \mathcal{W}(t,s)\,\mathcal{KT}\,g(s)\,ds \right\|_{X^1} \\
&\quad \leq (\Delta_1^2 + 1)\,(t + t^2)^{1/2}\,\|\psi\|_{X^2} + \sqrt{\frac{2}{\varepsilon_*}}\,(1 + e^{4\varepsilon_* t})^{1/2} C_{\mathcal{KT}}\,\sqrt{t}\,\|g\|_{\infty,1} \to 0, \quad t \to 0^+.
\end{aligned} \tag{C.9}$$

Next we are going to verify that h is continuous at $t_0 \in J$ such that $t_0 \in]0, a[$. W.l.o.g. consider $\eta > 0$ only, where η is so small that $t_0 + \eta \in J$. Fix $\delta \in]0, t_0[$. Then by (C.85), (C.78), (C.79), (C.80) and (B.42):

$$\begin{aligned}
&\|h(t_0+\eta) - h(t_0)\|_{X^1} \\
&\quad \leq \|[\mathcal{W}(t_0+\eta,0) - \mathcal{W}(t_0,0)]\psi\|_{X^1} + \left\| \int_{t_0}^{t_0+\eta} \mathcal{W}(t_0+\eta,s)\,\mathcal{KT}\,g(s)\,ds \right\|_{X^1} \\
&\qquad + \left\| \int_0^{t_0-\delta} [\mathcal{W}(t_0+\eta,s) - \mathcal{W}(t_0,s)]\,\mathcal{KT}\,g(s)\,ds \right\|_{X^1} \\
&\qquad + \left\| \int_{t_0-\delta}^{t_0} \mathcal{W}(t_0+\eta,s)\,\mathcal{KT}\,g(s)\,ds \right\|_{X^1} + \left\| \int_{t_0-\delta}^{t_0} \mathcal{W}(t_0,s)\,\mathcal{KT}\,g(s)\,ds \right\|_{X^1} \\
&\quad \leq 2\,(\Delta_1^2+1)\left[\frac{1}{\varepsilon_* t_0}\sqrt{\eta} + e^{2\varepsilon_* t_0}\,\eta\right]\|\psi\|_{X^0} \\
&\qquad + \sqrt{\frac{2}{\varepsilon_*}}\,(1 + e^{2\varepsilon_*\eta})\left(\int_{t_0}^{t_0+\eta} \|\mathcal{KT}\,g(s)\|_{X^0}^2\,ds\right)^{1/2} \\
&\qquad + \int_0^{t_0-\delta} \|[\mathcal{W}(t_0+\eta,s) - \mathcal{W}(t_0,s)]\,\mathcal{KT}\,g(s)\|_{X^1}\,ds \\
&\qquad + \frac{4}{\varepsilon_*^{1/2}}\,(1 + e^{2\varepsilon_*(\eta+\delta)})\left(\int_{t_0-\delta}^{t_0} \|\mathcal{KT}\,g(s)\|_{X^0}^2\,ds\right)^{1/2} \\
&\quad \leq 2\,(\Delta_1^2+1)\left[\frac{1}{\varepsilon_* t_0}\sqrt{\eta} + e^{2\varepsilon_* t_0}\,\eta\right]\|\psi\|_{X^0} \\
&\qquad + \sqrt{\frac{2}{\varepsilon_*}}\,(1 + e^{2\varepsilon_*\eta})\,C_{\mathcal{KT}}\left(\int_{t_0}^{t_0+\eta} \|g(s)\|_{X^0}^2\,ds\right)^{1/2} \\
&\qquad + 2\,(\Delta_1^2+1)\,C_{\mathcal{KT}} \int_0^{t_0-\delta} \left[\frac{1}{\varepsilon_*(t_0-s)}\sqrt{\eta} + \exp(2\varepsilon_*(t_0-s))\,\eta\right]\|g(s)\|_{X^0}\,ds \\
&\qquad + \frac{4}{\varepsilon_*^{1/2}}\,(1 + e^{2\varepsilon_*(\eta+\delta)})\,C_{\mathcal{KT}}\left(\int_{t_0-\delta}^{t_0} \|g(s)\|_{X^0}^2\,ds\right)^{1/2} \\
&\quad \leq 2\,(\Delta_1^2+1)\left[\frac{1}{\varepsilon_* t_0}\sqrt{\eta} + e^{2\varepsilon_* t_0}\,\eta\right]\|\psi\|_{X^0} + \sqrt{\frac{2}{\varepsilon_*}}\,(1 + e^{2\varepsilon_*\eta})\,C_{\mathcal{KT}}\,\|g\|_{\infty,1}\sqrt{\eta}
\end{aligned}$$

$$+\frac{\Delta_1^2+1}{\varepsilon_*}\,C_{\mathcal{KT}}\,\|g\|_{\infty,1}\left[2\ln\left(\frac{t_0}{\delta}\right)\sqrt{\eta}+e^{2\varepsilon_* t_0}\,\eta\right]+\frac{4}{\varepsilon_*^{1/2}}\,(1+e^{2\varepsilon_*(\eta+\delta)})\,C_{\mathcal{KT}}\,\|g\|_{\infty,1}\sqrt{\delta}. \tag{C.10}$$

So if we for instance set $\delta=\eta$, it follows that $\lim_{\eta\to 0^+}\|h(t_0+\eta)-h(t_0)\|_{X^1}=0$. Since a similar argument proves the continuity of h at $t_0=a$, in summary we have shown that $h\in C(J,X^1_{\rm odd})$. It remains to check the condition $\Phi(h(t))\le\lambda_*+2\varepsilon_*$ for $t\in J$. For this, owing to Lemma C.9 and by (C.9) for $t\in J$:

$$\begin{aligned}
&|\Phi(h(t))-\Phi(\psi)|\\
&\le 2C_\Phi(\|h(t)\|_{X^1}+\|\psi\|_{X^1})\|h(t)-\psi\|_{X^1}\\
&\le 22\,C_\Phi\,\|\psi\|_{X^1}\,\|h(t)-\psi\|_{X^1}\\
&\le 22\,C_\Phi\,\|\psi\|_{X^1}\left[(\Delta_1^2+1)\,(t+t^2)^{1/2}\,\|\psi\|_{X^2}+\sqrt{\frac{2}{\varepsilon_*}}\,(1+e^{4\varepsilon_* t})^{1/2}C_{\mathcal{KT}}\,\sqrt{t}\,\|g\|_{\infty,1}\right]\\
&\le 22\,C_\Phi\,\|\psi\|_{X^1}\left[\sqrt{2}\,(\Delta_1^2+1)\,\|\psi\|_{X^2}+10\sqrt{\frac{2}{\varepsilon_*}}\,(1+e^{4\varepsilon_*})^{1/2}C_{\mathcal{KT}}\,\|\psi\|_{X^1}\right]\sqrt{t}\\
&\le\frac{44\,C_\Phi}{\varepsilon_*^{1/2}}\,\|\psi\|_{X^1}\,\|\psi\|_{X^2}\,(\Delta_1^2+1+20\,C_{\mathcal{KT}})\,\sqrt{a}\\
&\le\varepsilon_*,
\end{aligned}$$

recalling the definition of a. Hence, we obtain

$$\Phi(h(t))\le|\Phi(h(t))-\Phi(\psi)|+\Phi(\psi)\le\varepsilon_*+\Phi(\psi)\le\lambda_*+2\varepsilon_*,\quad t\in J.$$

Altogether, so far we have verified that $F(\mathcal{G})\subset\mathcal{G}$. Next we will show that $F:\mathcal{G}\to\mathcal{G}$ is a contraction. Let $g_1,g_2\in\mathcal{G}$. Then $\Phi(g_1(t))\le\lambda_*+2\varepsilon_*$ and $\Phi(g_2(t))\le\lambda_*+2\varepsilon_*$ for $t\in J$ by the definition of $\mathcal{G}$. Furthermore, in the notation of Lemma C.18 below,

$$\begin{aligned}
\Lambda(t;g_1,g_2)&=2C_\Phi(\|g_1\|_{\infty,1}+\|g_2\|_{\infty,1})\int_0^t\|g_1(\tau)-g_2(\tau)\|_{X^1}\,d\tau\\
&\le 40\,C_\Phi\,\|\psi\|_{X^1}\int_0^t\|g_1(\tau)-g_2(\tau)\|_{X^1}\,d\tau\\
&\le 40\,C_\Phi\,\|\psi\|_{X^1}\,\|g_1-g_2\|_{\infty,1}\,a\\
&\le 800\,C_\Phi\,\|\psi\|^2_{X^1}\,a\\
&\le 1
\end{aligned}$$

by the choice of a. Thus, in particular

$$\Lambda(t;g_1,g_2)\,\exp(\Lambda(t;g_1,g_2))\le 120\,C_\Phi\,\|\psi\|_{X^1}\,\|g_1-g_2\|_{\infty,1}\,a.$$

Hence for $t \in J$ one deduces from this estimate, Lemma C.18, (C.75) for $\alpha = 1$ and (B.42) that

$$
\begin{aligned}
&\|(F(g_1))(t) - (F(g_2))(t)\|_{X^1} \\
&= \left\| \mathcal{W}(t,0;g_1)\psi + \int_0^t \mathcal{W}(t,s;g_1)\mathcal{KT}g_1(s)\,ds - \mathcal{W}(t,0;g_2)\psi - \int_0^t \mathcal{W}(t,s;g_2)\mathcal{KT}g_2(s)\,ds \right\|_{X^1} \\
&\le \|\mathcal{W}(t,0;g_1) - \mathcal{W}(t,0;g_2))\psi\|_{X^1} + \left\| \int_0^t [\mathcal{W}(t,s;g_1) - \mathcal{W}(t,s;g_2)]\mathcal{KT}g_2(s)\,ds \right\|_{X^1} \\
&\quad + \left\| \int_0^t \mathcal{W}(t,s;g_1)\mathcal{KT}(g_1(s) - g_2(s))\,ds \right\|_{X^1} \\
&\le 2\Lambda(t;g_1,g_2)\exp(\Lambda(t;g_1,g_2))\,(1+e^{2\varepsilon_* t})\,\|\psi\|_{X^1} \\
&\quad + \frac{2}{\varepsilon_*^{1/2}}\Lambda(t;g_1,g_2)\exp(\Lambda(t;g_1,g_2))\,(1+e^{2\varepsilon_* t})\left(\int_0^t \|\mathcal{KT}g_2(s)\|_{X^0}^2\,ds\right)^{1/2} \\
&\quad + \frac{2}{\varepsilon_*^{1/2}}(1+e^{2\varepsilon_* t})\left(\int_0^t \|\mathcal{KT}(g_1(s) - g_2(s))\|_{X^0}^2\,ds\right)^{1/2} \\
&\le 960\,C_\Phi\,\|\psi\|_{X^1}^2\,\|g_1 - g_2\|_{\infty,1}\,a + \frac{960\,C_\Phi C_{\mathcal{KT}}}{\varepsilon_*^{1/2}}\|\psi\|_{X^1}\,\|g_1 - g_2\|_{\infty,1}\left(\int_0^t \|g_2(s)\|_{X^0}^2\,ds\right)^{1/2} a \\
&\quad + \frac{8C_{\mathcal{KT}}}{\varepsilon_*^{1/2}}\left(\int_0^t \|g_1(s) - g_2(s)\|_{X^0}^2\,ds\right)^{1/2} \\
&\le 960\,C_\Phi\,\|\psi\|_{X^1}^2\,\|g_1 - g_2\|_{\infty,1}\,a + \frac{9600\,C_\Phi C_{\mathcal{KT}}}{\varepsilon_*^{1/2}}\|\psi\|_{X^1}^2\,\|g_1 - g_2\|_{\infty,1}\,a^{3/2} \\
&\quad + \frac{8C_{\mathcal{KT}}}{\varepsilon_*^{1/2}}\|g_1 - g_2\|_{\infty,1}\,a^{1/2} \\
&\le \frac{9600}{\varepsilon_*^{1/2}}\Big(C_\Phi\,\|\psi\|_{X^1}^2 + C_\Phi C_{\mathcal{KT}}\,\|\psi\|_{X^1}^2 + C_{\mathcal{KT}}\Big)\,\|g_1 - g_2\|_{\infty,1}\,a^{1/2},
\end{aligned}
$$

where we used in between that $2\varepsilon_* t \le 2\varepsilon_* a \le 2\varepsilon_* \le 1/2$. Thus, by the choice of a, we obtain $\|F(g_1) - F(g_2)\|_{\infty,1} \le \frac{1}{2}\|g_1 - g_2\|_{\infty,1}$, and the Banach fixed point theorem applies to yield the claim. □

Corollary C.3 *In the setting of Lemma C.2, the solution g hat the following additional properties:*

(a) $\|g(t) - g(s)\|_{X^1} \le C_4(g,a)(t-s)^{1/6}$ *for* $t, s \in J$, $t \ge s$,
(b) $g(t) \in X^2_{\mathrm{odd}}$ *for* $t \in J$ *and*

$$\|g(t)\|_{X^2} \le C_5(g,a), \quad t \in J,$$

where $C_4(g,a) > 0$ and $C_5(g,a) > 0$ are explicit constants (see below) that depend upon Δ_1, $a = \max J$, $\|\psi\|_{X^2}$, ε_*, $C_{\mathcal{KT}}$ and $\|g\|_{\infty,1}$.

(c) The function $J \ni t \mapsto g(t) \in X^2_{\mathrm{odd}}$ *is continuous in* $]0,a[$.
(d) The function $J \ni t \mapsto g(t) \in X^0_{\mathrm{odd}}$ *is differentiable at every* $t \in]0,a[$ *and its derivative is given by*

$$g'(t) = -Lg(t) + \Phi(g(t))\,g(t). \tag{C.11}$$

(e) *The function* $J \ni t \mapsto \Phi(g(t)) \in \mathbb{R}$ *is differentiable at every* $t \in]0, a[$ *and its derivative is given by*

$$\frac{d}{dt}\Phi(g(t)) = -2\,(\|Lg(t)\|_{X^0}^2 - \Phi(g(t))^2).$$

(f) *If* $\|\psi\|_{X^0} = 1$*, then* $\|g(t)\|_{X^0} = 1$ *for* $t \in J$ *and* $(g'(t), g(t))_{X^0} = 0$ *for* $t \in]0, a[$. *Furthermore,*

$$\Phi(g(t_1)) - \Phi(g(t_0)) = -2\int_{t_0}^{t_1} \|g'(t)\|_{X^0}^2\,dt \tag{C.12}$$

for $t_0, t_1 \in J$, $t_1 \geq t_0$. *In particular,* $J \ni t \mapsto \Phi(g(t)) \in \mathbb{R}$ *is monotone decreasing and* $\Phi(g(t)) \geq \lambda_*$ *for* $t \in J$.

Proof Since the solution g is a fixed point of F in the proof of Lemma C.2, one has $h = F(g) = g$. Hence, (C.9) and (C.10) for $\delta = \eta$ and $\eta \leq 1$ imply that

$$\|g(t) - \psi\|_{X^1} \leq C_1(g, a)\sqrt{t}, \quad (t \in J), \tag{C.13}$$

$$\|g(t+\eta) - g(t)\|_{X^1} \leq C_2(g, a)\left(\frac{1}{t} + 1 + \ln\left(\frac{t}{\eta}\right)\right)\sqrt{\eta},$$
$$(t, t+\eta \in J, t > 0, \eta > 0), \tag{C.14}$$

where

$$C_1(g, a) = (\Delta_1^2 + 1)\,(1 + \sqrt{a})\,\|\psi\|_{X^2} + \sqrt{\frac{2}{\varepsilon_*}}\,(1 + e^{4\varepsilon_* a})^{1/2} C_{\mathcal{KT}}\,\|g\|_{\infty,1},$$

$$C_2(g, a) = \frac{2}{\varepsilon_*}\,(\Delta_1^2 + 1)\,e^{2\varepsilon_* a}\,\|\psi\|_{X^0} + \frac{4}{\varepsilon_*}\Big[(\Delta_1^2 + 1)e^{2\varepsilon_* a} + 2(1 + e^{4\varepsilon_*})\Big]\,C_{\mathcal{KT}}\,\|g\|_{\infty,1}.$$

If $t \leq \eta^{1/3}$, then (C.13) leads to

$$\begin{aligned}
\|g(t+\eta) - g(t)\|_{X^1} &\leq \|g(t+\eta) - \psi\|_{X^1} + \|g(t) - \psi\|_{X^1} \\
&\leq C_1(g, a)(\sqrt{t+\eta} + \sqrt{t}) \\
&\leq 2C_1(g, a)\sqrt{\eta^{1/3} + \eta} \\
&\leq 2\sqrt{2}\,C_1(g, a)\,\eta^{1/6}.
\end{aligned}$$

On the other hand, if $t \geq \eta^{1/3}$, then by (C.14):

$$\begin{aligned}
\|g(t+\eta) - g(t)\|_{X^1} &\leq C_2(g, a)\left(\eta^{-1/3} + 1 + \ln\left(\frac{a}{\eta}\right)\right)\sqrt{\eta} \\
&\leq C_2(g, a)\left(2\eta^{-1/3} + \ln\left(\frac{a}{\eta}\right)\right)\sqrt{\eta} \\
&\leq C_2(g, a)\left(2 + |\ln a| + \frac{3}{e}\right)\eta^{1/6}
\end{aligned}$$

for $a = \max J$, where we also used that $3\,|\ln x|\,x \le \frac{3}{e}$ for $x \in [0, 1]$. In summary, we have shown that if $t, t + \eta \in J$, $t > 0$, $\eta \in]0, 1]$, then

$$\|g(t + \eta) - g(t)\|_{X^1} \le C_3(g, a)\, \eta^{1/6},$$

provided that we define

$$C_3(g, a) = 4C_1(g, a) + (4 + |\ln a|)C_2(g, a).$$

It follows that for $t, s \in J$, $t \ge s$, one always has the Hölder bound

$$\|g(t) - g(s)\|_{X^1} \le C_4(g, a)(t - s)^{1/6}$$

for

$$C_4(g, a) = C_3(g, a) + 2\|g\|_{\infty,1},$$

as is claimed in (a).

Owing to Remark B.2(b), this in particular implies that $\|g(t) - g(s)\|_{X^0} \le C_4(g, a)(t - s)^{1/6}$ and thus also

$$\|\mathcal{KT} g(t) - \mathcal{KT} g(s)\|_{X^0} \le C_{\mathcal{KT}} C_4(g, a)(t - s)^{1/6} \tag{C.15}$$

for $t, s \in J$ by (B.42). Hence, we obtain from (C.7), (C.68) and (C.76) with $A = C_{\mathcal{KT}} C_4(g, a)$, $\gamma = 1/6$, $\alpha = 2$, and (C.77) that for $t \in J$:

$$\begin{aligned}
\|g(t)\|_{X^2} &\le \|\mathcal{W}(t, 0)\psi\|_{X^2} + \left\| \int_0^t \mathcal{W}(t, s)\, \mathcal{KT}[g(s) - g(t)]\, ds \right\|_{X^2} \\
&\quad + \left\| \int_0^t \mathcal{W}(t, s)\, \mathcal{KT} g(t)\, ds \right\|_{X^2} \\
&\le e^{2\varepsilon_* t} \|\psi\|_{X^2} + 3\, C_{\mathcal{KT}}^2 C_4(g, a)^2 \left(\frac{4}{\varepsilon_*^2}\, t^{1/3} + \frac{1}{4\varepsilon_*}\, e^{4\varepsilon_* t}\, t^{4/3} \right) \\
&\quad + \frac{1}{\varepsilon_*}\, (2 + e^{2\varepsilon_* t})\, \|\mathcal{KT} g\|_{X^0} \\
&\le e^{2\varepsilon_* a} \|\psi\|_{X^2} + 3C_{\mathcal{KT}}^2 C_4(g, a)^2 \left(\frac{4}{\varepsilon_*^2}\, a^{1/3} + \frac{1}{4\varepsilon_*}\, e^{4\varepsilon_* a}\, a^{4/3} \right) \\
&\quad + \frac{C_{\mathcal{KT}}}{\varepsilon_*}\, (2 + e^{2\varepsilon_* a})\, \|g\|_{\infty,1} \\
&=: C_5(g, a),
\end{aligned}$$

which yields the asserted bound in (b).

In order to prove the continuity of g in X^2_{odd} as claimed in (c), fix $t_0 \in]0, a[$ and $\eta > 0$ sufficiently small. Then by (C.7):

$$\begin{aligned} g(t_0+\eta)-g(t_0) &= \mathcal{W}(t_0+\eta,0)\psi - \mathcal{W}(t_0,0)\psi \\ &\quad + \int_{t_0}^{t_0+\eta} \mathcal{W}(t_0+\eta,s)\,\mathcal{KT}\,g(s)\,ds \\ &\quad + \int_0^{t_0} [\mathcal{W}(t_0+\eta,s)-\mathcal{W}(t_0,s)]\,\mathcal{KT}\,g(s)\,ds \\ &= B_1+B_2+B_3, \end{aligned} \tag{C.16}$$

with B_j, $j=1,2,3$, denoting the three lines. We will bound each of the three terms individually and we will start with B_1. Here, it suffices to invoke (C.86), since by this estimate on has

$$\begin{aligned} \|B_1\|_{X^2} &= \|\mathcal{W}(t_0+\eta,0)\psi - \mathcal{W}(t_0,0)\psi\|_{X^2} \\ &\le 2\,(\Delta_1^2+1)\left[\frac{2}{\varepsilon_*^{3/2}t_0^{3/2}}\sqrt{\eta}+\exp(2\varepsilon_* t_0)\,\eta\right]\|\psi\|_{X^0}. \end{aligned} \tag{C.17}$$

To deal with B_2 in (C.16), we write this term as

$$\begin{aligned} B_2 &= \int_{t_0}^{t_0+\eta} \mathcal{W}(t_0+\eta,s)\,[\mathcal{KT}\,g(s)-\mathcal{KT}\,g(t_0+\eta)]\,ds + \left(\int_{t_0}^{t_0+\eta}\mathcal{W}(t_0+\eta,s)\,ds\right)\mathcal{KT}\,g(t_0+\eta) \\ &= B_{21}+B_{22}. \end{aligned} \tag{C.18}$$

First we consider B_{21}. Recall from (C.15) that $\|\mathcal{KT}\,g(t)-\mathcal{KT}\,g(s)\|_{X^0}\le C_{\mathcal{KT}}\,C_4(g,a)(t-s)^{1/6}$. Hence, (C.76) with $\alpha=2$ and $\gamma=1/6$ leads to

$$\begin{aligned} \|B_{21}\|_{X^2}^2 &= \left\|\int_{t_0}^{t_0+\eta}\mathcal{W}(t_0+\eta,s)\,[\mathcal{KT}\,g(t_0+\eta)-\mathcal{KT}\,g(s)]\,ds\right\|_{X^2}^2 \\ &\le 3\,C_{\mathcal{KT}}^2\,C_4(g,a)^2\left(\frac{4}{\varepsilon_*^2}+\frac{1}{4\varepsilon_*}e^{4\varepsilon_*\eta}\,\eta\right)\eta^{1/3} \\ &\le \frac{15}{\varepsilon_*^2}\,C_{\mathcal{KT}}^2\,C_4(g,a)^2\,\eta^{1/3}. \end{aligned} \tag{C.19}$$

What concerns B_{22} in (C.18), we have

$$\|B_{22}\|_{X^2}^2=16\pi^3\sum_{k\in\mathbb{Z}}(1+k^2)^2\iint_D dI\,d\ell\,\ell\,\frac{1}{|Q'(e)|}\left|\int_{t_0}^{t_0+\eta}\mathcal{W}_k(t_0+\eta,s)(I,\ell)\,ds\right|^2|(\mathcal{KT}\,g(t_0+\eta))_k(I,\ell)|^2.$$

Therefore, (C.64), (C.65) together with $1-e^{-x}\le\min\{1,x\}$ implies that

$$\|B_{22}\|_{X^2}^2 \le 32\pi^3\sum_{k=2}^{\infty}(1+k^2)^2\iint_D dI\,d\ell\,\ell\,\frac{1}{|Q'(e)|}\left[\int_{t_0}^{t_0+\eta}e^{-\varepsilon_* k^2(t_0+\eta-s)}\,ds\right]^2|(\mathcal{KT}\,g(t_0+\eta))_k(I,\ell)|^2$$

$$
\begin{aligned}
&+128\pi^3 \iint_D dI\, d\ell\, \ell\, \frac{1}{|Q'(e)|}\left[\int_{t_0}^{t_0+\eta} e^{2\varepsilon_*(t_0+\eta-s)}\, ds\right]^2 |(\mathcal{KT}g(t_0+\eta))_1(I,\ell)|^2 \\
&= \frac{32\pi^3}{\varepsilon_*^2}\sum_{k=2}^{\infty}\frac{(1+k^2)^2}{k^4}(1-e^{-\varepsilon_* k^2\eta})^2 \iint_D dI\, d\ell\, \ell\, \frac{1}{|Q'(e)|}|(\mathcal{KT}g(t_0+\eta))_k(I,\ell)|^2 \\
&\quad + \frac{32\pi^3}{\varepsilon_*^2}(e^{2\varepsilon_*\eta}-1)^2 \iint_D dI\, d\ell\, \ell\, \frac{1}{|Q'(e)|}|(\mathcal{KT}g(t_0+\eta))_1(I,\ell)|^2 \\
&\le \frac{64\pi^3}{\varepsilon_*^2}\sum_{k=2}^{\infty}\min\{1, k^4\eta^2\}\, \|(\mathcal{KT}g(t_0+\eta))_k\|^2_{L^2_{\frac{1}{|Q'|}}(D)} \\
&\quad + \frac{32\pi^3}{\varepsilon_*^2}(e^{2\varepsilon_*\eta}-1)^2\, \|(\mathcal{KT}g(t_0+\eta))_1\|^2_{L^2_{\frac{1}{|Q'|}}(D)}.
\end{aligned}
$$

Observing that, for all $t \in J$ and $k \in \mathbb{Z}$,

$$
\begin{aligned}
16\pi^3\|(\mathcal{KT}g(t))_k\|^2_{L^2_{\frac{1}{|Q'|}}(D)} &\le \|\mathcal{KT}g(t)\|^2_{X^0} \le C^2_{\mathcal{KT}}\|g(t)\|^2_{X^0} \\
&\le C^2_{\mathcal{KT}}\|g\|^2_{\infty,1} \le 100\, C^2_{\mathcal{KT}}\, \|\psi\|^2_{X^1},
\end{aligned} \tag{C.20}
$$

this leads to

$$
\begin{aligned}
\|B_{22}\|^2_{X^2} \le\ & \frac{32\pi^3}{\varepsilon_*^2}\sum_{k\ne 0}\min\{1, k^4\eta^2\}\, \|(\mathcal{KT}g(t_0+\eta))_k\|^2_{L^2_{\frac{1}{|Q'|}}(D)} \\
&+ \frac{200}{\varepsilon_*^2}\, C^2_{\mathcal{KT}}\, \|\psi\|^2_{X^1}\, (e^{2\varepsilon_*\eta}-1)^2.
\end{aligned} \tag{C.21}
$$

Thus if we go back to (C.18) and invoke (C.19) and (C.21), we have shown that

$$
\begin{aligned}
\|B_2\|_{X^2} &\le \|B_{21}\|_{X^2} + \|B_{22}\|_{X^2} \\
&\le \frac{5}{\varepsilon_*}C_{\mathcal{KT}}\, C_4(g,a)\, \eta^{1/6} + \frac{2}{\varepsilon_*}\left(16\pi^3\sum_{k\ne 0}\min\{1, k^4\eta^2\}\, \|(\mathcal{KT}g(t_0+\eta))_k\|^2_{L^2_{\frac{1}{|Q'|}}(D)}\right)^{1/2} \\
&\quad + \frac{15}{\varepsilon_*}C_{\mathcal{KT}}\, \|\psi\|_{X^1}\, (e^{2\varepsilon_*\eta}-1).
\end{aligned} \tag{C.22}
$$

Concerning B_3 in (C.16), we fix $\delta \in]0, \min\{1, t_0\}[$ and split this term further up into

$$
\begin{aligned}
B_3 =& \int_0^{t_0-\delta}[\mathcal{W}(t_0+\eta, s) - \mathcal{W}(t_0, s)]\, \mathcal{KT}g(s)\, ds \\
&+ \int_{t_0-\delta}^{t_0}[\mathcal{W}(t_0+\eta, s) - \mathcal{W}(t_0, s)]\, (\mathcal{KT}g(s) - \mathcal{KT}g(t_0))\, ds \\
&+ \left(\int_{t_0-\delta}^{t_0}[\mathcal{W}(t_0+\eta, s) - \mathcal{W}(t_0, s)]\, ds\right)\mathcal{KT}g(t_0) \\
=& B_{31} + B_{32} + B_{33}.
\end{aligned} \tag{C.23}
$$

For B_{31}, one has from (C.86) and (B.42):

$$\begin{aligned}
\|B_{31}\|_{X^2} &\le \int_0^{t_0-\delta} \|[\mathcal{W}(t_0+\eta,s)-\mathcal{W}(t_0,s)]\,\mathcal{KT}g(s)\|_{X^2}\,ds \\
&\le 2\,(\Delta_1^2+1)\int_0^{t_0-\delta}\left[\frac{2}{\varepsilon_*^{3/2}(t_0-s)^{3/2}}\sqrt{\eta}+\exp(2\varepsilon_*(t_0-s))\,\eta\right]\|\mathcal{KT}g(s)\|_{X^0}\,ds \\
&\le 2\,(\Delta_1^2+1)\,C_{\mathcal{KT}}\sqrt{\eta}\int_0^{t_0-\delta}\left[\frac{2}{\varepsilon_*^{3/2}(t_0-s)^{3/2}}+e^{2\varepsilon_* t_0}\right]\|g(s)\|_{X^0}\,ds \\
&\le 2\,(\Delta_1^2+1)\,C_{\mathcal{KT}}\sqrt{\eta}\,\|g\|_{\infty,1}\left[\frac{4}{\varepsilon_*^{3/2}}\,\delta^{-1/2}+e^{2\varepsilon_* t_0}\,t_0\right] \\
&\le \frac{8C_{\mathcal{KT}}}{\varepsilon_*^{3/2}}\,(\Delta_1^2+1)e^{2\varepsilon_* t_0}\,(1+t_0^{3/2})\,\|g\|_{\infty,1}\sqrt{\eta}\,\delta^{-1/2}.
\end{aligned} \tag{C.24}$$

In order to bound B_{32} from (C.23), we use (C.69). It follows that

$$\begin{aligned}
\|B_{32}\|_{X^2}^2 &= 16\pi^3\sum_{k\in\mathbb{Z}}(1+k^2)^2\iint_D dI\,d\ell\,\ell\,\frac{1}{|Q'(e)|}\left|\int_{t_0-\delta}^{t_0}[\mathcal{W}_k(t_0+\eta,s)(I,\ell)-\mathcal{W}_k(t_0,s)(I,\ell)]\right. \\
&\qquad\qquad \left.\times\,((\mathcal{KT}g(s))_k(I,\ell)-(\mathcal{KT}g(t_0))_k(I,\ell))\,ds\right|^2 \\
&\le 32\pi^3(\Delta_1^2+1)^2\sum_{k=2}^{\infty}(1+k^2)^2\iint_D dI\,d\ell\,\ell\,\frac{1}{|Q'(e)|}\left[\int_{t_0-\delta}^{t_0} ds\,e^{-\varepsilon_* k^2(t_0-s)}\right. \\
&\qquad\qquad \left.\times\,((\mathcal{KT}g(s))_k(I,\ell)-(\mathcal{KT}g(t_0))_k(I,\ell))\right]^2 \\
&+ 128\pi^3(\Delta_1^2+1)^2\iint_D dI\,d\ell\,\ell\,\frac{1}{|Q'(e)|}\left[\int_{t_0-\delta}^{t_0} ds\,e^{2\varepsilon_*(t_0-s)}\right. \\
&\qquad\qquad \left.\times\,((\mathcal{KT}g(s))_1(I,\ell)-(\mathcal{KT}g(t_0))_1(I,\ell))\right]^2.
\end{aligned}$$

To the right-hand side we apply Lemma C.13. It follows that

$$\begin{aligned}
\|B_{32}\|_{X^2}^2 &\le \frac{64\pi^3}{\varepsilon_*}(\Delta_1^2+1)^2\sum_{k=2}^{\infty}k^2\int_{t_0-\delta}^{t_0} e^{-\varepsilon_* k^2(t_0-s)}\,\|(\mathcal{KT}g(s))_k-(\mathcal{KT}g(t_0))_k\|^2_{L^2_{\frac{1}{|Q'|}}(D)}\,ds \\
&+\frac{32\pi^3}{\varepsilon_*}(\Delta_1^2+1)^2\,(e^{4\varepsilon_*\delta}-1)\int_{t_0-\delta}^{t_0}\|(\mathcal{KT}g(s))_1-(\mathcal{KT}g(t_0))_1\|^2_{L^2_{\frac{1}{|Q'|}}(D)}\,ds.
\end{aligned}$$

By Remark C.14 one has $k^2 e^{-\varepsilon_* k^2(t_0-s)} \le \frac{1}{\varepsilon_*(t_0-s)}$. Thus, from (B.42) and (a), it follows that

$$\begin{aligned}
\|B_{32}\|_{X^2}^2 &\le \frac{2}{\varepsilon_*^2}(\Delta_1^2+1)^2\int_{t_0-\delta}^{t_0}\frac{ds}{t_0-s}\,\|\mathcal{KT}g(s)-\mathcal{KT}g(t_0)\|_{X^0}^2\\
&\quad+\frac{1}{\varepsilon_*}(\Delta_1^2+1)^2(e^{4\varepsilon_*\delta}-1)\int_{t_0-\delta}^{t_0}\|\mathcal{KT}g(s)-\mathcal{KT}g(t_0)\|_{X^0}^2\,ds\\
&\le \frac{2}{\varepsilon_*^2}(\Delta_1^2+1)^2 C_{\mathcal{KT}}^2\,C_4(g,a)^2\int_{t_0-\delta}^{t_0}\frac{ds}{(t_0-s)^{2/3}}\\
&\quad+\frac{1}{\varepsilon_*}(\Delta_1^2+1)^2 C_{\mathcal{KT}}^2\,C_4(g,a)^2(e^{4\varepsilon_*\delta}-1)\int_{t_0-\delta}^{t_0}(t_0-s)^{1/3}\,ds\\
&= \frac{6}{\varepsilon_*^2}(\Delta_1^2+1)^2 C_{\mathcal{KT}}^2\,C_4(g,a)^2\delta^{1/3}+\frac{3}{4\varepsilon_*}(\Delta_1^2+1)^2 C_{\mathcal{KT}}^2\,C_4(g,a)^2(e^{4\varepsilon_*\delta}-1)\,\delta^{4/3}\\
&\le \frac{12}{\varepsilon_*^2}(\Delta_1^2+1)^2 C_{\mathcal{KT}}^2\,C_4(g,a)^2\,e^{4\varepsilon_*\delta}\,\delta^{1/3}.
\end{aligned}\tag{C.25}$$

Next we turn to B_{33} from (C.23). Here, we have

$$\begin{aligned}
\|B_{33}\|_{X^2}^2 = 16\pi^3\sum_{k\neq0}(1+k^2)^2\iint_D dI\,d\ell\,\ell\,\frac{1}{|Q'(e)|}\left|\int_{t_0-\delta}^{t_0}[\mathcal{W}_k(t_0+\eta,s)(I,\ell)-\mathcal{W}_k(t_0,s)(I,\ell)]\,ds\right|^2\\
\times|(\mathcal{KT}g(t_0))_k(I,\ell)|^2,
\end{aligned}$$

whence (C.69) leads to

$$\begin{aligned}
&\|B_{33}\|_{X^2}^2\\
&\le 32\pi^3(\Delta_1^2+1)^2\sum_{k=2}^{\infty}(1+k^2)^2\min\{1,k^4\eta^2\}\iint_D dI\,d\ell\,\ell\,\frac{1}{|Q'(e)|}\left[\int_{t_0-\delta}^{t_0}\exp(-\varepsilon_*k^2(t_0-s))\,ds\right]^2\\
&\qquad\times|(\mathcal{KT}g(t_0))_k(I,\ell)|^2\\
&\quad+128\pi^3(\Delta_1^2+1)^2\eta^2\iint_D dI\,d\ell\,\ell\,\frac{1}{|Q'(e)|}\left[\int_{t_0-\delta}^{t_0}\exp(2\varepsilon_*(t_0-s))\,ds\right]^2\\
&\qquad\times|(\mathcal{KT}g(t_0))_1(I,\ell)|^2\\
&\le \frac{64\pi^3}{\varepsilon_*^2}(\Delta_1^2+1)^2\sum_{k=2}^{\infty}\min\{1,k^4\eta^2\}\|(\mathcal{KT}g(t_0))_k\|_{L^2_{\frac{1}{|Q'|}}(D)}^2\\
&\quad+\frac{32\pi^3}{\varepsilon_*^2}(\Delta_1^2+1)^2\eta^2(e^{2\varepsilon_*\delta}-1)^2\|(\mathcal{KT}g(t_0))_1\|_{L^2_{\frac{1}{|Q'|}}(D)}^2\\
&\le \frac{32\pi^3}{\varepsilon_*^2}(\Delta_1^2+1)^2 e^{4\varepsilon_*\delta}\sum_{k\neq0}\min\{1,k^4\eta^2\}\|(\mathcal{KT}g(t_0))_k\|_{L^2_{\frac{1}{|Q'|}}(D)}^2.
\end{aligned}\tag{C.26}$$

Thus if we summarize (C.24), (C.25) and (C.26), it follows from (C.23) that

$$\begin{aligned}
\|B_3\|_{X^2} &\le \|B_{31}\|_{X^2}+\|B_{32}\|_{X^2}+\|B_{33}\|_{X^2}\\
&\le \frac{8C_{\mathcal{KT}}}{\varepsilon_*^{3/2}}(\Delta_1^2+1)e^{2\varepsilon_*t_0}(1+t_0^{3/2})\|g\|_{\infty,1}\sqrt{\eta}\,\delta^{-1/2}+\frac{4}{\varepsilon_*}(\Delta_1^2+1)C_{\mathcal{KT}}\,C_4(g,a)\,e^{2\varepsilon_*\delta}\,\delta^{1/6}\\
&\quad+\frac{2}{\varepsilon_*}(\Delta_1^2+1)\,e^{2\varepsilon_*\delta}\left(16\pi^3\sum_{k\neq0}\min\{1,k^4\eta^2\}\|(\mathcal{KT}g(t_0))_k\|_{L^2_{\frac{1}{|Q'|}}(D)}^2\right)^{1/2}.
\end{aligned}\tag{C.27}$$

Altogether, now we can go back to (C.16) and use (C.17), (C.22) and (C.27). In this way, we obtain

$$\begin{aligned}
&\|g(t_0+\eta)-g(t_0)\|_{X^2}\\
&\quad\le \|B_1\|_{X^2}+\|B_2\|_{X^2}+\|B_3\|_{X^2}\\
&\quad\le 2\,(\Delta_1^2+1)\left[\frac{2}{\varepsilon_*^{3/2}t_0^{3/2}}\sqrt{\eta}+\exp(2\varepsilon_* t_0)\,\eta\right]\|\psi\|_{X^0}+\frac{5}{\varepsilon_*}\,C_{\mathcal{KT}}\,C_4(g,a)\,\eta^{1/6}\\
&\qquad+\frac{2}{\varepsilon_*}\left(16\pi^3\sum_{k\neq 0}\min\{1,k^4\eta^2\}\,\|(\mathcal{KT}g(t_0+\eta))_k\|^2_{L^2_{\frac{1}{|Q'|}}(D)}\right)^{1/2}\\
&\qquad+\frac{15}{\varepsilon_*}\,C_{\mathcal{KT}}\,\|\psi\|_{X^1}\,(e^{2\varepsilon_*\eta}-1)\\
&\qquad+\frac{8C_{\mathcal{KT}}}{\varepsilon_*^{3/2}}\,(\Delta_1^2+1)e^{2\varepsilon_* t_0}\,(1+t_0^{3/2})\,\|g\|_{\infty,1}\,\sqrt{\eta}\,\delta^{-1/2}+\frac{4}{\varepsilon_*}\,(\Delta_1^2+1)\,C_{\mathcal{KT}}\,C_4(g,a)\,e^{2\varepsilon_*\delta}\,\delta^{1/6}\\
&\qquad+\frac{2}{\varepsilon_*}\,(\Delta_1^2+1)\,e^{2\varepsilon_*\delta}\left(16\pi^3\sum_{k\neq 0}\min\{1,k^4\eta^2\}\|(\mathcal{KT}g(t_0))_k\|^2_{L^2_{\frac{1}{|Q'|}}(D)}\right)^{1/2}.
\end{aligned}\tag{C.28}$$

Then the next step is to take, for instance, $\delta=\eta^{3/4}$. In addition, we have

$$\lim_{\eta\to 0^+}16\pi^3\sum_{k\neq 0}\min\{1,k^4\eta^2\}\,\|(\mathcal{KT}g(t_0+\chi(\eta)))_k\|^2_{L^2_{\frac{1}{|Q'|}}(D)}=0,\tag{C.29}$$

if $\lim_{\eta\to 0^+}\chi(\eta)=0$. This is a consequence of the generalized Lebesgue dominated convergence theorem (for sums). In fact, we have $\lim_{\eta\to 0^+}\min\{1,k^4\eta^2\}\,\|(\mathcal{KT}g(t_0+\chi(\eta)))_k\|^2_{L^2_{\frac{1}{|Q'|}}(D)}=0$ for every $k\in\mathbb{Z}\setminus\{0\}$, cf. (C.20). Furthermore, using (B.42), $\|g\|_{\infty,1}\le 10\,\|\psi\|_{X^1}$ and (a),

$$\lim_{\eta\to 0^+}\|(\mathcal{KT}g(t_0+\chi(\eta)))_k\|^2_{L^2_{\frac{1}{|Q'|}}(D)}=\|(\mathcal{KT}g(t_0))_k\|^2_{L^2_{\frac{1}{|Q'|}}(D)},\quad k\in\mathbb{Z}\setminus\{0\},$$

and similarly

$$\begin{aligned}
16\pi^3\sum_{k\neq 0}\|(\mathcal{KT}g(t_0+\eta))_k\|^2_{L^2_{\frac{1}{|Q'|}}(D)}&=\|\mathcal{KT}g(t_0+\eta)\|^2_{X^0}\\
\to\|\mathcal{KT}g(t_0)\|^2_{X^0}&=16\pi^3\sum_{k\neq 0}\|(\mathcal{KT}g(t_0))_k\|^2_{L^2_{\frac{1}{|Q'|}}(D)}
\end{aligned}$$

as $\eta\to 0^+$. This yields (C.29), and therefore (C.28) shows that $\|g(t_0+\eta)-g(t)\|_{X^2}=\mathrm{o}(1)$ as $\eta\to 0^+$, which means that $g: J\to X^2_{\mathrm{odd}}$ is continuous at t_0. To establish (d), let $d(t)=-Lg(t)+\Phi(g(t))g(t)$. Since $g(t)\in X^2_{\mathrm{odd}}$ by (b), Remark B.18(a) and (B.18) imply that $\mathcal{T}^2g(t)\in X^0_{\mathrm{odd}}$. As a consequence of $Lg(t)=-\mathcal{T}^2g(t)-\mathcal{KT}g(t)$, it follows from Remark B.18(b) and (B.42) that

$Lg(t) \in X^0_{\text{odd}}$. Hence, in particular $d(t) \in X^0_{\text{odd}}$. Now by (C.7) one has for $\eta > 0$ small enough:

$$\begin{aligned}
&g(t+\eta) - g(t) - \eta d(t) \\
&= \mathcal{W}(t+\eta,0)\psi + \int_0^{t+\eta} \mathcal{W}(t+\eta,s)\mathcal{K}\mathcal{T}g(s)\,ds - \mathcal{W}(t,0)\psi - \int_0^t \mathcal{W}(t,s)\mathcal{K}\mathcal{T}g(s)\,ds - \eta d(t) \\
&= \mathcal{W}(t+\eta,0)\psi - \mathcal{W}(t,0)\psi + \int_t^{t+\eta} \mathcal{W}(t+\eta,s)\mathcal{K}\mathcal{T}g(s)\,ds - \eta\mathcal{K}\mathcal{T}g(t) \\
&\quad + \int_0^t (\mathcal{W}(t+\eta,s) - \mathcal{W}(t,s))\mathcal{K}\mathcal{T}g(s)\,ds - \eta[\mathcal{T}^2 + \Phi(g(t))]g(t) \\
&= \mathcal{W}(t+\eta,0)\psi - \mathcal{W}(t,0)\psi + \int_t^{t+\eta} \mathcal{W}(t+\eta,s)\mathcal{K}\mathcal{T}g(s)\,ds - \eta\mathcal{K}\mathcal{T}g(t) \\
&\quad + \int_0^t (\mathcal{W}(t+\eta,s) - \mathcal{W}(t,s))\mathcal{K}\mathcal{T}g(s)\,ds \\
&\quad - \eta[\mathcal{T}^2 + \Phi(g(t))]\left(\mathcal{W}(t,0)\psi + \int_0^t \mathcal{W}(t,s)\mathcal{K}\mathcal{T}g(s)\,ds\right) \\
&= \mathcal{W}(t+\eta,0)\psi - \mathcal{W}(t,0)\psi - \eta[\mathcal{T}^2 + \Phi(g(t))]\mathcal{W}(t,0)\psi \\
&\quad + \int_t^{t+\eta} \mathcal{W}(t+\eta,s)\mathcal{K}\mathcal{T}g(s)\,ds - \eta\mathcal{K}\mathcal{T}g(t) \\
&\quad + \int_0^t \left(\mathcal{W}(t+\eta,s) - \mathcal{W}(t,s) - \eta[\mathcal{T}^2 + \Phi(g(t))]\mathcal{W}(t,s)\right)\mathcal{K}\mathcal{T}g(s)\,ds. \qquad \text{(C.30)}
\end{aligned}$$

We will write this relation as

$$g(t+\eta) - g(t) - \eta d(t) = A_1 + A_2 + A_3, \qquad \text{(C.31)}$$

with A_j, $j = 1, 2, 3$, denoting the three lines, and look at each A_j individually. Let $m_k(t) = m_k(t)(I,\ell) = k^2\omega_1^2(I,\ell) - \Phi(g(t))$. For A_1, one has by (C.70) for $s = 0$:

$$\begin{aligned}
\|A_1\|^2_{X^0} &= 16\pi^3 \sum_{k\neq 0} \iint_D dI\,d\ell\,\ell\,\frac{1}{|Q'(e)|}\Big|\mathcal{W}_k(t+\eta,0)(I,\ell) - \mathcal{W}_k(t,0)(I,\ell) \\
&\qquad + \eta\, m_k(t)(I,\ell)\,\mathcal{W}_k(t,0)(I,\ell)\Big|^2 |\psi_k(I,\ell)|^2 \\
&\le 32\pi^3\,(\Delta_1^2+1)^4\eta^2 \sum_{k=2}^{\infty}\left(|\Delta(t,\eta)| + k^2\min\{1,k^2\eta\}\right)^2 e^{-2\varepsilon_* k^2 t}\,\|\psi_k\|^2_{L^2_{\frac{1}{|Q'|}}(D)} \\
&\quad + 32\pi^3\,(\Delta_1^2+1)^4\eta^2\,(|\Delta(t,\eta)|+\eta)^2 e^{4\varepsilon_*(1+t)}\,\|\psi_1\|^2_{L^2_{\frac{1}{|Q'|}}(D)} \\
&\le 64\pi^3\,(\Delta_1^2+1)^4\eta^2 \sum_{k=2}^{\infty}\left(|\Delta(t,\eta)|^2 + k^4\min\{1,k^4\eta^2\}\right)\|\psi_k\|^2_{L^2_{\frac{1}{|Q'|}}(D)} \\
&\quad + (\Delta_1^2+1)^4\eta^2\,(|\Delta(t,\eta)|+\eta)^2 e^{4\varepsilon_*(1+t)}\,\|\psi\|^2_{X^0},
\end{aligned}$$

where

$$\Delta(t,\eta) = \frac{\exp(-\int_t^{t+\eta}[\Phi(g(t)) - \Phi(g(\tau))]\,d\tau) - 1}{\eta}. \tag{C.32}$$

As a consequence,

$$\begin{aligned}\|A_1\|_{X^0}^2 &\le 3(\Delta_1^2+1)^4\eta^2\,(|\Delta(t,\eta)|+\eta)^2 e^{4\varepsilon_*(1+a)}\,\|\psi\|_{X^0}^2\\ &\quad + 32\pi^3(\Delta_1^2+1)^4\eta^2\sum_{|k|\ge 2} k^4\,\min\{1,k^4\eta^2\}\,\|\psi_k\|^2_{L^2_{\frac{1}{|Q'|}}(D)}\end{aligned}$$

so that

$$\begin{aligned}\|A_1\|_{X^0} &\le 2(\Delta_1^2+1)^2\eta\,(|\Delta(t,\eta)|+\eta)\,e^{2\varepsilon_*(1+a)}\,\|\psi\|_{X^0}\\ &\quad + 2(\Delta_1^2+1)^2\eta\left(16\pi^3\sum_{k\neq 0} k^4\,\min\{1,k^4\eta^2\}\,\|\psi_k\|^2_{L^2_{\frac{1}{|Q'|}}(D)}\right)^{1/2}.\end{aligned} \tag{C.33}$$

To bound A_2 in (C.30), we decompose it as

$$\begin{aligned}A_2 &= \int_t^{t+\eta} \mathcal{W}(t+\eta,s)\,\mathcal{KT}[g(s)-g(t)]\,ds\\ &\quad + \left[\int_t^{t+\eta}\mathcal{W}(t+\eta,s)\,ds\right]\mathcal{KT}\,g(t) - \eta\,\mathcal{KT}\,g(t)\\ &= A_{21} + A_{22}.\end{aligned} \tag{C.34}$$

Then by (C.68), (B.42) and Remark B.2(b):

$$\begin{aligned}\|A_{21}\|_{X^0} &\le \int_t^{t+\eta} \|\mathcal{W}(t+\eta,s)\,\mathcal{KT}[g(s)-g(t)]\|_{X^0}\,ds\\ &\le \int_t^{t+\eta} e^{2\varepsilon_*(t+\eta-s)}\,\|\mathcal{KT}[g(s)-g(t)]\|_{X^0}\,ds\\ &\le C_{\mathcal{KT}}\,e^{2\varepsilon_*\eta}\int_t^{t+\eta}\|g(s)-g(t)\|_{X^0}\,ds\\ &\le C_{\mathcal{KT}}\,e^{2\varepsilon_*}\int_t^{t+\eta}\|g(s)-g(t)\|_{X^1}\,ds\\ &\le C_{\mathcal{KT}}\,e^{2\varepsilon_*}C_4(g,a)\int_t^{t+\eta}(s-t)^{1/6}\,ds\\ &\le C_{\mathcal{KT}}\,e^{2\varepsilon_*}C_4(g,a)\,\eta^{7/6}.\end{aligned} \tag{C.35}$$

With regard to A_{22} from (C.34), here one has

$$\begin{aligned}&\|A_{22}\|^2_{X0}\\ &= 16\pi^3\sum_{k\neq 0}\iint_D dI\,d\ell\,\ell\,\frac{1}{|Q'(e)|}\left[\int_t^{t+\eta}\mathcal{W}_k(t+\eta,s)(I,\ell)\,ds - \eta\right]^2 |(\mathcal{KT}\,g(t))_k(I,\ell)|^2\end{aligned}$$

$$\leq 16\pi^3 \sum_{k\neq 0} \iint_D dI\, d\ell\, \ell\, \frac{1}{|Q'(e)|} \Big(\int_t^{t+\eta} ds \,\Big|1 - \exp\Big(- \int_s^{t+\eta} m_k(\tau)(I,\ell)\, d\tau\Big)\Big|\Big)^2 |(\mathcal{KT} g(t))_k(I,\ell)|^2.$$

From (C.62) we know that $0 \leq m_k(\tau)(I,\ell) \leq \Delta_1^2 k^2$ for $|k| \geq 2$, so that $\int_s^{t+\eta} m_k(\tau)(I,\ell)\, d\tau \in [0, \Delta_1^2 k^2 (t+\eta-s)]$. If $|k| = 1$, then $-2\varepsilon_* \leq m_k(\tau)(I,\ell) \leq \Delta_1^2$ by (C.63) yields $\int_s^{t+\eta} m_k(\tau)(I,\ell)\, d\tau \in [-2\varepsilon_*(t+\eta-s), \Delta_1^2(t+\eta-s)] \subset [-2\varepsilon_*, \infty[$. Since $|1-e^{-x}| \leq e^{2\varepsilon_*}\min\{1,|x|\}$ for $x \in [-2\varepsilon_*, \infty[$, we deduce that

$$\begin{aligned}\Big|1 - \exp\Big(- \int_s^{t+\eta} m_k(\tau)(I,\ell)\, d\tau\Big)\Big| &\leq e^{2\varepsilon_*} \min\Big\{1, \Big|\int_s^{t+\eta} m_k(\tau)(I,\ell)\, d\tau\Big|\Big\} \\ &\leq e^{2\varepsilon_*}\min\{1, \Delta_1^2 k^2 (t+\eta-s)\}\end{aligned}$$

for all $k \in \mathbb{Z} \setminus \{0\}$. Hence, we may continue the above estimate as

$$\begin{aligned}&\|A_{22}\|_{X^0}^2 \\ &\leq 16\pi^3 e^{4\varepsilon_*} \sum_{k\neq 0} \iint_D dI\, d\ell\, \ell\, \frac{1}{|Q'(e)|}\Big(\int_t^{t+\eta} ds\, \min\{1, \Delta_1^2 k^2 (t+\eta-s)\}\Big)^2 |(\mathcal{KT} g(t))_k(I,\ell)|^2 \\ &\leq \eta^2\, 16\pi^3 e^{4\varepsilon_*} \sum_{k\neq 0} \iint_D dI\, d\ell\, \ell\, \frac{1}{|Q'(e)|} \min\{1, \Delta_1^4 k^4 \eta^2\}\, |(\mathcal{KT} g(t))_k(I,\ell)|^2 \\ &= \eta^2\, 16\pi^3 e^{4\varepsilon_*} \sum_{k\neq 0} \min\{1, \Delta_1^4 k^4 \eta^2\}\, \|(\mathcal{KT} g(t))_k\|^2_{L^2_{\frac{1}{|Q'|}}(D)} \\ &\leq \eta^2\, 16\pi^3 e^{4\varepsilon_*} (\Delta_1^4 + 1) \sum_{k\neq 0} \min\{1, k^4\eta^2\}\, \|(\mathcal{KT} g(t))_k\|^2_{L^2_{\frac{1}{|Q'|}}(D)}. \end{aligned} \tag{C.36}$$

Thus, we can go back to (C.34) and summarize (C.35) and (C.36). This yields

$$\begin{aligned}\|A_2\|_{X^0} &\leq \|A_{21}\|_{X^0} + \|A_{22}\|_{X^0} \\ &\leq C_{\mathcal{KT}}\, e^{2\varepsilon_*} C_4(g,a)\, \eta^{7/6} \\ &\quad + \eta\, e^{2\varepsilon_*} (\Delta_1^2 + 1) \Big(16\pi^3 \sum_{k\neq 0} \min\{1, k^4\eta^2\}\, \|(\mathcal{KT} g(t))_k\|^2_{L^2_{\frac{1}{|Q'|}}(D)}\Big)^{1/2}. \end{aligned} \tag{C.37}$$

Concerning A_3 in (C.30), this term is further split up into

$$\begin{aligned}A_3 &= \int_0^t \Big(\mathcal{W}(t+\eta, s) - \mathcal{W}(t,s) - \eta\,[\mathcal{T}^2 + \Phi(g(t))]\,\mathcal{W}(t,s)\Big)\, \mathcal{KT}[g(s) - g(t)]\, ds \\ &\quad + \Big[\int_0^t \Big(\mathcal{W}(t+\eta, s) - \mathcal{W}(t,s) - \eta\,[\mathcal{T}^2 + \Phi(g(t))]\,\mathcal{W}(t,s)\Big)\, ds\Big]\, \mathcal{KT} g(t) \\ &= A_{31} + A_{32}. \end{aligned} \tag{C.38}$$

Then we obtain from (C.70):

$$
\begin{aligned}
\|A_{31}\|^2_{X^0} &= 16\pi^3 \sum_{k\neq 0} \iint\limits_D dI\, d\ell\, \ell\, \frac{1}{|Q'(e)|} \bigg| \int_0^t \Big(\mathcal{W}_k(t+\eta, s)(I,\ell) - \mathcal{W}_k(t,s)(I,\ell) \\
&\qquad + \eta\, m_k(t)(I,\ell)\, \mathcal{W}_k(t,s)(I,\ell) \Big) \\
&\qquad \times [(\mathcal{KT} g(s))_k(I,\ell) - (\mathcal{KT} g(t))_k(I,\ell)]\, ds \bigg|^2 \\
&\le 32\pi^3 (\Delta_1^2+1)^4 \eta^2 \sum_{k=2}^{\infty} \Big(|\Delta(t,\eta)| + k^2 \min\{1, k^2\eta\} \Big)^2 \\
&\qquad \times \iint\limits_D dI\, d\ell\, \ell\, \frac{1}{|Q'(e)|} \bigg[\int_0^t \exp(-\varepsilon_* k^2 (t-s)) \\
&\qquad \times |(\mathcal{KT} g(s))_k(I,\ell) - (\mathcal{KT} g(t))_k(I,\ell)|\, ds \bigg]^2 \\
&\quad + 32\pi^3 (\Delta_1^2+1)^4 e^{4\varepsilon_*} \eta^2 (|\Delta(t,\eta)| + \eta)^2 \\
&\qquad \times \iint\limits_D dI\, d\ell\, \ell\, \frac{1}{|Q'(e)|} \bigg[\int_0^t \exp(2\varepsilon_* (t-s)) \\
&\qquad \times |(\mathcal{KT} g(s))_1(I,\ell) - (\mathcal{KT} g(t))_1(I,\ell)|\, ds \bigg]^2
\end{aligned}
\tag{C.39}
$$

for $\Delta(t,\eta)$ from (C.32). To the right-hand side we apply Lemma C.13. This yields

$$
\begin{aligned}
&\|A_{31}\|^2_{X^0} \\
&\le \frac{32\pi^3}{\varepsilon_*} (\Delta_1^2+1)^4 \eta^2 \sum_{k=2}^{\infty} \frac{1}{k^2} \Big(|\Delta(t,\eta)| + k^2 \min\{1, k^2\eta\} \Big)^2 \\
&\qquad \times \int_0^t ds\, e^{-\varepsilon_* k^2 (t-s)} \|(\mathcal{KT} g(s))_k - (\mathcal{KT} g(t))_k\|^2_{L^2_{\frac{1}{|Q'|}}(D)} \\
&\quad + \frac{8\pi^3}{\varepsilon_*} (\Delta_1^2+1)^4 e^{4\varepsilon_*} \eta^2 (|\Delta(t,\eta)| + \eta)^2 \\
&\qquad \times (e^{4\varepsilon_* t} - 1) \int_0^t \|(\mathcal{KT} g(s))_1 - (\mathcal{KT} g(t))_1\|^2_{L^2_{\frac{1}{|Q'|}}(D)}\, ds \\
&\le \frac{2}{\varepsilon_*} (\Delta_1^2+1)^4 \eta^2 |\Delta(t,\eta)|^2 \int_0^t \|\mathcal{KT} g(s) - \mathcal{KT} g(t)\|^2_{X^0}\, ds \\
&\quad + \frac{64\pi^3}{\varepsilon_*} (\Delta_1^2+1)^4 \eta^2 \sum_{k=2}^{\infty} k^2 \min\{1, k^4\eta^2\} \\
&\qquad \times \int_0^t ds\, e^{-\varepsilon_* k^2 (t-s)} \|(\mathcal{KT} g(s))_k - (\mathcal{KT} g(t))_k\|^2_{L^2_{\frac{1}{|Q'|}}(D)} \\
&\quad + \frac{1}{4\varepsilon_*} (\Delta_1^2+1)^4 e^{4\varepsilon_*} \eta^2 (|\Delta(t,\eta)| + \eta)^2
\end{aligned}
$$

$$\times\,(e^{4\varepsilon_* t}-1)\int_0^t \|\mathcal{KT}g(s)-\mathcal{KT}g(t)\|_{X^0}^2\,ds$$
$$\le \frac{4}{\varepsilon_*}(\Delta_1^2+1)^4\,e^{4\varepsilon_*(1+t)}\,C_{\mathcal{KT}}^2\,\eta^2\,(|\Delta(t,\eta)|+\eta)^2\int_0^t \|g(s)-g(t)\|_{X^0}^2\,ds$$
$$+\frac{32\pi^3}{\varepsilon_*}(\Delta_1^2+1)^4\,\eta^4\sum_{0<|k|\le\eta^{-1/2}} k^4$$
$$\times\int_0^t ds\, e^{-\varepsilon_* k^2(t-s)}\,\|(\mathcal{KT}g(s))_k-(\mathcal{KT}g(t))_k\|^2_{L^2_{\frac{1}{|Q'|}}(D)}$$
$$+\frac{32\pi^3}{\varepsilon_*}(\Delta_1^2+1)^4\,\eta^2\sum_{|k|\ge\eta^{-1/2}} k^2$$
$$\times\int_0^t ds\, e^{-\varepsilon_* k^2(t-s)}\,\|(\mathcal{KT}g(s))_k-(\mathcal{KT}g(t))_k\|^2_{L^2_{\frac{1}{|Q'|}}(D)}$$
$$\le \frac{16}{\varepsilon_*}(\Delta_1^2+1)^4\,e^{4\varepsilon_*(1+a)}\,C_{\mathcal{KT}}^2\,\|g\|_{\infty,1}^2\,a\,\eta^2(|\Delta(t,\eta)|+\eta)^2$$
$$+\frac{64\pi^3}{\varepsilon_*}(\Delta_1^2+1)^4\,\eta^{13/6}\sum_{k\neq0}|k|^{7/3}$$
$$\times\int_0^t ds\, e^{-\varepsilon_* k^2(t-s)}\,\|(\mathcal{KT}g(s))_k-(\mathcal{KT}g(t))_k\|^2_{L^2_{\frac{1}{|Q'|}}(D)}.$$

From Remark C.14 one has $|k|^{7/3}e^{-\varepsilon_* k^2(t-s)}\le\frac{4}{\varepsilon_*^{7/6}(t-s)^{7/6}}$. As a consequence, by (a),

$$\|A_{31}\|_{X^0}^2 \le \frac{16}{\varepsilon_*}(\Delta_1^2+1)^4\,e^{4\varepsilon_*(1+a)}\,C_{\mathcal{KT}}^2\,\|g\|_{\infty,1}^2\,a\,\eta^2(|\Delta(t,\eta)|+\eta)^2$$
$$+\frac{256\pi^3}{\varepsilon_*^{13/6}}(\Delta_1^2+1)^4\,\eta^{13/6}\sum_{k\neq0}\int_0^t\frac{ds}{(t-s)^{7/6}}\|(\mathcal{KT}g(s))_k-(\mathcal{KT}g(t))_k\|^2_{L^2_{\frac{1}{|Q'|}}(D)}$$
$$\le \frac{16}{\varepsilon_*}(\Delta_1^2+1)^4\,e^{4\varepsilon_*(1+a)}\,C_{\mathcal{KT}}^2\,\|g\|_{\infty,1}^2\,a\,\eta^2(|\Delta(t,\eta)|+\eta)^2$$
$$+\frac{8}{\varepsilon_*^{13/6}}(\Delta_1^2+1)^4\,\eta^{13/6}\int_0^t\frac{ds}{(t-s)^{7/6}}\|\mathcal{KT}g(s)-\mathcal{KT}g(t)\|_{X^0}^2$$
$$\le \frac{16}{\varepsilon_*}(\Delta_1^2+1)^4\,e^{4\varepsilon_*(1+a)}\,C_{\mathcal{KT}}^2\,\|g\|_{\infty,1}^2\,a\,\eta^2(|\Delta(t,\eta)|+\eta)^2$$
$$+\frac{8}{\varepsilon_*^{13/6}}(\Delta_1^2+1)^4\,C_{\mathcal{KT}}^2\,C_4(g,a)^2\,\eta^{13/6}\int_0^t\frac{ds}{(t-s)^{7/6}}(t-s)^{1/3}$$
$$\le \frac{16}{\varepsilon_*}(\Delta_1^2+1)^4\,e^{4\varepsilon_*(1+a)}\,C_{\mathcal{KT}}^2\,\|g\|_{\infty,1}^2\,a\,\eta^2(|\Delta(t,\eta)|+\eta)^2$$
$$+\frac{48}{\varepsilon_*^{13/6}}(\Delta_1^2+1)^4\,C_{\mathcal{KT}}^2\,C_4(g,a)^2\,\eta^{13/6}\,a^{1/6}. \tag{C.40}$$

Next we turn to A_{32} in (C.38). Analogously to (C.39) one gets

$$\|A_{32}\|_{X^0}^2$$
$$\le 32\pi^3(\Delta_1^2+1)^4\,\eta^2\sum_{k=2}^{\infty}\Big(|\Delta(t,\eta)|+k^2\min\{1,k^2\eta\}\Big)^2$$

$$\times \iint\limits_{D} dI\, d\ell\, \ell \frac{1}{|Q'(e)|}\Big[\int_0^t \exp(-\varepsilon_* k^2(t-s))\, |(\mathcal{KT} g(t))_k(I,\ell)|\, ds\Big]^2$$
$$+32\pi^3 (\Delta_1^2+1)^4 e^{4\varepsilon_*} \eta^2 (|\Delta(t,\eta)|+\eta)^2$$
$$\times \iint\limits_{D} dI\, d\ell\, \ell \frac{1}{|Q'(e)|}\Big[\int_0^t \exp(2\varepsilon_*(t-s))\, |(\mathcal{KT} g(t))_1(I,\ell)|\, ds\Big]^2.$$

Thus integrating out $\int_0^t ds$, we see that

$$\begin{aligned}
\|A_{32}\|_{X^0}^2 &\le \frac{32\pi^3}{\varepsilon_*^2}(\Delta_1^2+1)^4 \eta^2 \sum_{k=2}^{\infty}\Big(|\Delta(t,\eta)|+\min\{1,k^2\eta\}\Big)^2 \|(\mathcal{KT} g(t))_k\|^2_{L^2_{\frac{1}{|Q'|}}(D)} \\
&\quad + \frac{8\pi^3}{\varepsilon_*^2}(\Delta_1^2+1)^4 e^{4\varepsilon_*(1+t)} \eta^2 (|\Delta(t,\eta)|+\eta)^2 \|(\mathcal{KT} g(t))_1\|^2_{L^2_{\frac{1}{|Q'|}}(D)} \\
&\le \frac{2}{\varepsilon_*^2}(\Delta_1^2+1)^4 \eta^2 |\Delta(t,\eta)|^2 \|\mathcal{KT} g(t)\|^2_{X^0} \\
&\quad + \frac{64\pi^3}{\varepsilon_*^2}(\Delta_1^2+1)^4 \eta^2 \sum_{k=2}^{\infty} \min\{1,k^4\eta^2\} \|(\mathcal{KT} g(t))_k\|^2_{L^2_{\frac{1}{|Q'|}}(D)} \\
&\quad + \frac{1}{4\varepsilon_*^2}(\Delta_1^2+1)^4 e^{4\varepsilon_*(1+a)} \eta^2 (|\Delta(t,\eta)|+\eta)^2 \|\mathcal{KT} g(t)\|^2_{X^0} \\
&\le \frac{400}{\varepsilon_*^2}(\Delta_1^2+1)^4 e^{4\varepsilon_*(1+a)} C_{\mathcal{KT}}^2 \|\psi\|^2_{X^1} \eta^2 (|\Delta(t,\eta)|+\eta)^2 \\
&\quad + \frac{64\pi^3}{\varepsilon_*^2}(\Delta_1^2+1)^4 \eta^2 \sum_{k=2}^{\infty} \min\{1,k^4\eta^2\} \|(\mathcal{KT} g(t))_k\|^2_{L^2_{\frac{1}{|Q'|}}(D)}.
\end{aligned} \tag{C.41}$$

Hence if we use (C.40) and (C.41) in (C.38), it follows that

$$\begin{aligned}
\|A_3\|_{X^0} &\le \|A_{31}\|_{X^0} + \|A_{32}\|_{X^0} \\
&\le \frac{60}{\varepsilon_*}(\Delta_1^2+1)^2 e^{2\varepsilon_*(1+a)} C_{\mathcal{KT}} \|\psi\|_{X^1} (1+a^{1/2})\, \eta\, (|\Delta(t,\eta)|+\eta) \\
&\quad + \frac{7}{\varepsilon_*^{13/12}} (\Delta_1^2+1)^2 C_{\mathcal{KT}}\, C_4(g,a)\, a^{1/12} \eta^{13/12} \\
&\quad + \frac{2}{\varepsilon_*}(\Delta_1^2+1)^2 \eta \Big(16\pi^3 \sum_{k\neq 0} \min\{1,k^4\eta^2\} \|(\mathcal{KT} g(t))_k\|^2_{L^2_{\frac{1}{|Q'|}}(D)}\Big)^{1/2}.
\end{aligned} \tag{C.42}$$

Therefore altogether from (C.31), (C.33), (C.37) and (C.42), we see that

$$\begin{aligned}
&\|g(t+\eta) - g(t) - \eta\, d(t)\|_{X^0} \\
&\quad \le \|A_1\|_{X^0} + \|A_2\|_{X^0} + \|A_3\|_{X^0}
\end{aligned}$$

$$
\begin{aligned}
&\leq \frac{62}{\varepsilon_*}(\Delta_1^2+1)^2\, e^{2\varepsilon_*(1+a)}\,(C_{\mathcal{KT}}+1)\,\|\psi\|_{X^1}\,(1+a^{1/2})\,\eta\,(|\Delta(t,\eta)|+\eta)\\
&+C_{\mathcal{KT}}\,e^{2\varepsilon_*}C_4(g,a)\,\eta^{7/6}+\frac{7}{\varepsilon_*^{13/12}}\,(\Delta_1^2+1)^2\,C_{\mathcal{KT}}\,C_4(g,a)\,a^{1/12}\,\eta^{13/12}\\
&+2(\Delta_1^2+1)^2\eta\left(16\pi^3\sum_{k\neq 0}k^4\,\min\{1,k^4\eta^2\}\,\|\psi_k\|^2_{L^2_{\frac{1}{|Q'|}}(D)}\right)^{1/2}\\
&+\frac{4}{\varepsilon_*}\,e^{2\varepsilon_*}\,(\Delta_1^2+1)^2\,\eta\left(16\pi^3\sum_{k\neq 0}\min\{1,k^4\eta^2\}\,\|(\mathcal{KT}g(t))_k\|^2_{L^2_{\frac{1}{|Q'|}}(D)}\right)^{1/2}.
\end{aligned}
\tag{C.43}
$$

Now $g \in C(J, X^1_{\mathrm{odd}})$ together with Lemma C.9 implies that $J \ni t \mapsto \Phi(g(t))$ is continuous; in fact this function is even Hölder continuous by (a). It follows that

$$
\lim_{\eta\to 0^+}\Delta(t,\eta)=\lim_{\eta\to 0^+}\frac{\exp(-\int_t^{t+\eta}[\Phi(g(t))-\Phi(g(\tau))]\,d\tau)-1}{\eta}=0.
$$

As in (C.29) we also have (recalling that $\psi \in X^2_{\mathrm{odd}}$) that

$$
\begin{aligned}
&\lim_{\eta\to 0^+}16\pi^3\sum_{k\neq 0}k^4\min\{1,k^4\eta^2\}\,\|\psi_k\|^2_{L^2_{\frac{1}{|Q'|}}(D)}=0,\\
&\lim_{\eta\to 0^+}16\pi^3\sum_{k\neq 0}\min\{1,k^4\eta^2\}\,\|(\mathcal{KT}g(t))_k\|^2_{L^2_{\frac{1}{|Q'|}}(D)}=0.
\end{aligned}
$$

Hence, (C.43) yields $\eta^{-1}\|g(t+\eta)-g(t)-\eta\, d(t)\|_{X^0}=\mathrm{o}(1)$ as $\eta \to 0^+$, which means that g is differentiable at t and $g'(t)=d(t)$.

To show (e), since $g(t) \in X^2_{\mathrm{odd}}$, (B.44) and (C.6) imply that $\Phi(g(t)) = (Lg(t), g(t))_{X^0}$. Hence, by Corollary B.19, one has for $\eta > 0$ small enough:

$$
\begin{aligned}
\Phi(g(t+\eta))-\Phi(g(t)) &= (Lg(t+\eta),g(t+\eta))_{X^0}-(Lg(t),g(t))_{X^0}\\
&=(Lg(t+\eta)-Lg(t),g(t+\eta))_{X^0}+(Lg(t),g(t+\eta)-g(t))_{X^0}\\
&=(g(t+\eta)-g(t),Lg(t+\eta))_{X^0}+(Lg(t),g(t+\eta)-g(t))_{X^0}\\
&=(g(t+\eta)-g(t)-\eta g'(t),Lg(t+\eta))_{X^0}+\eta\,(g'(t),Lg(t+\eta))_{X^0}\\
&\quad+(Lg(t),g(t+\eta)-g(t)-\eta g'(t))_{X^0}+\eta\,(Lg(t),g'(t))_{X^0}\\
&=\eta\,(g'(t),Lg(t+\eta))_{X^0}+\eta\,(Lg(t),g'(t))_{X^0}\\
&\quad+(Lg(t),g(t+\eta)-g(t)-\eta g'(t))_{X^0}\\
&\quad+(g(t+\eta)-g(t)-\eta g'(t),Lg(t+\eta))_{X^0}.
\end{aligned}
\tag{C.44}
$$

From (B.18) and (B.42), we deduce

$$
\begin{aligned}
\|Lg(t+\eta)-Lg(t)\|_{X^0} &\leq \|\mathcal{T}^2g(t+\eta)-\mathcal{T}^2g(t)\|_{X^0}+\|\mathcal{KT}g(t+\eta)-\mathcal{KT}g(t)\|_{X^0}\\
&\leq \Delta_1^2\,\|g(t+\eta)-g(t)\|_{X^2}+C_{\mathcal{KT}}\,\|g(t+\eta)-g(t)\|_{X^0}
\end{aligned}
$$

$$\leq (\Delta_1^2 + C_{\mathcal{K}\mathcal{T}})\, \|g(t+\eta) - g(t)\|_{X^2}.$$

Thus $\lim_{\eta\to 0^+} Lg(t+\eta) = Lg(t)$ in X^0 by (c). Furthermore, from (d) we know that

$$\eta^{-1}\|g(t+\eta) - g(t) - \eta g'(t)\|_{X^0} \to 0, \quad \eta \to 0^+.$$

As a consequence, (C.44) together with (C.11) yields

$$\begin{aligned}
\lim_{\eta\to 0^+} \frac{\Phi(g(t+\eta)) - \Phi(g(t))}{\eta} &= (g'(t), Lg(t))_{X^0} + (Lg(t), g'(t))_{X^0} \\
&= 2\,(Lg(t), -Lg(t) + \Phi(g(t))\, g(t))_{X^0} \\
&= -2\,(\|Lg(t)\|_{X^0}^2 - \Phi(g(t))^2).
\end{aligned}$$

Finally, we turn to the proof of (f) so that $\|\psi\|_{X^0} = 1$ is added as a hypothesis. Denote $\varphi(t) = \|g(t)\|_{X^0}^2$. Then $\varphi(0) = 1$, and moreover by (d):

$$\begin{aligned}
\varphi'(t) &= 2\,(g(t), g'(t))_{X^0} \\
&= 2\Big[-(g(t), Lg(t))_{X^0} + \Phi(g(t))\,(g(t), g(t))_{X^0}\Big] \\
&= 2\Phi(g(t))\,(\varphi(t) - 1).
\end{aligned}$$

Therefore $\varphi(t) = 1$ for $t \in J$ due to uniqueness, and in particular $(g(t), g'(t))_{X^0} = \frac{1}{2}\varphi'(t) = 0$ for $t \in]0, a[$. Furthermore,

$$\begin{aligned}
\|g'(t)\|_{X^0}^2 &= \| - Lg(t) + \Phi(g(t))g(t)\|_{X^0}^2 \\
&= \|Lg(t)\|_{X^0}^2 - 2\Phi(g(t))\,(Lg(t), g(t))_{X^0} + \Phi(g(t))^2\, \|g(t)\|_{X^0}^2 \\
&= \|Lg(t)\|_{X^0}^2 - \Phi(g(t))^2,
\end{aligned}$$

and (C.12) is obtained from (e) upon integration. Lastly, Lemma C.10 and $\|g(t)\|_{X^0} = 1$ imply that $\Phi(g(t)) \geq \lambda_* $ for $t \in J$. □

Theorem C.4 *Let $\psi \in X^2_{\text{odd}}$ be such that $\|\psi\|_{X^0} = 1$ and $\Phi(\psi) \leq \lambda_* + \varepsilon_*$. Then there exists a continuous solution $g : [0, \infty[\to X^1_{\text{odd}}$ of (C.7) that has all the additional properties as listed in Corollary C.3(a)–(f), where (a) and (b) are valid on every finite time interval* $[0, a]$.

Proof Denote

$$\begin{aligned}
T = \sup\Big\{a > 0 : &\text{ there exists a solution } g \in C([0, a], X^1_{\text{odd}}) \text{ of } (C.7) \\
&\text{such that } \Phi(g(t)) \leq \lambda_* + 2\varepsilon_* \text{ for } t \in [0, a]\Big\}.
\end{aligned}$$

By Lemma C.2 and Corollary C.3 one has $\{\ldots\} \neq \emptyset$, and hence $T > 0$. Let $g, h \in C([0, a], X^1_{\text{odd}})$ be two solutions of (C.7) that are defined on a common time interval

$[0, a]$ such that $\Phi(g(t)) \le \lambda_* + 2\varepsilon_*$ and $\Phi(h(t)) \le \lambda_* + 2\varepsilon_*$ for all $t \in [0, a]$. Also $\|g\|_{\infty,1} < \infty$ and $\|h\|_{\infty,1} < \infty$, where $\|g\|_{\infty,1} = \max\{\|g(t)\|_{X^1} : t \in [0, a]\}$. Hence for $t \in [0, a]$ one deduces from Lemma C.18 and (C.75) for $\alpha = 1$ that

$$
\begin{aligned}
&\|g(t) - h(t)\|_{X^1} \\
&= \left\| \mathcal{W}(t, 0; g)\psi + \int_0^t \mathcal{W}(t, s; g)\mathcal{K}\mathcal{T} g(s)\, ds - \mathcal{W}(t, 0; h)\psi - \int_0^t \mathcal{W}(t, s; h)\mathcal{K}\mathcal{T} h(s)\, ds \right\|_{X^1} \\
&\le \|(\mathcal{W}(t, 0; g) - \mathcal{W}(t, 0; h))\psi\|_{X^1} + \left\| \int_0^t [\mathcal{W}(t, s; g) - \mathcal{W}(t, s; h)]\mathcal{K}\mathcal{T} h(s)\, ds \right\|_{X^1} \\
&\quad + \left\| \int_0^t \mathcal{W}(t, s; g)\mathcal{K}\mathcal{T}(g(s) - h(s))\, ds \right\|_{X^1} \\
&\le 2\Lambda(t; g, h) \exp(\Lambda(t; g, h))\,(1 + e^{2\varepsilon_* t})\,\|\psi\|_{X^1} \\
&\quad + \frac{2}{\varepsilon_*^{1/2}} \Lambda(t; g, h) \exp(\Lambda(t; g, h))\,(1 + e^{2\varepsilon_* t}) \left(\int_0^t \|\mathcal{K}\mathcal{T} h(s)\|_{X^0}^2\, ds \right)^{1/2} \\
&\quad + \frac{2}{\varepsilon_*^{1/2}} (1 + e^{2\varepsilon_* t}) \left(\int_0^t \|\mathcal{K}\mathcal{T}(g(s) - h(s))\|_{X^0}^2\, ds \right)^{1/2},
\end{aligned}
$$

where

$$
\Lambda(t; g, h) = 2C_\Phi(\|g\|_{\infty,1} + \|h\|_{\infty,1}) \int_0^t \|g(\tau) - h(\tau)\|_{X^1}\, d\tau \le C \int_0^t \|g(\tau) - h(\tau)\|_{X^1}\, d\tau \le C, \tag{C.45}
$$

with constants denoted by $C > 0$ being allowed to depend upon C_Φ, $C_{\mathcal{K}\mathcal{T}}$, ε_*, $\|\psi\|_{X^1}$, $\|g\|_{\infty,1}$, $\|h\|_{\infty,1}$ and $a < \infty$. Thus, it follows from (C.45) and (B.42) that

$$
\|g(t) - h(t)\|_{X^1} \le C \int_0^t \|g(\tau) - h(\tau)\|_{X^1}\, d\tau + C\left(\int_0^t \|g(s) - h(s)\|_{X^0}^2\, ds \right)^{1/2}.
$$

Squaring this relation, one obtains by using Hölder's inequality that

$$
\begin{aligned}
\|g(t) - h(t)\|_{X^1}^2 &\le C\left(\int_0^t \|g(\tau) - h(\tau)\|_{X^1}\, d\tau \right)^2 + C \int_0^t \|g(s) - h(s)\|_{X^1}^2\, ds \\
&\le C \int_0^t \|g(s) - h(s)\|_{X^1}^2\, ds.
\end{aligned}
$$

As a consequence, Gronwall's inequality yields $g(t) = h(t)$ for $t \in [0, a]$. From this uniqueness, one obtains a maximal and continuous solution $g : [0, T[\to X^1_{\text{odd}}$. Since the arguments from Corollary C.3 can be applied on any compact subinterval $[0, a]$ of $[0, T[$, we even know that $g(t) \in X^2_{\text{odd}}$ and $\|g(t)\|_{X^0} = 1$ for $t \in [0, T[$, and in addition $t \mapsto \Phi(g(t))$ is monotone decreasing on $[0, T[$; in particular, this implies that $\Phi(g(t)) \le \Phi(g(0)) = \Phi(\psi) \le \lambda_* + \varepsilon_*$ for $t \in [0, T[$.

Suppose now that $T < \infty$. First we are going to show that in this case $\psi_* = \lim_{t\to T} g(t)$ does exist in $X^1_{\rm odd}$. Let $(t_j) \subset [0, T[$ be a sequence such that $\lim_{j\to\infty} t_j = T$. W.l.o.g. we consider t_i and t_j such that $t_i > t_j \geq T/2$. Then for $\delta > 0$ small by (C.7), (C.85), (C.78), (C.79), (B.42) and (C.85):

$$\begin{aligned}
\|g(t_i) - g(t_j)\|_{X^1} &\leq \|(\mathcal{W}(t_i, 0) - \mathcal{W}(t_j, 0))\psi\|_{X^1} + \left\| \int_{t_j}^{t_i} \mathcal{W}(t_i, s)\,\mathcal{KT} g(s)\, ds \right\|_{X^1} \\
&\quad + \left\| \int_0^{t_j-\delta} (\mathcal{W}(t_i, s) - \mathcal{W}(t_j, s))\,\mathcal{KT} g(s)\, ds \right\|_{X^1} \\
&\quad + \left\| \int_{t_j-\delta}^{t_j} \mathcal{W}(t_i, s)\,\mathcal{KT} g(s)\, ds \right\|_{X^1} + \left\| \int_{t_j-\delta}^{t_j} \mathcal{W}(t_j, s)\,\mathcal{KT} g(s)\, ds \right\|_{X^1} \\
&\leq 2\,(\Delta_1^2 + 1) \left[\frac{1}{\varepsilon_* t_j} \sqrt{t_i - t_j} + \exp(2\varepsilon_* t_j)\,(t_i - t_j) \right] \|\psi\|_{X^0} \\
&\quad + \frac{2}{\varepsilon_*^{1/2}} (1 + e^{2\varepsilon_*(t_i - t_j)}) \left(\int_{t_j}^{t_i} \|\mathcal{KT} g(s)\|_{X^0}^2\, ds \right)^{1/2} \\
&\quad + \int_0^{t_j-\delta} \|(\mathcal{W}(t_i, s) - \mathcal{W}(t_j, s))\,\mathcal{KT} g(s)\|_{X^1}\, ds \\
&\quad + \frac{4}{\varepsilon_*^{1/2}} (1 + e^{2\varepsilon_*(t_i - t_j + \delta)}) \left(\int_{t_j-\delta}^{t_j} \|\mathcal{KT} g(s)\|_{X^0}^2\, ds \right)^{1/2} \\
&\leq 4\,(\Delta_1^2 + 1) \left[\frac{1}{\varepsilon_* T} \sqrt{t_i - t_j} + e^{2\varepsilon_* T}\,(t_i - t_j) \right] \|\psi\|_{X^0} \\
&\quad + \frac{2}{\varepsilon_*^{1/2}} (1 + e^{2\varepsilon_* T})\, C_{\mathcal{KT}} \left(\int_{t_j}^{t_i} \|g(s)\|_{X^0}^2\, ds \right)^{1/2} \\
&\quad + 2\,(\Delta_1^2 + 1)\, C_{\mathcal{KT}} \int_0^{t_j-\delta} \left[\frac{1}{\varepsilon_*(t_j - s)} \sqrt{t_i - t_j} \right. \\
&\qquad\qquad \left. + \exp(2\varepsilon_*(t_j - s))\,(t_i - t_j) \right] \|g(s)\|_{X^0}\, ds \\
&\quad + \frac{4}{\varepsilon_*^{1/2}} (1 + e^{2\varepsilon_*(T+1)})\, C_{\mathcal{KT}} \left(\int_{t_j-\delta}^{t_j} \|g(s)\|_{X^0}^2\, ds \right)^{1/2} \\
&= 4\,(\Delta_1^2 + 1) \left[\frac{1}{\varepsilon_* T} \sqrt{t_i - t_j} + e^{2\varepsilon_* T}\,(t_i - t_j) \right] \|\psi\|_{X^0} \\
&\quad + \frac{2}{\varepsilon_*^{1/2}} (1 + e^{2\varepsilon_* T})\, C_{\mathcal{KT}} \sqrt{t_i - t_j} + \frac{4}{\varepsilon_*^{1/2}} (1 + e^{2\varepsilon_*(T+1)})\, C_{\mathcal{KT}} \sqrt{\delta} \\
&\quad + 2\,(\Delta_1^2 + 1)\, C_{\mathcal{KT}} \int_0^{t_j-\delta} \left[\frac{1}{\varepsilon_*(t_j - s)} \sqrt{t_i - t_j} \right. \\
&\qquad\qquad \left. + \exp(2\varepsilon_*(t_j - s))\,(t_i - t_j) \right] ds \\
&\leq 4\,(\Delta_1^2 + 1) \left[\frac{1}{\varepsilon_* T} \sqrt{t_i - t_j} + e^{2\varepsilon_* T}\,(t_i - t_j) \right] \|\psi\|_{X^0} \\
&\quad + \frac{4}{\varepsilon_*^{1/2}} (1 + e^{2\varepsilon_*(T+1)})\, C_{\mathcal{KT}} \left(\sqrt{t_i - t_j} + \sqrt{\delta} \right) \\
&\quad + 2\,(\Delta_1^2 + 1)\, C_{\mathcal{KT}} \left[\ln\left(\frac{T}{\delta} \right) \sqrt{t_i - t_j} + \frac{1}{2\varepsilon_*} e^{2\varepsilon_* T}\,(t_i - t_j) \right].
\end{aligned} \tag{C.46}$$

So if we set for instance $\delta = \sqrt{t_i - t_j}$, this estimate proves that $\psi_* = \lim_{t\to T} g(t)$ exists in $X^1_{\rm odd}$, and the function

$$g_* : [0, T] \to X^1_{\rm odd}, \quad g_*(t) = \begin{cases} g(t) & : \; t \in [0, T[\\ \psi_* & : \; t = T \end{cases},$$

is continuous. Since $\Phi : X^1_{\rm odd} \to \mathbb{R}$ is continuous by Lemma C.9, one has

$$\Phi(\psi_*) = \lim_{t\to T} \Phi(g(t)) \le \lambda_* + \varepsilon_*.$$

Similarly, we also find that $\|\psi_*\|_{X^0} = \lim_{t\to T} \|g(t)\|_{X^0} = 1$. The estimate which leads to (C.46) also yields that

$$\begin{aligned} \|g(t) - g(s)\|_{X^1} &\le 4\,(\Delta_1^2 + 1)\left[\frac{1}{\varepsilon_* T}\sqrt{t-s} + e^{2\varepsilon_* T}\,(t-s)\right]\|\psi\|_{X^0} \\ &\quad + \frac{4}{\varepsilon_*^{1/2}}\,(1 + e^{2\varepsilon_*(T+1)})\,C_{\mathcal{K}\mathcal{T}}\,(\sqrt{t-s} + \sqrt{\delta}) \\ &\quad + 2\,(\Delta_1^2 + 1)\,C_{\mathcal{K}\mathcal{T}}\left[\ln\left(\frac{T}{\delta}\right)\sqrt{t-s} + \frac{1}{2\varepsilon_*}\,e^{2\varepsilon_* T}\,(t-s)\right] \end{aligned}$$

for $t, s \in [0, T[$, $t > s \ge T/2$. If $\delta = \sqrt{t-s} \le 1$, then $-x\ln x^{3/2} \le \frac{3}{2e}$ for $x \in [0, 1]$ shows that

$$\begin{aligned} \|g(t) - g(s)\|_{X^1} &\le C\delta + C\sqrt{\delta} + C\ln\left(\frac{T}{\delta}\right)\delta \\ &\le C\delta + C\sqrt{\delta} + C\delta^{1/3} \\ &\le C_*\delta^{1/3} \\ &= C_*(t-s)^{1/6}, \end{aligned}$$

where $C_* > 0$ depend upon Δ_1, $C_{\mathcal{K}\mathcal{T}}$, ε_*, T, $\|\psi\|_{X^0}$. On the other hand, if $\delta = \sqrt{t-s} \ge 1$, then

$$\|g(t) - g(s)\|_{X^1} \le 2\,\|g_*\|_{\infty,1} \le 2\,\|g_*\|_{\infty,1}\,(t-s)^{1/6}$$

for $\|g_*\|_{\infty,1} = \max\{\|g_*(t)\|_{X^1} : t \in [0, T]\} < \infty$. In summary, we have verified that

$$\|g(t) - g(s)\|_{X^1} \le C_{**}(t-s)^{1/6}, \quad C_{**} = C_* + 2\,\|g_*\|_{\infty,1}, \tag{C.47}$$

for $t, s \in [0, T]$, $t > s \ge T/2$; note that is was not possible to derive this estimate directly from Corollary C.3(a), since a bound on $\|g(t)\|_{X^1}$ for $t \in [0, a]$ enters the constants $C_1(g, a)$ and $C_2(g, a)$, and a priori we do not know that $\sup\{\|g(t)\|_{X^1} : t \in [0, T[\} < \infty$. From (C.47) it follows that $\psi_* \in X^2_{\rm odd}$: by (C.7), (C.68), (C.81), (C.76) for $A = C_{\mathcal{K}\mathcal{T}} C_{**}$, $\gamma = 1/6$, $\alpha = 2$ and (C.77) one has

$$\begin{aligned}
\|\psi_*\|_{X^2} &\le \|\mathcal{W}(T,0)\psi\|_{X^2} + \left\| \int_0^{T/2} \mathcal{W}(T,s)\,\mathcal{KT}g(s)\,ds \right\|_{X^2} \\
&\quad + \left\| \int_{T/2}^{T} \mathcal{W}(T,s)\,\mathcal{KT}(g(s)-\psi_*)\,ds \right\|_{X^2} + \left\| \left(\int_{T/2}^{T} \mathcal{W}(T,s)\,ds \right) \mathcal{KT}\psi_* \right\|_{X^2} \\
&\le e^{2\varepsilon_* T}\|\psi\|_{X^2} + \frac{4}{\varepsilon_*}\left(\frac{1}{\sqrt{T}} + e^{2\varepsilon_* T}\right)\left(\int_0^{T/2} \|\mathcal{KT}g(s)\|_{X^0}^2\,ds \right)^{1/2} \\
&\quad + \frac{4C_{\mathcal{KT}}C_{**}}{\varepsilon_*}\,(T/2)^{1/6} + \frac{C_{\mathcal{KT}}C_{**}}{\varepsilon_*^{1/2}}\,e^{\varepsilon_* T}\,(T/2)^{2/3} \\
&\quad + \frac{1}{\varepsilon_*}\,(2+e^{\varepsilon_* T})\,\|\mathcal{KT}\psi_*\|_{X^0} \\
&< \infty;
\end{aligned}$$

recall that $\|\psi_*\|_{X^0} = 1$ and observe $\int_0^{T/2} \|\mathcal{KT}g(s)\|_{X^0}^2\,ds \le C_{\mathcal{KT}}^2 \int_0^{T/2} \|g(s)\|_{X^0}^2\,ds = C_{\mathcal{KT}}^2(T/2)$. Therefore, we have shown that in fact $\psi_* \in X^2_{\text{odd}}$ is verified.

Now consider the evolution equation

$$h(t) = \mathcal{W}(T+t,T)\psi_* + \int_0^t \mathcal{W}(T+t,T+s)\,\mathcal{KT}h(s)\,ds. \tag{C.48}$$

Owing to $\psi_* \in X^2_{\text{odd}}$ and $\Phi(\psi_*) \le \lambda_* + \varepsilon_*$, a fixed point argument analogous to the proof of Lemma C.2 can be employed to show that there is $\delta > 0$ and a continuous solution $h : [0,\delta] \to X^1_{\text{odd}}$ of (C.48) such that $\Phi(h(t)) \le \lambda_* + 2\varepsilon_*$ for $t \in [0,\delta]$. Define

$$\tilde{g} : [0, T+\delta] \to X^1_{\text{odd}}, \quad \tilde{g}(t) = \begin{cases} g_*(t) & : \ t \in [0,T] \\ h(t-T) & : \ t \in]T, T+\delta] \end{cases}.$$

Then $\mathcal{W}(T,T) = \text{id}$ implies that $\tilde{g}$ is continuous, and furthermore $\Phi(\tilde{g}(t)) \le \lambda_* + 2\varepsilon_*$ for $t \in [0, T+\delta]$. If $t \in [0,T[$, then

$$\tilde{g}(t) = g(t) = \mathcal{W}(t,0)\psi + \int_0^t \mathcal{W}(t,s)\,\mathcal{KT}g(s)\,ds = \mathcal{W}(t,0)\psi + \int_0^t \mathcal{W}(t,s)\,\mathcal{KT}\tilde{g}(s)\,ds$$

by (C.7) for g. On the other hand, if $t \in]T, T+\delta]$, then owing to (C.7) at $t = T$ and (C.3):

$$\begin{aligned}
\tilde{g}(t) &= h(t-T) \\
&= \mathcal{W}(t,T)\psi_* + \int_0^{t-T} \mathcal{W}(t,T+s)\,\mathcal{KT}h(s)\,ds \\
&= \mathcal{W}(t,T)\left[\mathcal{W}(T,0)\psi + \int_0^T \mathcal{W}(T,s)\,\mathcal{KT}g(s)\,ds\right] + \int_T^t \mathcal{W}(t,\tau)\,\mathcal{KT}h(\tau-T)\,d\tau \\
&= \mathcal{W}(t,0)\psi + \int_0^T \mathcal{W}(t,s)\,\mathcal{KT}g(s)\,ds + \int_T^t \mathcal{W}(t,\tau)\,\mathcal{KT}\tilde{g}(\tau)\,d\tau \\
&= \mathcal{W}(t,0)\psi + \int_0^t \mathcal{W}(t,s)\,\mathcal{KT}\tilde{g}(s)\,ds.
\end{aligned}$$

This proves that in fact $\tilde{g}$ is a solution of (C.7) on $[0, T+\delta]$, which however contradicts the definition of T. Therefore, we must have $T = \infty$ and the claims follow. □

C.4 Compactness

The next result is well-known in principle; see Remark B.2(a) for the definition of the space $L^2_{\text{sph}, \frac{1}{|Q'|}}(K) = X^0$.

Lemma C.5 *The linear operator*

$$\widetilde{\mathcal{K}} : L^2_{\text{sph}, \frac{1}{|Q'|}}(K) \to L^2_{\text{sph}, \frac{1}{|Q'|}}(K), \quad \widetilde{\mathcal{K}}g = |Q'(e_Q)|^{1/2}\, p_r\, U_g'(r),$$

is compact. Furthermore, $|Q'(e_Q)|^{1/2}\widetilde{\mathcal{K}}g = \mathcal{K}g$ for $\mathcal{K}$ from (1.15) and $(\mathcal{K}g)_k = |Q'(e_Q)|^{1/2}(\widetilde{\mathcal{K}}g)_k$ for the Fourier coefficients.

Proof We closely follow [48, Lemma 2.2]. Let $(g_j) \subset L^2_{\text{sph}, \frac{1}{|Q'|}}(K)$ be bounded. The associated densities ρ_{g_j} have compact support $\operatorname{supp} \rho_{g_j} \subset \{x \in \mathbb{R}^3 : |x| \le r_Q\}$. Furthermore, if we denote by $\max_K |v| < \infty$ the maximal value of $|v|$ for some $(x, v) \in K$, then using Hölder's inequality, (Q3) and $e \in [U_Q(0), e_0]$:

$$\begin{aligned}
|\rho_{g_j}(x)| &= \left| \int_{\max_K |v|} g_j(x, v)\, dv \right| \\
&\le \left(\int_{\max_K |v|} |Q'(e_Q)|\, dv \right)^{1/2} \left(\int_{\mathbb{R}^3} \frac{1}{|Q'(e_Q)|} |g_j(x, v)|^2\, dv \right)^{1/2} \\
&\le C(\max_K |v|)^{3/2} \sup\{|Q'(e)| : e \in [U_Q(0), e_0]\}^{1/2} \left(\int_{\max_K |v|} \frac{1}{|Q'(e_Q)|} |g_j(x, v)|^2\, dv \right)^{1/2} \\
&\le C \left(\int_{\max_K |v|} \frac{1}{|Q'(e_Q)|} |g_j(x, v)|^2\, dv \right)^{1/2}.
\end{aligned}$$

Hence, we obtain

$$\int_{\mathbb{R}^3} |\rho_{g_j}(x)|^2\, dx \le C \iint_K \frac{1}{|Q'(e_Q)|} |g_j(x, v)|^2\, dx\, dv = C \|g_j\|^2_{L^2_{\text{sph}, \frac{1}{|Q'|}}(K)} \le C, \tag{C.49}$$

independently of j. Thus, $(\rho_{g_j}) \subset L^2(\mathbb{R}^3)$ is bounded, and therefore also

$$\|\nabla(\nabla U_{g_j})\|_{L^2(\mathbb{R}^3)} \le C \|\nabla(\nabla \Delta^{-1} \rho_{g_j})\|_{L^2(\mathbb{R}^3)} \le C \|\rho_{g_j}\|_{L^2(\mathbb{R}^3)} \le C.$$

In addition, by the Hardy-Littlewood-Sobolev inequality, we have

$$\|\nabla U_{g_j}\|_{L^2(\mathbb{R}^3)} \le C\|\nabla\Delta^{-1}\rho_{g_j}\|_{L^2(\mathbb{R}^3)} \le C\|\rho_{g_j}\|_{L^{6/5}(\mathbb{R}^3)} \le C\|\rho_{g_j}\|_{L^2(\mathbb{R}^3)} \le C, \tag{C.50}$$

using (C.49) and the compact support of the ρ_{g_j}. Accordingly, we have shown that $(\nabla U_{g_j}) \subset H^1(\mathbb{R}^3)$ is bounded, which means in particular that the sequence has a convergent subsequence in any $L^2(\overline{B_R(0)})$, $R > 0$, where $B_R(0) = \{x \in \mathbb{R}^3 : |x| < R\}$. Next, if $|x| \ge 2r_Q$ and $|y - x| \le r_Q$, then $|y| \ge |x| - r_Q \ge |x|/2$. From (1.14), we find that for $|x| \ge 2r_Q$:

$$|\nabla U_{g_j}(x)| \le \int_{|y-x|\le r_Q} \frac{|\rho_{g_j}(y-x)|}{|y|^2}\,dy \le \frac{4}{|x|^2}\,\|\rho_{g_j}\|_{L^1(\mathbb{R}^3)} \le \frac{C}{|x|^2}.$$

Since $\int_{|x|\ge R} \frac{dx}{|x|^4} \le CR^{-1}$, due to the local compactness it is now straightforward to prove that $(\nabla U_{g_j}) \subset L^2(\mathbb{R}^3)$ has a strong limit. This yields

$$\begin{aligned}
\|\widetilde{\mathcal{K}}g_j - \widetilde{\mathcal{K}}g_k\|^2_{L^2_{\mathrm{sph},\frac{1}{|Q'|}}(K)} &= \iint_K p_r^2\,|U'_{g_j}(r) - U'_{g_k}(r)|^2\,dx\,dv \\
&\le (\max_K |v|)^5 \int_{\mathbb{R}^3} |\nabla U_{g_j}(x) - \nabla U_{g_k}(x)|^2\,dx \\
&\to 0, \quad j,k \to \infty,
\end{aligned}$$

which shows that $(\widetilde{\mathcal{K}}g_j) \subset L^2_{\mathrm{sph},\frac{1}{|Q'|}}(K)$ has a strongly convergent subsequence. Next, $\widetilde{\mathcal{K}}$ is a bounded operator, since

$$\begin{aligned}
\|\widetilde{\mathcal{K}}g\|^2_{L^2_{\mathrm{sph},\frac{1}{|Q'|}}(K)} &= \iint_K p_r^2\,|U'_g(r)|^2\,dx\,dv \\
&\le C(\max_K |v|)^5\,\|\nabla U_g\|^2_{L^2(\mathbb{R}^3)} \le C\|\rho_g\|^2_{L^2(\mathbb{R}^3)} \le C\|g\|^2_{L^2_{\mathrm{sph},\frac{1}{|Q'|}}(K)},
\end{aligned}$$

cf. (C.50) and (C.49). The relation $|Q'(e_Q)|^{1/2}\widetilde{\mathcal{K}}g = \mathcal{K}g$ is obvious and for $(\mathcal{K}g)_k = |Q'(e_Q)|^{1/2}(\widetilde{\mathcal{K}}g)_k$ for $k \in \mathbb{Z}$ it suffices to remark that $e = e_Q$ is independent of θ. □

Corollary C.6 *The linear operator*

$$\mathcal{K} : L^2_{\mathrm{sph},\frac{1}{|Q'|}}(K) \to L^2_{\mathrm{sph},\frac{1}{|Q'|}}(K), \quad \mathcal{K}g = |Q'(e_Q)|\,p_r\,U'_g(r),$$

is compact.

Proof Since $e_Q \in [e_0, U_Q(0)]$, we have $|Q'(e_Q)| \le C$. Due to Lemma C.5 we therefore obtain $|\mathcal{K}g| = |Q'(e_Q)|^{1/2}|\widetilde{\mathcal{K}}g| \le C^{1/2}|\widetilde{\mathcal{K}}g|$, which implies that $\mathcal{K}$ is bounded and compact. □

Lemma C.7 *Suppose that $\lambda_* + 3\varepsilon_* \le \delta_1^2$ is satisfied. Let $(\psi_j) \subset X^2_{\mathrm{odd}}$ and $(t_j) \subset]0, \infty[$ be sequences such that $\|\psi_j\|_{X^0} = 1$ and $\Phi(\psi_j) \le \lambda_* + \varepsilon_*$ for $j \in \mathbb{N}$, and furthermore $t_j \to \infty$ as $j \to \infty$. Denote by g_j the solution of (C.7) with initial data $g_j(0) = \psi_j$. Then*

$$\{g_j(t_j) : j \in \mathbb{N}\} \subset X^0$$

is relatively compact.

Proof Owing to Theorem C.4 one has $\|g_j(t)\|_{X^0} = 1$ for all $j \in \mathbb{N}$ and $t \in [0, \infty[$. Moreover,

$$\lambda_* \le \Phi(g_j(t)) \le \Phi(g_j(0)) = \Phi(\psi_j) \le \lambda_* + \varepsilon_* \tag{C.51}$$

for $j \in \mathbb{N}$ and $t \in [0, \infty[$. Therefore by (C.6), and also using (B.42),

$$\begin{aligned}\|\mathcal{T} g_j(t)\|^2_{X^0} &= \Phi(g_j(t)) + (\mathcal{K}\mathcal{T} g_j(t), g_j(t))_{X^0} \\ &\le \lambda_* + \varepsilon_* + \|\mathcal{K}\mathcal{T} g_j(t)\|_{X^0} \, \|g_j(t)\|_{X^0} \\ &\le \lambda_* + \varepsilon_* + C_{\mathcal{K}\mathcal{T}},\end{aligned}$$

which says that $\{\mathcal{T} g_j(t) : j \in \mathbb{N}, t \in [0, \infty[\} \subset X^0$ is bounded. Hence, according to Lemma C.5, the set

$$\{\widetilde{\mathcal{K}}\mathcal{T} g_j(t) : j \in \mathbb{N}, t \in [0, \infty[\} \subset X^0 = L^2_{\mathrm{sph}, \frac{1}{|Q'|}}(K)$$

is relatively compact. By (C.7),

$$g_j(t_j) = \mathcal{W}(t_j, 0)\psi_j + \int_0^{t_j} \mathcal{W}(t_j, s)\,\mathcal{K}\mathcal{T} g_j(s)\, ds, \quad j \in \mathbb{N}.$$

For the initial data terms, (C.72) yields

$$\begin{aligned}\|\mathcal{W}(t_j, 0)\psi_j\|^2_{X^0} &= 16\pi^3 \sum_{k\neq 0} \iint_D dI\, d\ell\, \ell\, \frac{1}{|Q'(e)|} |\mathcal{W}_k(t_j, 0)(I, \ell)(\psi_j)_k(I, \ell)|^2 \\ &\le 16\pi^3 \sum_{k\neq 0} \exp(-2\varepsilon_* k^2 t_j) \iint_D dI\, d\ell\, \ell\, \frac{1}{|Q'(e)|}\, |(\psi_j)_k(I, \ell)|^2 \\ &\le 16\pi^3 \exp(-2\varepsilon_* t_j) \sum_{k\neq 0} \iint_D dI\, d\ell\, \ell\, \frac{1}{|Q'(e)|}\, |(\psi_j)_k(I, \ell)|^2 \\ &= \exp(-2\varepsilon_* t_j)\, \|\psi_j\|^2_{X^0} \to 0, \quad j \to \infty.\end{aligned}$$

Hence, it suffices to establish that

$$B = \left\{ \int_0^{t_j} \mathcal{W}(t_j, s)\,\mathcal{K}\mathcal{T} g_j(s)\, ds : j \in \mathbb{N} \right\} \subset X^0$$

is relatively compact, where $\mathcal{W}(t_j, s) = \mathcal{W}(t_j, s; g_j)$. For this we are going to use Remark B.2(d) and we introduce

$$S_j(N) = 16\pi^3 \sum_{|k|\geq N} \left\| \int_0^{t_j} \mathcal{W}_k(t_j, s)\,(\mathcal{KT} g_j(s))_k\, ds \right\|^2_{L^2_{\frac{1}{|Q'|}}(D)}, \quad j, N \in \mathbb{N}.$$

Then (C.72), Lemma C.13, (B.42) and $\|g_j(s)\|^2_{X^0} = 1$ lead to

$$\begin{aligned}
S_j(N) &= 16\pi^3 \sum_{|k|\geq N} \iint_D dI\, d\ell\, \ell \frac{1}{|Q'(e)|} \left| \int_0^{t_j} \mathcal{W}_k(t_j, s)(I, \ell)\,(\mathcal{KT} g_j(s))_k(I, \ell)\, ds \right|^2 \\
&\leq \frac{16\pi^3}{\varepsilon_*} \sum_{|k|\geq N} \frac{1}{k^2} \int_0^{t_j} ds\, e^{-\varepsilon_* k^2 (t_j - s)}\, \|(\mathcal{KT} g_j(s))_k\|^2_{L^2_{\frac{1}{|Q'|}}(D)} \\
&\leq \frac{1}{\varepsilon_*} \sum_{|k|\geq N} \frac{1}{k^2} \int_0^{t_j} ds\, e^{-\varepsilon_* k^2 (t_j - s)}\, \|\mathcal{KT} g_j(s)\|^2_{X^0} \\
&\leq \frac{1}{\varepsilon_*} C^2_{\mathcal{KT}} \sum_{|k|\geq N} \frac{1}{k^2} \int_0^{t_j} e^{-\varepsilon_* k^2 (t_j - s)}\, ds \\
&\leq \frac{1}{\varepsilon_*^2} C^2_{\mathcal{KT}} \sum_{|k|\geq N} \frac{1}{k^4}.
\end{aligned}$$

Taking $N = 1$ shows that

$$\sup_{j\in\mathbb{N}} \left\| \int_0^{t_j} \mathcal{W}(t_j, s)\, \mathcal{KT} g_j(s)\, ds \right\|^2_{X^0} = \sup_{j\in\mathbb{N}} S_j(1) < \infty,$$

so that $B \subset X^0$ is bounded. In addition,

$$\sup_{j\in\mathbb{N}} \sum_{|k|\geq N} \left\| \int_0^{t_j} \mathcal{W}_k(t_j, s)\,(\mathcal{KT} g_j(s))_k\, ds \right\|^2_{L^2_{\frac{1}{|Q'|}}(D)} \leq \frac{1}{16\pi^3 \varepsilon_*^2} C^2_{\mathcal{KT}} \sum_{|k|\geq N} \frac{1}{k^4}$$

will be smaller than ε^2, if $N = N(\varepsilon)$ is taken sufficiently large. It remains to validate (iii) from Remark B.2(d), i.e., the fact that each

$$B_k = \left\{ \int_0^{t_j} \mathcal{W}_k(t_j, s)\,(\mathcal{KT} g_j(s))_k\, ds : j \in \mathbb{N} \right\} \subset L^2_{\frac{1}{|Q'|}}(D)$$

is relatively compact. Since $(\mathcal{KT} g_j(s))_k = |Q'(e_Q)|^{1/2} (\widetilde{\mathcal{KT}} g_j(s))_k$ by Lemma C.5, this is equivalent to the statement that each

$$\tilde{B}_k = \left\{ \int_0^{t_j} \mathcal{W}_k(t_j, s)\, (\widetilde{\mathcal{K}}\mathcal{T} g_j(s))_k\, ds \,:\, j \in \mathbb{N} \right\} \subset L^2(D)$$

is relatively compact. As every projection $\pi_k : X^0 \to L^2_{\frac{1}{|Q'|}}(D)$, $\pi_k g = g_k$, is continuous, the above discussion implies that

$$\tilde{C}_k = \{(\widetilde{\mathcal{K}}\mathcal{T} g_j(t))_k : j \in \mathbb{N}, t \in [0, \infty[\} \subset L^2_{\frac{1}{|Q'|}}(D)$$

is relatively compact. Due to (Q3) and $e \in [U_Q(0), e_0]$, we have

$$\begin{aligned}
\|\phi\|^2_{L^2(D)} &= \iint\limits_D dI\, d\ell\, \ell\, \frac{|Q'(e)|}{|Q'(e)|}\, |\phi(I, \ell)|^2 \\
&\le \sup\{|Q'(e)| : e \in [U_Q(0), e_0]\}^{1/2} \iint\limits_D dI\, d\ell\, \ell\, \frac{1}{|Q'(e)|}\, |\phi(I, \ell)|^2 \\
&\le C \|\phi\|^2_{L^2_{\frac{1}{|Q'|}}(D)}.
\end{aligned}$$

In other words, the identity map $I : L^2_{\frac{1}{|Q'|}}(D) \to L^2(D)$ is continuous, which in turn yields that also

$$\tilde{C}_k = \{(\widetilde{\mathcal{K}}\mathcal{T} g_j(t))_k : j \in \mathbb{N}, t \in [0, \infty[\} \subset L^2(D) \tag{C.52}$$

is relatively compact. Next we write

$$\begin{aligned}
\mathcal{W}_k(t, s)(I, \ell) &= \exp\Big(- \int_s^t [k^2 \omega_1^2(I, \ell) - \Phi(g_j(\tau))]\, d\tau \Big) \\
&= \exp\Big(- \int_s^t [\delta_1^2 - \varepsilon_* - \Phi(g_j(\tau))]\, d\tau \Big)\, e^{-(t-s)\varphi(I, \ell)}
\end{aligned}$$

for $\varphi(I, \ell) = k^2 \omega_1^2(I, \ell) + \varepsilon_* - \delta_1^2$; then $\varepsilon_* \le \varphi(I, \ell) \le k^2 \Delta_1^2 + \varepsilon_*$ for $(I, \ell) \in D$. Consider the probability measure

$$d\mu_j = \alpha_j\, \mathbf{1}_{[0, t_j]}(s)\, e^{-\int_s^{t_j} [\delta_1^2 - \varepsilon_* - \Phi(g_j(\tau))]\, d\tau}\, ds, \quad \alpha_j = \left(\int_0^{t_j} e^{-\int_s^{t_j} [\delta_1^2 - \varepsilon_* - \Phi(g_j(\tau))]\, d\tau}\, ds \right)^{-1}.$$

This is well-defined: according to (C.51) we have

$$\delta_1^2 \ge \delta_1^2 - \lambda_* - \varepsilon_* \ge \delta_1^2 - \varepsilon_* - \Phi(g_j(\tau)) \ge \delta_1^2 - \lambda_* - 2\varepsilon_* \ge \varepsilon_*, \tag{C.53}$$

which shows that

$$\frac{1}{2\delta_1^2} \le \frac{1}{\delta_1^2}(1 - e^{-\delta_1^2 t_j}) \le \alpha_j^{-1} \le \frac{1}{\varepsilon_*}(1 - e^{-\delta_1^2 t_j}) \le \frac{1}{\varepsilon_*},$$

at least if j is sufficiently large (as we may assume). By means of $d\mu_j$ we may write the elements $\phi_j = \int_0^{t_j} \mathcal{W}_k(t_j, s)\,(\widetilde{\mathcal{K}\mathcal{T}}\,g_j(s))_k\,ds$ of $\tilde{B}_k$ at a point $(I, \ell) \in D$ as

$$\begin{aligned}\phi_j(I, \ell) &= \int_0^{t_j} e^{-\int_s^{t_j}[\delta_1^2 - \varepsilon_* - \Phi(g_j(\tau))]\,d\tau}\, e^{-(t_j - s)\varphi(I,\ell)}\,(\widetilde{\mathcal{K}\mathcal{T}}\,g_j(s))_k(I, \ell)\,ds \\ &= \int_{\mathbb{R}} \alpha_j^{-1} e^{-(t_j - s)\varphi(I,\ell)}\,(\widetilde{\mathcal{K}\mathcal{T}}\,g_j(s))_k(I, \ell)\,d\mu_j(s).\end{aligned}$$

Due to Lemma C.19, we thus have

$$\begin{aligned}\phi_j &\in \overline{\mathrm{co}}\left\{\alpha_j^{-1} e^{-(t_j - s)\varphi}\,(\widetilde{\mathcal{K}\mathcal{T}}\,g_j(s))_k : s \in [0, t_j]\right\} \\ &\subset \overline{\mathrm{co}}\left\{\beta e^{-\sigma\varphi}\chi : \beta \in \left[\frac{1}{2\delta_1^2}, \frac{1}{\varepsilon_*}\right], \sigma \in [0, t_j], \chi \in \tilde{C}_k\right\} \subset L^2(D),\end{aligned}$$

cf. (C.52). As a consequence,

$$\tilde{B}_k \subset \overline{\mathrm{co}}\left\{\beta e^{-\sigma\varphi}\chi : \beta \in \left[\frac{1}{2\delta_1^2}, \frac{1}{\varepsilon_*}\right], \sigma \in [0, \infty[, \chi \in \tilde{C}_k\right\} \subset L^2(D). \qquad \text{(C.54)}$$

The set $S = \{\ldots\} \subset L^2(D)$ is relatively compact. To see this, let $(\beta_i) \subset [\frac{1}{2\delta_1^2}, \frac{1}{\varepsilon_*}]$, $(\sigma_i) \subset [0, \infty[$ and $(\chi_i) \subset \tilde{C}_k$ be sequences. By passing to subsequences, if necessary, we may assume that $\beta_i \to \beta_0 \in [\frac{1}{2\delta_1^2}, \frac{1}{\varepsilon_*}]$, $\sigma_i \to \sigma_0 \in [0, \infty]$ and $\chi_i \to \chi_0$ in $L^2(D)$, the latter by the relative compactness of $\tilde{C}_k$. Case 1: $\sigma_0 = \infty$. Then

$$\|\beta_i\, e^{-\sigma_i\varphi}\chi_i\|_{L^2(D)} \le \beta_i\, e^{-\sigma_i\varepsilon_*}\|\chi_i\|_{L^2(D)} \le C e^{-\sigma_i\varepsilon_*} \to 0, \quad i \to \infty,$$

so that $\beta_i\, e^{-\sigma_i\varphi}\chi_i \to 0$ in $L^2(D)$. Case 2: $\sigma_0 < \infty$. Here, we have

$$\begin{aligned}\|\beta_i\, e^{-\sigma_i\varphi}\chi_i - \beta_0\, e^{-\sigma_0\varphi}\chi_0\|_{L^2(D)} &\le \|e^{-\sigma_i\varphi}\beta_i\chi_i - e^{-\sigma_0\varphi}\beta_i\chi_i\|_{L^2(D)} \\ &\quad + \|e^{-\sigma_0\varphi}\beta_i\chi_i - e^{-\sigma_0\varphi}\beta_0\chi_0\|_{L^2(D)} \\ &\le \sup_{(I,\ell)\in D} |e^{-\sigma_i\varphi(I,\ell)} - e^{-\sigma_0\varphi(I,\ell)}|\, \|\beta_i\chi_i\|_{L^2(D)} \\ &\quad + e^{-\sigma_0\varepsilon_*}\|\beta_i\chi_i - \beta_0\chi_0\|_{L^2(D)}.\end{aligned}$$

So if we furthermore use the bound

$$|e^{-\sigma_i y} - e^{-\sigma_0 y}| \le |1 - e^{-(\sigma_0 - \sigma_i)y}| \le |\sigma_i - \sigma_0|\, e^{|\sigma_0 - \sigma_i|(k^2\Delta_1^2 + \varepsilon_*)}$$

for $y \in [\varepsilon_*, k^2\Delta_1^2 + \varepsilon_*]$, it follows that $\beta_i\, e^{-\sigma_i\varphi}\chi_i \to \beta_0\, e^{-\sigma_0\varphi}\chi_0$ in $L^2(D)$. Therefore $S \subset L^2(D)$ is relatively compact, whence so is $\overline{\mathrm{co}}\, S$, which implies by (C.54) that also $\tilde{B}_k \subset L^2(D)$ is relatively compact. □

Theorem C.8 *Suppose that* $\lambda_* < \delta_1^2$. *Then* λ_* *is an eigenvalue of* L.

Proof Let $\varepsilon_* > 0$ be so small that $\lambda_* + 3\varepsilon_* \le \delta_1^2$ and $\varepsilon_* \le \min\{\frac{1}{4}, \frac{\delta_1^2}{2}\}$ are satisfied. From (C.58) in Lemma C.10, we know that

$$\lambda_* = \inf\{\Phi(g) : g \in X^{00}_{\mathrm{odd}},\, \|g\|_{X^0} = 1\}.$$

Hence for every $j \in \mathbb{N}$ we can fix a function $\psi_j \in X^{00}_{\mathrm{odd}} \subset X^2_{\mathrm{odd}}$ such that $\|\psi_j\|_{X^0} = 1$ and $\Phi(\psi_j) \le \lambda_* + 1/j$. Then if $j \ge 1/\varepsilon_*$ one has $\Phi(\psi_j) \le \lambda_* + 1/j \le \lambda_* + \varepsilon_*$, which we assume for simplicity to hold for $j \ge 1$. Denote by g_j the solution of (C.7) with initial data $g_j(0) = \psi_j$. According to Theorem C.4 (i.e., Corollary C.3), one then has $g_j(t) \in X^2_{\mathrm{odd}}$ and $\|g_j(t)\|_{X^0} = 1$ as well as

$$g_j'(t) = -Lg_j(t) + \Phi(g_j(t))\, g_j(t) \tag{C.55}$$

and

$$\Phi(g_j(t)) - \Phi(g_j(s)) = -2\int_s^t \|g_j'(\tau)\|^2_{X^0}\, d\tau \tag{C.56}$$

for $j \in \mathbb{N}$ and $t, s \in [0, \infty[$, $t \ge s$. Since $\Phi(g_j(\cdot))$ is monotone decreasing and

$$\Phi(g_j(t)) \ge \lambda_* \|g_j(t)\|^2_{X^0} = \lambda_*,$$

the limit $l_j = \lim_{t\to\infty} \Phi(g_j(t))$ does exist. Hence, (C.56) leads to

$$\int_0^\infty \|g_j'(t)\|^2_{X^0}\, d\tau = \frac{1}{2}\,(\Phi(\psi_j) - l_j) < \infty$$

for every $j \in \mathbb{N}$. As $]0, \infty[\ni t \mapsto g_j(t) \in X^2_{\mathrm{odd}}$ is continuous, also the derivative $]0, \infty[\ni t \mapsto g_j'(t) = -Lg_j(t) + \Phi(g_j(t))\, g_j(t) \in X^0_{\mathrm{odd}}$ is continuous. As a consequence, there must be a time $t_j \ge j$ such that

$$\|g_j'(t_j)\|_{X^0} \le \frac{1}{j}. \tag{C.57}$$

Then $t_j \to \infty$ as $j \to \infty$ and

$$\lambda_* \le \Phi(g_j(t_j)) \le \Phi(g_j(0)) = \Phi(\psi_j) \le \lambda_* + \frac{1}{j},$$

which implies that $\lim_{j\to\infty} \Phi(g_j(t_j)) = \lambda_*$; in particular, the sequence $(g_j(t_j)) \subset X^0$ is a minimal sequence for λ_*. Using Lemma C.7, by passing to a subsequence (which is not relabeled), we may furthermore assume that $g_j(t_j) \to \varphi_*$ in X^0 as $j \to \infty$, for some function $\varphi_* \in X^0_{\text{odd}}$. Owing to $\|g_j(t_j)\|_{X^0} = 1$ we also have $\|\varphi_*\|_{X^0} = 1$. Next we take the inner product of (C.55) at $t = t_j$ with an odd function $h \in X^{00}$ to obtain from Corollary B.19 that

$$\begin{aligned} -(g_j(t_j), Lh)_{X^0} &= -(Lg_j(t_j), h)_{X^0} \\ &= (g_j'(t_j), h)_{X^0} - \Phi(g_j(t_j))\,(g_j(t_j), h)_{X^0}. \end{aligned}$$

Recalling (C.57) we can pass to the limit $j \to \infty$ and it follows that

$$(\varphi_*, Lh)_{X^0} = \lambda_*(\varphi_*, h)_{X^0}$$

for any $h \in X^{00}$ that is odd. Then Lemma C.11 implies that $\varphi_* \in X^2_{\text{odd}}$ and $L\varphi_* = \lambda_*\varphi_*$, which completes the proof. □

C.5 Some Technical Lemmas

Lemma C.9 *There is a $C_\Phi > 0$ such that for $g, h \in X^1_{\text{odd}}$ we have*

$$|\Phi(g)-\Phi(h)| \le C_\Phi\Big[(\|g\|_{X^1}+\|h\|_{X^1})\,\|g-h\|_{X^1}+(\|g\|_{X^0}+\|h\|_{X^0})\,\|g-h\|_{X^0}\Big].$$

In particular,

$$|\Phi(g) - \Phi(h)| \le 2C_\Phi(\|g\|_{X^1} + \|h\|_{X^1})\,\|g-h\|_{X^1}.$$

Proof By means of Lemma B.8(a) and (B.42) we estimate

$$\begin{aligned} |\Phi(g) - \Phi(h)| &= \Big| \|\mathcal{T}g\|^2_{X^0} - (\mathcal{K}\mathcal{T}g, g)_{X^0} - \|\mathcal{T}h\|^2_{X^0} + (\mathcal{K}\mathcal{T}h, h)_{X^0} \Big| \\ &\le (\|\mathcal{T}g\|_{X^0} + \|\mathcal{T}h\|_{X^0})\|\mathcal{T}(g-h)\|_{X^0} \\ &\quad + |(\mathcal{K}\mathcal{T}g, g-h)_{X^0}| + |(\mathcal{K}\mathcal{T}(g-h), h)_{X^0}| \\ &\le \Delta_1^2\,(\|g\|_{X^1} + \|h\|_{X^1})\,\|g-h\|_{X^1} \\ &\quad + C_{\mathcal{K}\mathcal{T}}\,(\|g\|_{X^0} + \|h\|_{X^0})\,\|g-h\|_{X^0}. \end{aligned}$$

Thus, we can define $C_\Phi = \Delta_1^2 + C_{\mathcal{K}\mathcal{T}}$. To obtain the second bound one just has to apply Remark B.2(b). □

Lemma C.10 *Let*

$$\hat\lambda = \inf\{\Phi(u) : u \in X^1_{\text{odd}}, \|u\|_{X^0} = 1\}.$$

Then $\hat\lambda = \lambda_*$, and hence $\Phi(u) \ge \lambda_*\|u\|^2_{X^0}$ for $u \in X^1_{\text{odd}}$. Moreover,

$$\lambda_* = \inf\{\Phi(u) : u \in X^{00}_{\mathrm{odd}}, \|u\|_{X^0} = 1\}. \tag{C.58}$$

Proof Let $\varepsilon > 0$. By definition there is a function $u \in X^2_{\mathrm{odd}}$ such that $\|u\|_{X^0} = 1$ and $(Lu, u)_{X^0} \le \lambda_* + \varepsilon$. Using (B.44) from Lemma B.19 we get $\lambda_* + \varepsilon \ge (Lu, u)_{X^0} = \Phi(u) \ge \hat{\lambda}$, so that $\lambda_* \ge \hat{\lambda}$. Conversely, for $\varepsilon > 0$ there is $u \in X^1_{\mathrm{odd}}$ such that $\|u\|_{X^0} = 1$ and $\Phi(u) \le \hat{\lambda} + \varepsilon$. If we write $u = \sum_{k\in\mathbb{Z}} u_k\, e^{ik\theta}$ and define $u^{(N)} = \sum_{|k|\le N} u_k\, e^{ik\theta}$ as in (B.5) from Remark B.2(c), then $u^{(N)} \in X^{00}$ is odd and $\|u^{(N)} - u\|_{X^1} \to 0$ as $N \to \infty$. Since in particular $u^{(N)} \in X^2_{\mathrm{odd}}$, the definition of λ_* and (B.44) imply that $\Phi(u^{(N)}) = (Lu^{(N)}, u^{(N)})_{X^0} \ge \lambda_* \|u^{(N)}\|^2_{X^0}$. Owing to Lemma C.9 we have $\Phi(u^{(N)}) \to \Phi(u)$ as $N \to \infty$. Thus, passing to the limit we infer that $\hat{\lambda} + \varepsilon \ge \Phi(u) \ge \lambda_* \|u\|^2_{X^0} = \lambda_*$ for every $\varepsilon > 0$. To establish (C.58), let $\tilde{\lambda} = \inf\{\Phi(u) : u \in X^{00}_{\mathrm{odd}}, \|u\|_{X^0} = 1\}$. Since $X^{00}_{\mathrm{odd}} \subset X^1_{\mathrm{odd}}$ we have $\tilde{\lambda} \ge \lambda_*$. To verify the converse, let $\varepsilon > 0$. Then there is a function $u \in X^1_{\mathrm{odd}}$ such that $\|u\|_{X^0} = 1$ as well as $\Phi(u) \le \lambda_* + \varepsilon$. Let the associated $u^{(N)} \in X^{00}_{\mathrm{odd}}$ be defined as above. Then $\delta_N = \|u^{(N)} - u\|_{X^1} \to 0$ as $N \to \infty$ and also

$$\|u^{(N)}\|^2_{X^\alpha} = 16\pi^3 \sum_{|k|\le N} (1+k^2)^\alpha \,\|u_k\|^2_{L^2_{\frac{1}{|Q'|}}(D)} \le 16\pi^3 \sum_{k\in\mathbb{Z}} (1+k^2)^\alpha \,\|u_k\|^2_{L^2_{\frac{1}{|Q'|}}(D)} = \|u\|^2_{X^\alpha} \tag{C.59}$$

for $\alpha = 0, 1$. Next let $A > 0$. Then $Au^{(N)} \in X^{00}_{\mathrm{odd}}$, and from Lemma C.9 together with the preceding estimate we get

$$\begin{aligned} |\Phi(Au^{(N)}) - \Phi(u)| &\le |\Phi(Au^{(N)}) - \Phi(u^{(N)})| + |\Phi(u^{(N)}) - \Phi(u)| \\ &\le 2C_\Phi(A+1)|A-1|\|u^{(N)}\|^2_{X^1} \\ &\quad + 2C_\Phi(\|u^{(N)}\|_{X^1} + \|u\|_{X^1})\|u^{(N)} - u\|_{X^1} \\ &\le 2C_\Phi(A+1)|A-1|\, \|u\|^2_{X^1} + 4C_\Phi \|u\|_{X^1}\delta_N. \end{aligned} \tag{C.60}$$

Let

$$\eta = \min\left\{\frac{1}{2}, \frac{\varepsilon}{12C_\Phi\, \|u\|^2_{X^1}}\right\}.$$

Owing to $\|u\|_{X^0} = 1$ and $u^{(N)} \to u$ in X^1 as $N \to \infty$, we may fix an $N \in \mathbb{N}$ large enough such that both conditions

$$\delta_N \le \frac{\varepsilon}{4C_\Phi \|u\|_{X^1}} \quad \text{and} \quad \|u^{(N)}\|_{X^0} \ge 1 - \eta$$

are verified; then $1 - \eta \le \|u^{(N)}\|_{X^0} \le 1$ holds, by (C.59). Now we take $A = \|u^{(N)}\|^{-1}_{X^0}$ to obtain $Au^{(N)} \in X^{00}_{\mathrm{odd}}$ and $\|Au^{(N)}\|_{X^0} = 1$. In addition,

$$|A - 1| = A\,|1 - \|u^{(N)}\|_{X^0}| \le A\eta \le 2\eta,$$

and also $A + 1 \leq 3$. Going back to (C.60), it follows that

$$\begin{aligned}|\Phi(Au^{(N)}) - \Phi(u)| &\leq 2C_\Phi(A+1)|A-1|\,\|u\|_{X^1}^2 + 4C_\Phi\|u\|_{X^1}\delta_N \\ &\leq 12C_\Phi\,\|u\|_{X^1}^2\eta + 4C_\Phi\|u\|_{X^1}\delta_N \\ &\leq 2\varepsilon.\end{aligned}$$

By definition, this shows that

$$\tilde{\lambda} \leq \Phi(Au^{(N)}) \leq \Phi(u) + |\Phi(Au^{(N)}) - \Phi(u)| \leq \Phi(u) + 2\varepsilon \leq \lambda_* + 3\varepsilon$$

is verified for all $\varepsilon > 0$. Hence $\tilde{\lambda} \leq \lambda_*$, and consequently $\tilde{\lambda} = \lambda_*$. □

The next result says that a weak solution to $Lu = \lambda u$ is also a strong solution.

Lemma C.11 *Let $\lambda > 0$ and $u \in X^0_{\text{odd}}$ be such that $(u, Lh)_{X^0} = \lambda(u, h)_{X^0}$ for all $h \in X^{00}$ that are odd. Then $u \in X^2_{\text{odd}}$ and $Lu = \lambda u$.*

Proof By assumption and by Lemma B.15(b), we have

$$(u, -\mathcal{T}^2 h)_{X^0} = \lambda(u, h)_{X^0} + (u, \mathcal{K}\mathcal{T}h)_{X^0} = \lambda(u, h)_{X^0} + (\mathcal{K}\mathcal{T}u, h)_{X^0}.$$

Therefore, (B.21) and (B.42) lead to

$$\begin{aligned}&\left|16\pi^3 \sum_{k\neq 0} \iint_D dI\,d\ell\,\ell\,\frac{1}{|Q'(e)|}\,k^2\omega_1^2\,\overline{u_k}\,h_k\right| \\ &= |(u, -\mathcal{T}^2 h)_{X^0}| \leq \lambda\,|(u, h)_{X^0}| + |(\mathcal{K}\mathcal{T}u, h)_{X^0}| \leq (\lambda + C_{\mathcal{K}\mathcal{T}})\,\|u\|_{X^0}\|h\|_{X^0}\end{aligned} \tag{C.61}$$

for all $h \in X^{00}$ that are odd. For fixed $N \in \mathbb{N}$ we apply this estimate to h given by $h_k = k^2 u_k$ for $0 \leq |k| \leq N$ and $h_k = 0$ for $|k| \geq N + 1$, i.e., $h = \sum_{0<|k|\leq N} k^2 u_k\, e^{ik\theta}$. Then (C.61) implies that

$$\begin{aligned}16\pi^3\delta_1^2 \sum_{0<|k|\leq N} k^4\,\|u_k\|^2_{L^2_{\frac{1}{|Q'|}}(D)} &= 16\pi^3\delta_1^2 \sum_{0<|k|\leq N} \iint_D dI\,d\ell\,\ell\,\frac{1}{|Q'(e)|}\,k^4\,|u_k|^2 \\ &\leq 16\pi^3 \sum_{0<|k|\leq N} \iint_D dI\,d\ell\,\ell\,\frac{1}{|Q'(e)|}\,k^4\omega_1^2\,|u_k|^2 \\ &\leq (\lambda + C_{\mathcal{K}\mathcal{T}})\,\|u\|_{X^0}\|h\|_{X^0}.\end{aligned}$$

On the other hand,

$$\|h\|^2_{X^0} = 16\pi^3 \sum_{k\in\mathbb{Z}} \|h_k\|^2_{L^2_{\frac{1}{|Q'|}}(D)} = 16\pi^3 \sum_{0<|k|\leq N} k^4\,\|u_k\|^2_{L^2_{\frac{1}{|Q'|}}(D)},$$

and this in turn leads to

$$
\begin{aligned}
16\pi^3 \sum_{|k|\le N} (1+k^2)^2 \|u_k\|^2_{L^2_{\frac{1}{|Q'|}}(D)} &= 16\pi^3 \sum_{0<|k|\le N} (1+k^2)^2 \|u_k\|^2_{L^2_{\frac{1}{|Q'|}}(D)} \\
&\le 64\pi^3 \sum_{0<|k|\le N} k^4 \|u_k\|^2_{L^2_{\frac{1}{|Q'|}}(D)} \\
&\le \frac{4}{\delta_1^4} (\lambda + C_{\mathcal{K}\mathcal{T}})^2 \|u\|^2_{X^0}
\end{aligned}
$$

for every $N \in \mathbb{N}$. As $N \to \infty$, it follows that $u \in X^2_{\text{odd}}$. Hence, $(Lu, h)_{X^0} = (u, Lh)_{X^0} = \lambda(u, h)_{X^0}$ for all $h \in X^{00}$ that are odd. Since $X^{00} \subset X^0$ is dense according to Remark B.2(c), we deduce that $Lu = \lambda u$. □

Lemma C.12 *Let $h : J \to X^1_{\text{odd}}$ be continuous and such that $\Phi(h(t)) \le \lambda_* + 2\varepsilon_*$ for $t \in J$. Denote $\mathcal{W}(t,s) = \mathcal{W}(t,s;h)$ and $m_k(t) = m_k(t)(I,\ell) = k^2\omega_1^2(I,\ell) - \Phi(h(t))$. Then*

$$\varepsilon_* k^2 \le m_k(t)(I,\ell) \le \Delta_1^2 k^2, \quad k \in \mathbb{Z} \setminus \{-1,0,1\}, \quad t \in J, \tag{C.62}$$

and

$$-2\varepsilon_* \le m_k(t)(I,\ell) \le \Delta_1^2, \quad k = \pm 1, \quad t \in J. \tag{C.63}$$

In particular,

$$|\mathcal{W}_k(t,s)(I,\ell)| \le \exp(-\varepsilon_* k^2 (t-s)), \quad k \in \mathbb{Z} \setminus \{-1,0,1\}, \quad t,s \in J, \ t \ge s, \tag{C.64}$$

$$|\mathcal{W}_k(t,s)(I,\ell)| \le \exp(2\varepsilon_* (t-s)), \quad k = \pm 1, \quad t,s \in J, \ t \ge s, \tag{C.65}$$

$$\int_\tau^t |\mathcal{W}_k(t,s)(I,\ell)|\, ds \le \frac{1}{\varepsilon_* k^2}, \quad k \in \mathbb{Z} \setminus \{-1,0,1\}, \quad t,\tau \in J, \ t \ge \tau, \tag{C.66}$$

$$\int_\tau^t |\mathcal{W}_k(t,s)(I,\ell)|\, ds \le \frac{1}{2\varepsilon_*} e^{2\varepsilon_*(t-\tau)}, \quad k = \pm 1, \quad t,\tau \in J, \ t \ge \tau, \tag{C.67}$$

$$\|\mathcal{W}(t,s) g\|_{X^\alpha} \le e^{2\varepsilon_*(t-s)} \|g\|_{X^\alpha}, \quad t,s \in J, \ t \ge s, \quad g \in X^\alpha_{\text{odd}}. \tag{C.68}$$

Furthermore, one has

$$
\begin{aligned}
&|\mathcal{W}_k(t+\eta,s)(I,\ell) - \mathcal{W}_k(t,s)(I,\ell)| \\
&\le (\Delta_1^2 + 1) \min\{1, k^2\eta\} \times \begin{cases} \exp(-\varepsilon_* k^2 (t-s)) & : \ k \in \mathbb{Z} \setminus \{-1,0,1\} \\ \exp(2\varepsilon_*(t-s)) & : \ k = \pm 1 \end{cases},
\end{aligned}
\tag{C.69}
$$

and moreover

$$
\begin{aligned}
&|\mathcal{W}_k(t+\eta,s)(I,\ell) - \mathcal{W}_k(t,s)(I,\ell) + \eta\, m_k(t)(I,\ell)\, \mathcal{W}_k(t,s)(I,\ell)| \\
&\le (\Delta_1^2+1)^2 \eta \left(\left| \frac{\exp(-\int_t^{t+\eta} [\Phi(h(t)) - \Phi(h(\tau))]\, d\tau) - 1}{\eta} \right| + k^2 \min\{1, k^2\eta\} \right) \\
&\quad \times \begin{cases} \exp(-\varepsilon_* k^2 (t-s)) & : \ k \in \mathbb{Z} \setminus \{-1,0,1\} \\ \exp(2\varepsilon_*) \exp(2\varepsilon_*(t-s)) & : \ k = \pm 1 \end{cases}
\end{aligned}
\tag{C.70}
$$

for $t+\eta, t, s \in J$, $\eta \in]0,1]$, $t \geq s$.

If we assume that even $\lambda_* + 3\varepsilon_* \leq \delta_1^2$ *holds, then (C.62), (C.63) and (C.64), (C.65) can be sharpened to*

$$\varepsilon_* k^2 \leq m_k(t)(I,\ell) \leq \Delta_1^2 k^2, \quad k \in \mathbb{Z} \setminus \{0\}, \quad t \in J, \tag{C.71}$$

and

$$|\mathcal{W}_k(t,s)(I,\ell)| \leq \exp(-\varepsilon_* k^2 (t-s)), \quad k \in \mathbb{Z} \setminus \{0\}, \quad t,s \in J, \ t \geq s. \tag{C.72}$$

Proof Since $\delta_1 = \inf \omega_1$ and $\lambda_* \leq \delta_1^2$ by Lemma 3.18, we have for $|k| \geq 2$

$$m_k(t)(I,\ell) \geq k^2\delta_1^2 - \lambda_* - 2\varepsilon_* \geq (k^2-1)\delta_1^2 - 2\varepsilon_* \geq \frac{1}{2}(k^2+2)\delta_1^2 - 2\varepsilon_* \geq \varepsilon_* k^2,$$

which yields the lower bound in (C.62). If $|k| = 1$, then

$$m_k(t)(I,\ell) \geq \delta_1^2 - \lambda_* - 2\varepsilon_* \geq -2\varepsilon_*$$

is the best we can get without assuming that $\lambda_* < \delta_1^2$. If, however, $\lambda_* + 3\varepsilon_* \leq \delta_1^2$ is verified as is the case for (C.71) and (C.72), then $|k| = 1$ entails

$$m_k(t)(I,\ell) \geq \delta_1^2 - \lambda_* - 2\varepsilon_* \geq \varepsilon_*$$

also in this case. Next, using Lemma C.10, we have in particular that $\Phi(h(t)) \geq 0$. Therefore, also $m_k(t)(I,\ell) \leq \Delta_1^2 k^2$ by Theorem 3.5 for all $k \in \mathbb{Z} \setminus \{0\}$. As $\mathcal{W}_k(t,s)(I,\ell) = \exp(-\int_s^t m_k(\tau)(I,\ell)\,d\tau)$, (C.64)-(C.67) are a direct consequence of (C.62) and (C.63), and also (C.71) and (C.72) are verified. Concerning (C.68), we have $|\mathcal{W}_k(t,s)(I,\ell)| \leq \exp(2\varepsilon_*(t-s))$ for all $k \in \mathbb{Z} \setminus \{0\}$. It follows that

$$\begin{aligned}
\|\mathcal{W}(t,s)g\|_{X^\alpha}^2 &= 16\pi^3 \sum_{k\in\mathbb{Z}} (1+k^2)^\alpha \, \|(\mathcal{W}(t,s)g)_k\|^2_{L^2_{\frac{1}{|Q'|}}(D)} \\
&= 16\pi^3 \sum_{k\in\mathbb{Z}} (1+k^2)^\alpha \, \|\mathcal{W}_k(t,s)g_k\|^2_{L^2_{\frac{1}{|Q'|}}(D)} \\
&\leq 16\pi^3 e^{4\varepsilon_*(t-s)} \sum_{k\in\mathbb{Z}} (1+k^2)^\alpha \, \|g_k\|^2_{L^2_{\frac{1}{|Q'|}}(D)} \\
&= e^{4\varepsilon_*(t-s)} \|g\|^2_{X^\alpha}.
\end{aligned}$$

To establish (C.69), one has

$$\begin{aligned}
&|\mathcal{W}_k(t+\eta,s)(I,\ell) - \mathcal{W}_k(t,s)(I,\ell)| \\
&= \left| \exp\left(-\int_s^{t+\eta} m_k(\tau)(I,\ell)\,d\tau\right) - \exp\left(-\int_s^t m_k(\tau)(I,\ell)\,d\tau\right) \right|
\end{aligned}$$

$$
\begin{aligned}
&= \left[1 - \exp\left(-\int_t^{t+\eta} m_k(\tau)(I,\ell)\,d\tau\right)\right]\exp\left(-\int_s^t m_k(\tau)(I,\ell)\,d\tau\right)\\
&\le \min\left\{1, \int_t^{t+\eta} m_k(\tau)(I,\ell)\,d\tau\right\} \times \begin{cases} \exp(-\varepsilon_* k^2(t-s)) & : \ k \in \mathbb{Z}\setminus\{-1,0,1\} \\ \exp(2\varepsilon_*(t-s)) & : \ k = \pm 1 \end{cases}\\
&\le (\Delta_1^2 + 1)\min\{1, k^2\eta\} \times \begin{cases} \exp(-\varepsilon_* k^2(t-s)) & : \ k \in \mathbb{Z}\setminus\{-1,0,1\} \\ \exp(2\varepsilon_*(t-s)) & : \ k = \pm 1 \end{cases},
\end{aligned}
$$

where we used the bound $1 - e^{-x} \le \min\{1, x\}$ for $x \ge 0$ together with (C.62) and (C.63). Finally, by definition,

$$
\begin{aligned}
&\mathcal{W}_k(t+\eta, s)(I,\ell) - \mathcal{W}_k(t,s)(I,\ell) + \eta\, m_k(t)(I,\ell)\,\mathcal{W}_k(t,s)(I,\ell)\\
&= \exp\left(-\int_s^{t+\eta} m_k(\tau)(I,\ell)\,d\tau\right) - \exp\left(-\int_s^{t} m_k(\tau)(I,\ell)\,d\tau\right)\\
&\quad + \eta\, m_k(t)(I,\ell)\exp\left(-\int_s^t m_k(\tau)(I,\ell)\,d\tau\right)\\
&= \left[\exp\left(-\int_t^{t+\eta} m_k(\tau)(I,\ell)\,d\tau\right) - 1 + \eta\, m_k(t)(I,\ell)\right]\exp\left(-\int_s^t m_k(\tau)(I,\ell)\,d\tau\right)\\
&= \eta\left[\frac{\exp\left(-\int_t^{t+\eta}[m_k(\tau)(I,\ell) - m_k(t)(I,\ell)]\,d\tau\right)\exp(-\eta\, m_k(t)(I,\ell)) - 1}{\eta}\right.\\
&\qquad \left. + m_k(t)(I,\ell)\right]\mathcal{W}_k(t,s)(I,\ell)\\
&= \eta\,\frac{\exp\left(-\int_t^{t+\eta}[\Phi(h(t)) - \Phi(h(\tau))]\,d\tau\right) - 1}{\eta}\exp(-\eta\, m_k(t)(I,\ell))\,\mathcal{W}_k(t,s)(I,\ell)\\
&\quad + \eta\left[\frac{\exp(-\eta\, m_k(t)(I,\ell)) - 1 + \eta\, m_k(t)(I,\ell)}{\eta}\right]\mathcal{W}_k(t,s)(I,\ell).
\end{aligned}
$$

Thus owing to (C.62)–(C.65):

$$
\begin{aligned}
&|\mathcal{W}_k(t+\eta, s)(I,\ell) - \mathcal{W}_k(t,s)(I,\ell) + \eta\, m_k(t)(I,\ell)\,\mathcal{W}_k(t,s)(I,\ell)|\\
&\le \eta\left|\frac{\exp(-\int_t^{t+\eta}[\Phi(h(t)) - \Phi(h(\tau))]\,d\tau) - 1}{\eta}\right|\\
&\qquad \times \begin{cases} \exp(-\varepsilon_* k^2\eta)\exp(-\varepsilon_* k^2(t-s)) & : \ k \in \mathbb{Z}\setminus\{-1,0,1\} \\ \exp(2\varepsilon_*\eta)\exp(2\varepsilon_*(t-s)) & : \ k = \pm 1 \end{cases}\\
&\quad + \eta\, m_k(t)(I,\ell)\, f(\eta\, m_k(t)(I,\ell)) \times \begin{cases} \exp(-\varepsilon_* k^2(t-s)) & : \ k \in \mathbb{Z}\setminus\{-1,0,1\} \\ \exp(2\varepsilon_*(t-s)) & : \ k = \pm 1 \end{cases},
\end{aligned}
\tag{C.73}
$$

where $f(x) = \frac{e^{-x}-1+x}{x}$ for $x \ge 0$. Now $f(x) = x\int_0^1 (1-\tau)e^{-\tau x}\,d\tau$ shows that always $f(x) \le x\int_0^1 (1-\tau)\,d\tau = x/2$. On the other hand, $f'(x) \ge 0$ and $\lim_{x\to\infty} f(x) = 1$ also yields $f(x) \le 1$, so that $0 \le f(x) \le \min\{1, x\}$. Since $m_k(t)(I,\ell) \le \Delta_1^2 k^2$ in all cases, it follows from (C.73) that

$$
\begin{aligned}
&|\mathcal{W}_k(t+\eta,s)(I,\ell)-\mathcal{W}_k(t,s)(I,\ell)+\eta\, m_k(t)(I,\ell)\,\mathcal{W}_k(t,s)(I,\ell)|\\
&\quad\le \eta\left|\frac{\exp(-\int_t^{t+\eta}[\Phi(h(t))-\Phi(h(\tau))]\,d\tau)-1}{\eta}\right|\\
&\qquad\times\begin{cases}\exp(-\varepsilon_* k^2(t-s)) & : \ k\in\mathbb{Z}\setminus\{-1,0,1\}\\ \exp(2\varepsilon_*)\,\exp(2\varepsilon_*(t-s)) & : \ k=\pm 1\end{cases}\\
&\quad+\Delta_1^2\,(\Delta_1^2+1)\,k^2\eta\,\min\{1,k^2\eta\}\times\begin{cases}\exp(-\varepsilon_* k^2(t-s)) & : \ k\in\mathbb{Z}\setminus\{-1,0,1\}\\ \exp(2\varepsilon_*(t-s)) & : \ k=\pm 1\end{cases},
\end{aligned}
$$

which implies (C.70). □

Lemma C.13 *Let $t_1, t_2, t_3 \in J$ be such that $t_1 \le t_2 \le t_3$. In addition, let $g = g(s,\lambda)$ denote an X^0_{odd}-valued continuous function depending upon a parameter λ (which itself is also allowed to depend upon t_1, t_2, t_3). Then for $k \in \mathbb{Z}\setminus\{0\}$:*

$$
\begin{aligned}
&\iint_D dI\,d\ell\,\ell\,\frac{1}{|Q'(e)|}\left[\int_{t_1}^{t_2}\exp(-\varepsilon_* k^2(t_3-s))\,|g_k(s,\lambda)(I,\ell)|\,ds\right]^2\\
&\quad\le\frac{1}{\varepsilon_* k^2}\,e^{\varepsilon_* k^2(t_2-2t_3)}\,\min\{1,\varepsilon_* k^2(t_2-t_1)\}\int_{t_1}^{t_2}ds\,e^{\varepsilon_* k^2 s}\,\|g_k(s,\lambda)\|^2_{L^2_{\frac{1}{|Q'|}}(D)}.
\end{aligned}
$$

Similarly, if $\phi = \phi(s,\lambda) \in L^2_{\frac{1}{|Q'|}}(D)$, then

$$
\begin{aligned}
&\iint_D dI\,d\ell\,\ell\,\frac{1}{|Q'(e)|}\left[\int_{t_1}^{t_2}\exp(2\varepsilon_*(t_3-s))\,|\phi(s,\lambda)(I,\ell)|\,ds\right]^2\\
&\quad\le\frac{1}{4\varepsilon_*}\,e^{4\varepsilon_* t_3}\,(e^{-4\varepsilon_* t_1}-e^{-4\varepsilon_* t_2})\int_{t_1}^{t_2}\|\phi(s,\lambda)\|^2_{L^2_{\frac{1}{|Q'|}}(D)}\,ds.
\end{aligned}
\tag{C.74}
$$

Proof Let $A = A(t_1,t_2,t_3,k,\lambda)$ denote the expression on the left-hand side. First we apply Minkowski's inequality [87, p. 271]. This yields

$$
\begin{aligned}
A&\le\left[\int_{t_1}^{t_2}ds\left(\iint_D dI\,d\ell\,\ell\,\frac{1}{|Q'(e)|}\,\exp(-2\varepsilon_* k^2(t_3-s))\,|g_k(s,\lambda)(I,\ell)|^2\right)^{1/2}\right]^2\\
&=e^{-2\varepsilon_* k^2 t_3}\left[\int_{t_1}^{t_2}ds\,e^{\varepsilon_* k^2 s}\,\|g_k(s,\lambda)\|_{L^2_{\frac{1}{|Q'|}}(D)}\right]^2.
\end{aligned}
$$

Next we will make use of Jensen's inequality for the convex function $f(x) = x^2$ and the probability measure $d\mu = \sigma\mathbf{1}_{[t_1,t_2]}(s)\,e^{\varepsilon_* k^2 s}\,ds$, where $\sigma = \varepsilon_* k^2(e^{\varepsilon_* k^2 t_2} - e^{\varepsilon_* k^2 t_1})^{-1}$. It follows that

$$
\begin{aligned}
A &= e^{-2\varepsilon_* k^2 t_3}\,\sigma^{-2}\Big[\int d\mu\,\|g_k(s,\lambda)\|_{L^2_{\frac{1}{|Q'|}}(D)}\Big]^2\\
&\le e^{-2\varepsilon_* k^2 t_3}\,\sigma^{-2}\int d\mu\,\|g_k(s,\lambda)\|^2_{L^2_{\frac{1}{|Q'|}}(D)}\\
&= \frac{1}{\varepsilon_* k^2}\,e^{-2\varepsilon_* k^2 t_3}\,(e^{\varepsilon_* k^2 t_2}-e^{\varepsilon_* k^2 t_1})\int_{t_1}^{t_2} ds\,e^{\varepsilon_* k^2 s}\,\|g_k(s,\lambda)\|^2_{L^2_{\frac{1}{|Q'|}}(D)}\\
&\le \frac{1}{\varepsilon_* k^2}\,e^{\varepsilon_* k^2 (t_2-2t_3)}\,\min\{1,\varepsilon_* k^2(t_2-t_1)\}\int_{t_1}^{t_2} ds\,e^{\varepsilon_* k^2 s}\,\|g_k(s,\lambda)\|^2_{L^2_{\frac{1}{|Q'|}}(D)},
\end{aligned}
$$

using the estimate $1-e^{-x}\le x$ for $x\ge 0$ for the last step. To establish (C.74) we proceed as above to get

$$
A \le e^{4\varepsilon_* t_3}\Big[\int_{t_1}^{t_2} ds\,e^{-2\varepsilon_* s}\,\|\phi(s,\lambda)\|_{L^2_{\frac{1}{|Q'|}}(D)}\Big]^2,
$$

where A stands for the left-hand side in (C.74). It remains to apply the Hölder inequality in s. □

The following elementary observation will be used at several places.

Remark C.14 Let $\gamma>0$, $\varepsilon_*>0$, and $t\ge s\ge 0$. The function $f(x)=x^\gamma e^{-\varepsilon_*(t-s)x}$ for $x>0$ attains its maximal value

$$
\Big(\frac{\gamma}{\varepsilon_*(t-s)}\Big)^\gamma e^{-\gamma}\le\Big(\frac{\gamma}{\varepsilon_*(t-s)}\Big)^\gamma
$$

at $x=\frac{\gamma}{\varepsilon_*(t_0-s)}$. ◇

Lemma C.15 *Let $h: J\to X^1_{\mathrm{odd}}$ be continuous and such that $\Phi(h(t))\le\lambda_*+2\varepsilon_*$ for $t\in J$. Denote $\mathcal{W}(t,s)=\mathcal{W}(t,s;h)$.*

(a) If $\alpha\in[1,2[$, then

$$
\Big\|\int_0^t \mathcal{W}(t,s)g(s)\,ds\Big\|^2_{X^\alpha}\le\frac{2^\alpha}{\varepsilon_*^\alpha}\int_0^t\Big[\frac{1}{(t-s)^{\alpha-1}}+e^{4\varepsilon_* t}\Big]\|g(s)\|^2_{X^0}\,ds,\quad\text{(C.75)}
$$

(b) if $\tau\in J$ and $\|g(t)-g(s)\|_{X^0}\le A(t-s)^\gamma$ for $t,s\in J$ such that $t\ge s\ge\tau$ and $\alpha<2(\gamma+1)$, then

$$
\begin{aligned}
\Big\|\int_\tau^t \mathcal{W}(t,s)[g(t)-g(s)]\,ds\Big\|^2_{X^\alpha}&\le\frac{2^\alpha A^2}{(2(\gamma+1)-\alpha)\varepsilon_*^\alpha}(t-\tau)^{2(\gamma+1)-\alpha}\\
&\quad+\frac{2^\alpha A^2}{4(2\gamma+1)\varepsilon_*}e^{4\varepsilon_*(t-\tau)}(t-\tau)^{2\gamma+1},
\end{aligned}
\quad\text{(C.76)}
$$

(c) in addition one has for $0 \le \tau \le t$

$$\left\| \int_\tau^t \mathcal{W}(t,s)g(t)\,ds \right\|_{X^2} \le \frac{1}{\varepsilon_*}\left(2 + e^{2\varepsilon_*(t-\tau)}\right) \|g(t)\|_{X^0}, \tag{C.77}$$

(d) if $h > 0$ *and* $t + h \in J$*, then*

$$\left\| \int_t^{t+h} \mathcal{W}(t+h,s)g(s)\,ds \right\|_{X^1}^2 \le \frac{2}{\varepsilon_*}\left(1 + e^{4\varepsilon_* h}\right) \int_t^{t+h} \|g(s)\|_{X^0}^2\,ds, \tag{C.78}$$

(e) if $h, \delta > 0$, $t + h, t - \delta \in J$*, then*

$$\left\| \int_{t-\delta}^{t} \mathcal{W}(t+h,s)g(s)\,ds \right\|_{X^1}^2 \le \frac{2}{\varepsilon_*}\left(1 + e^{4\varepsilon_*(h+\delta)}\right) \int_{t-\delta}^{t} \|g(s)\|_{X^0}^2\,ds, \tag{C.79}$$

(f) if $\delta > 0$ *and* $t - \delta \in J$*, then*

$$\left\| \int_{t-\delta}^{t} \mathcal{W}(t,s)g(s)\,ds \right\|_{X^1}^2 \le \frac{2}{\varepsilon_*}\left(1 + e^{4\varepsilon_*\delta}\right) \int_{t-\delta}^{t} \|g(s)\|_{X^0}^2\,ds, \tag{C.80}$$

(g) and moreover

$$\left\| \int_0^{t/2} \mathcal{W}(t,s)g(s)\,ds \right\|_{X^2} \le \frac{4}{\varepsilon_*}\left(\frac{1}{\sqrt{t}} + e^{2\varepsilon_* t}\right)\left(\int_0^{t/2} \|g(s)\|_{X^0}^2\,ds\right)^{1/2}, \tag{C.81}$$

for $t \in J$ *and* $g : J \to X^0_{\rm odd}$ *continuous.*

Proof We start out somewhat more generally and fix $t_1, t_2, t_3 \in J$ such that $t_1 \le t_2 \le t_3$. Let $g = g(s, \lambda)$ denote an $X^0_{\rm odd}$-valued continuous function depending upon a parameter λ (which itself is also allowed to depend upon t_1, t_2, t_3). Then by (C.62) and (C.63):

$$\begin{aligned}
&\left\| \int_{t_1}^{t_2} \mathcal{W}(t_3,s)g(s,\lambda)\,ds \right\|_{X^\alpha}^2 \\
&= 16\pi^3 \sum_{k\in\mathbb{Z}} (1+k^2)^\alpha \left\| \int_{t_1}^{t_2} \mathcal{W}_k(t_3,s)g_k(s,\lambda)\,ds \right\|_{L^2_{\frac{1}{|Q'|}}(D)}^2 \\
&= 16\pi^3 \sum_{k\in\mathbb{Z}} (1+k^2)^\alpha \iint_D dI\,d\ell\,\ell\,\frac{1}{|Q'(e)|} \left| \int_{t_1}^{t_2} \mathcal{W}_k(t_3,s)(I,\ell)\,g_k(s,\lambda)(I,\ell)\,ds \right|^2 \\
&\le 32\pi^3\, 2^\alpha \sum_{k=2}^{\infty} k^{2\alpha} \iint_D dI\,d\ell\,\ell\,\frac{1}{|Q'(e)|} \left[\int_{t_1}^{t_2} \exp(-\varepsilon_* k^2(t_3-s))\,|g_k(s,\lambda)(I,\ell)|\,ds \right]^2 \\
&\quad + 32\pi^3\, 2^\alpha \iint_D dI\,d\ell\,\ell\,\frac{1}{|Q'(e)|} \left[\int_{t_1}^{t_2} \exp(2\varepsilon_*(t_3-s))\,|g_1(s,\lambda)(I,\ell)|\,ds \right]^2;
\end{aligned}$$

recall that $g_{-k} = -g_k$. Hence, from Lemma C.13, we deduce that

$$
\begin{aligned}
\left\| \int_{t_1}^{t_2} \mathcal{W}(t_3, s) g(s, \lambda)\, ds \right\|_{X^\alpha}^2 \leq\ & \frac{32\pi^3 2^\alpha}{\varepsilon_*} \sum_{k=2}^{\infty} k^{2\alpha-2} e^{\varepsilon_* k^2 (t_2 - 2t_3)} \min\{1, \varepsilon_* k^2 (t_2 - t_1)\} \\
& \times \int_{t_1}^{t_2} ds\, e^{\varepsilon_* k^2 s} \|g_k(s, \lambda)\|_{L^2_{\frac{1}{|Q'|}}(D)}^2 \\
& + \frac{32\pi^3 2^\alpha}{4\varepsilon_*} e^{4\varepsilon_* t_3} (e^{-4\varepsilon_* t_1} - e^{-4\varepsilon_* t_2}) \int_{t_1}^{t_2} \|g_1(s, \lambda)\|_{L^2_{\frac{1}{|Q'|}}(D)}^2 ds.
\end{aligned} \tag{C.82}
$$

Now to verify (C.75), we take $t_1 = \tau$, $t_2 = t_3 = t$, $g(s, \lambda) = g(s, t)$, and $\alpha \in [1, 2[$. Hence,

$$
\begin{aligned}
\left\| \int_{\tau}^{t} \mathcal{W}(t, s) g(s, t)\, ds \right\|_{X^\alpha}^2 \leq\ & \frac{32\pi^3 2^\alpha}{\varepsilon_*} \sum_{k=2}^{\infty} k^{2\alpha-2} \int_{\tau}^{t} ds\, e^{-\varepsilon_* k^2 (t-s)} \|g_k(s, t)\|_{L^2_{\frac{1}{|Q'|}}(D)}^2 \\
& + \frac{32\pi^3 2^\alpha}{4\varepsilon_*} e^{4\varepsilon_* t} (e^{-4\varepsilon_* \tau} - e^{-4\varepsilon_* t}) \int_{\tau}^{t} \|g_1(s, t)\|_{L^2_{\frac{1}{|Q'|}}(D)}^2 ds.
\end{aligned}
$$

From Remark C.14 one has $k^{2\alpha-2} e^{-\varepsilon_* k^2 (t-s)} \leq (\frac{1}{\varepsilon_*(t-s)})^{\alpha-1}$. As a consequence,

$$
\begin{aligned}
\left\| \int_{\tau}^{t} \mathcal{W}(t, s) g(s, t)\, ds \right\|_{X^\alpha}^2 \leq\ & \frac{32\pi^3 2^\alpha}{\varepsilon_*^\alpha} \sum_{k=2}^{\infty} \int_{\tau}^{t} ds\, \frac{1}{(t-s)^{\alpha-1}} \|g_k(s, t)\|_{L^2_{\frac{1}{|Q'|}}(D)}^2 \\
& + \frac{32\pi^3 2^\alpha}{4\varepsilon_*} e^{4\varepsilon_* t} (e^{-4\varepsilon_* \tau} - e^{-4\varepsilon_* t}) \int_{\tau}^{t} \|g_1(s, t)\|_{L^2_{\frac{1}{|Q'|}}(D)}^2 ds \\
\leq\ & \frac{2^\alpha}{\varepsilon_*^\alpha} \int_{\tau}^{t} ds\, \frac{1}{(t-s)^{\alpha-1}} \|g(s, t)\|_{X^0}^2 \\
& + \frac{2^\alpha}{4\varepsilon_*} e^{4\varepsilon_* t} (e^{-4\varepsilon_* \tau} - e^{-4\varepsilon_* t}) \int_{\tau}^{t} \|g(s, t)\|_{X^0}^2 ds.
\end{aligned} \tag{C.83}
$$

Thus if we take $\tau = 0$ and $g(s, t) = g(s)$, then we obtain (C.75) for $\alpha \in [1, 2[$. To prove (C.76), let $g(s, t) = g(t) - g(s)$. Then by assumption

$$
\|g(s, t)\|_{X^0} = \|g(t) - g(s)\|_{X^0} \leq A(t-s)^\gamma, \quad t \geq s \geq \tau,
$$

and accordingly (C.83) leads to

$$
\begin{aligned}
\left\| \int_{\tau}^{t} \mathcal{W}(t, s) g(s, t)\, ds \right\|_{X^\alpha}^2 \leq\ & \frac{2^\alpha A^2}{\varepsilon_*^\alpha} \int_{\tau}^{t} (t-s)^{2\gamma-\alpha+1} ds \\
& + \frac{2^\alpha A^2}{4\varepsilon_*} e^{4\varepsilon_* t} (e^{-4\varepsilon_* \tau} - e^{-4\varepsilon_* t}) \int_{\tau}^{t} (t-s)^{2\gamma} ds
\end{aligned}
$$

$$\leq \frac{2^\alpha A^2}{(2(\gamma+1)-\alpha)\varepsilon_*^\alpha}(t-\tau)^{2(\gamma+1)-\alpha} + \frac{2^\alpha A^2}{4(2\gamma+1)\varepsilon_*} e^{4\varepsilon_*(t-\tau)}(t-\tau)^{2\gamma+1}.$$

Next we turn to (C.77), where the argument is similar, but more direct. Here by (C.66) and (C.67):

$$\begin{aligned}
&\left\| \int_\tau^t \mathcal{W}(t,s)g(t)\,ds \right\|_{X^2}^2 \\
&= 16\pi^3 \sum_{k\in\mathbb{Z}} (1+k^2)^2 \iint_D dI\,d\ell\,\ell\,\frac{1}{|Q'(e)|}\,|g_k(t)(I,\ell)|^2 \left| \int_\tau^t \mathcal{W}_k(t,s)(I,\ell)\,ds \right|^2 \\
&\leq 32\pi^3 \Bigg(\sum_{k=2}^\infty (1+k^2)^2 \iint_D dI\,d\ell\,\ell\,\frac{1}{|Q'(e)|}\,|g_k(t)(I,\ell)|^2\,\frac{1}{\varepsilon_*^2 k^4} \\
&\qquad + 4\iint_D dI\,d\ell\,\ell\,\frac{1}{|Q'(e)|}\,|g_1(t)(I,\ell)|^2\,\frac{1}{4\varepsilon_*^2}\,e^{4\varepsilon_*(t-\tau)} \Bigg) \\
&\leq \frac{1}{\varepsilon_*^2}\left(4+e^{4\varepsilon_*(t-\tau)}\right)\|g(t)\|_{X^0}^2.
\end{aligned}$$

What concerns (C.78), due to (C.82) with $t_1 = t$, $t_2 = t+h$, $t_3 = t+h$, $g(s,\lambda) = g(s)$, and $\alpha = 1$ it follows that

$$\begin{aligned}
\left\| \int_t^{t+h} \mathcal{W}(t+h,s)g(s)\,ds \right\|_{X^1}^2 &\leq \frac{64\pi^3}{\varepsilon_*} \sum_{k=2}^\infty \int_t^{t+h} ds\, e^{-\varepsilon_* k^2 (t+h-s)}\,\|g_k(s)\|^2_{L^2_{\frac{1}{|Q'|}}(D)} \\
&\quad + \frac{64\pi^3}{4\varepsilon_*} e^{4\varepsilon_*(t+h)}\left(e^{-4\varepsilon_* t} - e^{-4\varepsilon_*(t+h)}\right) \int_t^{t+h} \|g_1(s)\|^2_{L^2_{\frac{1}{|Q'|}}(D)}\,ds \\
&\leq \frac{64\pi^3}{\varepsilon_*} \sum_{k=2}^\infty \int_t^{t+h} ds\,\|g_k(s)\|^2_{L^2_{\frac{1}{|Q'|}}(D)} \\
&\quad + \frac{64\pi^3}{4\varepsilon_*} e^{4\varepsilon_* h} \int_t^{t+h} \|g_1(s)\|^2_{L^2_{\frac{1}{|Q'|}}(D)}\,ds \\
&\leq \frac{2}{\varepsilon_*}\left(1+e^{4\varepsilon_* h}\right) \int_t^{t+h} \|g(s)\|^2_{X^0}\,ds.
\end{aligned}$$

To establish (C.79), one takes $t_1 = t-\delta$, $t_2 = t$, $t_3 = t+h$, $g(s,\lambda) = g(s)$, and $\alpha = 1$ in (C.82). This yields

$$
\begin{aligned}
\left\| \int_{t-\delta}^{t} \mathcal{W}(t+h,s)g(s)\,ds \right\|_{X^1}^2 &\le \frac{64\pi^3}{\varepsilon_*} \sum_{k=2}^{\infty} \int_{t-\delta}^{t} ds\, e^{-\varepsilon_* k^2 (t+2h-s)} \|g_k(s)\|^2_{L^2_{\frac{1}{|Q'|}}(D)} \\
&\quad + \frac{64\pi^3}{4\varepsilon_*} e^{4\varepsilon_* h} (e^{4\varepsilon_* \delta} - 1) \int_{t-\delta}^{t} \|g_1(s)\|^2_{L^2_{\frac{1}{|Q'|}}(D)}\, ds \\
&\le \frac{2}{\varepsilon_*} (1 + e^{4\varepsilon_*(h+\delta)}) \int_{t-\delta}^{t} \|g(s)\|^2_{X^0}\, ds.
\end{aligned}
$$

To show (C.80), we specialize (C.82) to $t_1 = t - \delta$, $t_2 = t_3 = t$, $g(s, \lambda) = g(s)$, and $\alpha = 1$. In this way, we obtain

$$
\begin{aligned}
\left\| \int_{t-\delta}^{t} \mathcal{W}(t,s)g(s)\,ds \right\|_{X^1}^2 &\le \frac{64\pi^3}{\varepsilon_*} \sum_{k=2}^{\infty} \int_{t-\delta}^{t} ds\, e^{-\varepsilon_* k^2 (t-s)} \|g_k(s)\|^2_{L^2_{\frac{1}{|Q'|}}(D)} \\
&\quad + \frac{64\pi^3}{4\varepsilon_*} (e^{4\varepsilon_* \delta} - 1) \int_{t-\delta}^{t} \|g_1(s)\|^2_{L^2_{\frac{1}{|Q'|}}(D)}\, ds \\
&\le \frac{2}{\varepsilon_*} (1 + e^{4\varepsilon_*\delta}) \int_{t-\delta}^{t} \|g(s)\|^2_{X^0}\, ds.
\end{aligned}
$$

Finally to verify (C.81), we use (C.82) with $t_1 = 0$, $t_2 = t/2$, $t_3 = t$, $g(s, \lambda) = g(s)$, and $\alpha = 2$. It follows that

$$
\begin{aligned}
\left\| \int_{0}^{t/2} \mathcal{W}(t,s)g(s)\,ds \right\|_{X^2}^2 &\le \frac{128\pi^3}{\varepsilon_*} \sum_{k=2}^{\infty} k^2\, e^{-\varepsilon_* k^2 t/2} \int_{0}^{t/2} ds\, e^{-\varepsilon_* k^2 (t-s)} \|g_k(s)\|^2_{L^2_{\frac{1}{|Q'|}}(D)} \\
&\quad + \frac{128\pi^3}{4\varepsilon_*} e^{4\varepsilon_* t} (1 - e^{-4\varepsilon_* t/2}) \int_{0}^{t/2} \|g_1(s)\|^2_{L^2_{\frac{1}{|Q'|}}(D)}\, ds.
\end{aligned}
$$

From Remark C.14 one has $k^2 e^{-\varepsilon_* k^2 t/2} \le \frac{2}{\varepsilon_* t}$. Hence, the claim follows and the proof is complete. □

Lemma C.16 *Let $h : J \to X^1_{\text{odd}}$ be continuous and such that $\Phi(h(t)) \le \lambda_* + 2\varepsilon_*$ for $t \in J$. Denote $\mathcal{W}(t,s) = \mathcal{W}(t,s;h)$. Then*

$$
\|(\mathcal{W}(t,0) - \mathcal{W}(0,0))g\|_{X^1} \le (\Delta_1^2 + 1)\,(t + t^2)^{1/2}\, \|g\|_{X^2} \tag{C.84}
$$

for $t \in J$ and $g \in X^2_{\text{odd}}$. In addition,

$$
\|(\mathcal{W}(t+h,s) - \mathcal{W}(t,s))g\|_{X^1} \le 2\,(\Delta_1^2 + 1) \left[\frac{1}{\varepsilon_*(t-s)} \sqrt{h} + \exp(2\varepsilon_*(t-s))\, h \right] \|g\|_{X^0} \tag{C.85}
$$

and

$$
\|(\mathcal{W}(t+h,s) - \mathcal{W}(t,s))g\|_{X^2} \le 2\,(\Delta_1^2 + 1) \left[\frac{2}{\varepsilon_*^{3/2}(t-s)^{3/2}} \sqrt{h} + \exp(2\varepsilon_*(t-s))\, h \right] \|g\|_{X^0} \tag{C.86}
$$

for $t > s$, $h > 0$, $s, t, t+h \in J$, *and* $g \in X^0_{\mathrm{odd}}$.

Proof By definition,

$$
\begin{aligned}
&\|(\mathcal{W}(t,s) - \mathcal{W}(\tau,s))g\|^2_{X^\alpha} \\
&= 16\pi^3 \sum_{k\in\mathbb{Z}} (1+k^2)^\alpha \, \|(\mathcal{W}_k(t,s) - \mathcal{W}_k(\tau,s))g_k\|^2_{L^2_{\frac{1}{|Q'|}}(D)} \\
&= 16\pi^3 \sum_{k\in\mathbb{Z}} (1+k^2)^\alpha \iint_D dI\,d\ell\,\ell \, \frac{1}{|Q'(e)|} \left| (\mathcal{W}_k(t,s)(I,\ell) - \mathcal{W}_k(\tau,s)(I,\ell))\, g_k(I,\ell) \right|^2 .
\end{aligned}
\tag{C.87}
$$

Write $m_k(\sigma) = m_k(\sigma)(I,\ell) = k^2\omega_1^2(I,\ell) - \Phi(h(\sigma))$ as before. Then by (C.62) for $|k| \geq 2$:

$$
\begin{aligned}
|\mathcal{W}_k(t,s)(I,\ell) - \mathcal{W}_k(\tau,s)(I,\ell)| &= \exp\left(- \int_s^\tau m_k(\sigma)\,d\sigma \right) \left| \exp\left(- \int_\tau^t m_k(\sigma)\,d\sigma \right) - 1 \right| \\
&\leq \exp(-\varepsilon_* k^2(\tau - s)) \left[1 - \exp\left(- \int_\tau^t m_k(\sigma)\,d\sigma \right) \right],
\end{aligned}
$$

where we have dropped the arguments (I, ℓ) on the right-hand side. Similarly, if $k = \pm 1$, then (C.63) yields

$$
|\mathcal{W}_k(t,s)(I,\ell) - \mathcal{W}_k(\tau,s)(I,\ell)| \leq \exp(2\varepsilon_*(\tau - s)) \left[1 - \exp\left(- \int_\tau^t m_k(\sigma)\,d\sigma \right) \right].
$$

Since always $m_k(\sigma) \leq \Delta_1^2 k^2$, we obtain the bound

$$
\begin{aligned}
|\mathcal{W}_k(t,s)(I,\ell) - \mathcal{W}_k(\tau,s)(I,\ell)| &\leq \exp(-\varepsilon_* k^2(\tau - s)) \left[1 - \exp(-\Delta_1^2 k^2 (t - \tau)) \right] \\
&\leq \exp(-\varepsilon_* k^2(\tau - s)) \min\{1, \Delta_1^2 k^2 (t-\tau)\}, \quad |k| \geq 2,
\end{aligned}
$$

where in the last step we have used that $1 - e^{-x} \leq x$ for $x \geq 0$. In the same way,

$$
|\mathcal{W}_k(t,s)(I,\ell) - \mathcal{W}_k(\tau,s)(I,\ell)| \leq \Delta_1^2 \exp(2\varepsilon_*(\tau - s))\,(t - \tau), \quad k = \pm 1.
$$

Thus, we deduce from (C.87):

$$
\begin{aligned}
&\|(\mathcal{W}(t,s) - \mathcal{W}(\tau,s))g\|^2_{X^\alpha} \\
&\leq 32\pi^3 \sum_{k=2}^{\infty} (1+k^2)^\alpha \iint_D dI\,d\ell\,\ell \, \frac{1}{|Q'(e)|} \left| (\mathcal{W}_k(t,s)(I,\ell) - \mathcal{W}_k(\tau,s)(I,\ell))\, g_k(I,\ell) \right|^2 \\
&\quad + 32\pi^3\, 2^\alpha \iint_D dI\,d\ell\,\ell \, \frac{1}{|Q'(e)|} \left| (\mathcal{W}_1(t,s)(I,\ell) - \mathcal{W}_1(\tau,s)(I,\ell))\, g_1(I,\ell) \right|^2
\end{aligned}
$$

$$\leq 32\pi^3\, 2^\alpha \sum_{k=2}^{\infty} k^{2\alpha} \exp(-2\varepsilon_* k^2(\tau - s)) \min\{1, \Delta_1^4 k^4 (t-\tau)^2\} \, \|g_k\|^2_{L^2_{\frac{1}{|Q'|}}(D)}$$
$$+ 32\pi^3\, 2^\alpha \Delta_1^4 \exp(4\varepsilon_*(\tau - s))\, (t-\tau)^2 \, \|g_1\|^2_{L^2_{\frac{1}{|Q'|}}(D)}. \tag{C.88}$$

To establish (C.84)–(C.86), it follows from (C.88) for $\alpha \in \{1, 2\}$ that

$$\|(\mathcal{W}(t+h, s) - \mathcal{W}(t, s))g\|^2_{X^\alpha}$$
$$\leq 32\pi^3\, 2^\alpha (\Delta_1^4 + 1) \sum_{k=2}^{\infty} k^{2\alpha} \exp(-2\varepsilon_* k^2(t-s)) \min\{1, k^4 h^2\} \, \|g_k\|^2_{L^2_{\frac{1}{|Q'|}}(D)}$$
$$+ 32\pi^3\, 2^\alpha \Delta_1^4 \exp(4\varepsilon_*(t-s))\, h^2 \, \|g_1\|^2_{L^2_{\frac{1}{|Q'|}}(D)}$$
$$\leq 32\pi^3\, 2^\alpha (\Delta_1^4 + 1) \left[\sum_{k=2}^{h^{-1/2}} k^{2\alpha+4} h^2 \exp(-2\varepsilon_* k^2(t-s)) \, \|g_k\|^2_{L^2_{\frac{1}{|Q'|}}(D)} + \sum_{k=h^{-1/2}}^{\infty} k^{2\alpha} \exp(-2\varepsilon_* k^2(t-s)) \, \|g_k\|^2_{L^2_{\frac{1}{|Q'|}}(D)} \right]$$
$$+ 32\pi^3\, 2^\alpha \Delta_1^4 \exp(4\varepsilon_*(t-s))\, h^2 \, \|g_1\|^2_{L^2_{\frac{1}{|Q'|}}(D)}$$
$$\leq 32\pi^3\, 2^\alpha (\Delta_1^4 + 1) \left[\sum_{k=2}^{h^{-1/2}} k^{2\alpha+2} h \exp(-2\varepsilon_* k^2(t-s)) \, \|g_k\|^2_{L^2_{\frac{1}{|Q'|}}(D)} + \sum_{k=h^{-1/2}}^{\infty} k^{2\alpha+2} h \exp(-2\varepsilon_* k^2(t-s)) \, \|g_k\|^2_{L^2_{\frac{1}{|Q'|}}(D)} \right]$$
$$+ 32\pi^3\, 2^\alpha \Delta_1^4 \exp(4\varepsilon_*(t-s))\, h^2 \, \|g_1\|^2_{L^2_{\frac{1}{|Q'|}}(D)}$$
$$\leq 32\pi^3\, 2^\alpha (\Delta_1^4 + 1)\, h \sum_{k=2}^{\infty} k^{2\alpha+2} \exp(-2\varepsilon_* k^2(t-s)) \, \|g_k\|^2_{L^2_{\frac{1}{|Q'|}}(D)}$$
$$+ 32\pi^3\, 2^\alpha \Delta_1^4 \exp(4\varepsilon_*(t-s))\, h^2 \, \|g_1\|^2_{L^2_{\frac{1}{|Q'|}}(D)}. \tag{C.89}$$

So if we take $\alpha = 1$, set $s = t = 0$ and replace h by t, then we obtain (C.84). Regarding (C.85) and (C.86), from Remark C.14 one has $k^{2\alpha+2} \exp(-2\varepsilon_* k^2(t-s)) \leq (\frac{\alpha+1}{2\varepsilon_*(t-s)})^{\alpha+1}$. Thus, we obtain by means of (C.89) that

$$\|(\mathcal{W}(t+h, s) - \mathcal{W}(t, s))g\|^2_{X^\alpha}$$
$$\leq 32\pi^3\, 2^\alpha (\Delta_1^4 + 1) \left(\frac{\alpha+1}{2\varepsilon_*(t-s)} \right)^{\alpha+1} h \sum_{k=2}^{\infty} \|g_k\|^2_{L^2_{\frac{1}{|Q'|}}(D)}$$

$$+ 32\pi^3\, 2^\alpha \Delta_1^4 \exp(4\varepsilon_*(t-s))\, h^2\, \|g_1\|^2_{L^2_{\frac{1}{|Q'|}}(D)}$$

$$\leq 2^\alpha (\Delta_1^4 + 1) \left[\left(\frac{\alpha+1}{2\varepsilon_*(t-s)} \right)^{\alpha+1} h + \exp(4\varepsilon_*(t-s))\, h^2 \right] \|g\|^2_{X^0},$$

which proves both (C.85) and (C.86). □

Lemma C.17 *Let $h_1, h_2 : J \to X^1_{\text{odd}}$ be continuous and such that $\Phi(h_1(t)) \leq \lambda_* + 2\varepsilon_*$ and $\Phi(h_2(t)) \leq \lambda_* + 2\varepsilon_*$ for $t \in J$. Then for $t, s \in J$ such that $t \geq s$:*

$$|\mathcal{W}_k(t,s;h_1)(I,\ell) - \mathcal{W}_k(t,s;h_2)(I,\ell)| \leq \Lambda(t;h_1,h_2) \exp(\Lambda(t;h_1,h_2)) \exp(-\varepsilon_* k^2 (t-s)) \tag{C.90}$$

for $k \in \mathbb{Z} \setminus \{-1, 0, 1\}$ and

$$|\mathcal{W}_k(t,s;h_1)(I,\ell) - \mathcal{W}_k(t,s;h_2)(I,\ell)| \leq \Lambda(t;h_1,h_2) \exp(\Lambda(t;h_1,h_2)) \exp(2\varepsilon_*(t-s)) \tag{C.91}$$

for $k = \pm 1$. Here,

$$\Lambda(t;h_1,h_2) = 2C_\Phi(\|h_1\|_{\infty,1} + \|h_2\|_{\infty,1}) \int_0^t \|h_1(\tau) - h_2(\tau)\|_{X^1}\, d\tau$$

with $\|g\|_{\infty,1} = \max\{\|g(t)\|_{X^1} : t \in J\}$.

Proof By definition and by means of Lemma C.12:

$$\begin{aligned}
&|\mathcal{W}_k(t,s;h_1)(I,\ell) - \mathcal{W}_k(t,s;h_2)(I,\ell)| \\
&= \left| \exp\left(-\int_s^t [k^2\omega_1^2(I,\ell) - \Phi(h_1(\tau))]\, d\tau \right) - \exp\left(-\int_s^t [k^2\omega_1^2(I,\ell) - \Phi(h_2(\tau))]\, d\tau \right) \right| \\
&= \left| 1 - \exp\left(-\int_s^t [\Phi(h_1(\tau)) - \Phi(h_2(\tau))]\, d\tau \right) \right| \exp\left(-\int_s^t [k^2\omega_1^2(I,\ell) - \Phi(h_1(\tau))]\, d\tau \right) \\
&\leq \left| 1 - \exp\left(-\int_s^t [\Phi(h_1(\tau)) - \Phi(h_2(\tau))]\, d\tau \right) \right| \times \begin{cases} \exp(-\varepsilon_* k^2 (t-s)) & : \ k \in \mathbb{Z} \setminus \{-1,0,1\} \\ \exp(2\varepsilon_*(t-s)) & : \ k = \pm 1 \end{cases}.
\end{aligned} \tag{C.92}$$

Now Lemma C.9 yields

$$\begin{aligned}
|\Phi(h_1(\tau)) - \Phi(h_2(\tau))| &\leq 2C_\Phi(\|h_1(\tau)\|_{X^1} + \|h_2(\tau)\|_{X^1}) \|h_1(\tau) - h_2(\tau)\|_{X^1} \\
&\leq 2C_\Phi(\|h_1\|_{\infty,1} + \|h_2\|_{\infty,1}) \|h_1(\tau) - h_2(\tau)\|_{X^1},
\end{aligned}$$

and consequently

$$\left| -\int_s^t [\Phi(h_1(\tau)) - \Phi(h_2(\tau))]\, d\tau \right| \leq 2C_\Phi(\|h_1\|_{\infty,1} + \|h_2\|_{\infty,1}) \int_0^t \|h_1(\tau) - h_2(\tau)\|_{X^1}\, d\tau.$$

Therefore, the inequality $|1 - e^x| \leq |x|e^{|x|}$ for $x \in \mathbb{R}$ in conjunction with (C.92) leads to (C.90) and (C.91). □

Lemma C.18 *Let $h_1, h_2 : J \to X^1_{\text{odd}}$ be continuous and such that $\Phi(h_1(t)) \leq \lambda_* + 2\varepsilon_*$ and $\Phi(h_2(t)) \leq \lambda_* + 2\varepsilon_*$ for $t \in J$.*

(a) If $t \in J$ and $g : J \to X^0_{\text{odd}}$ is continuous, then

$$\left\| \int_0^t [\mathcal{W}(t,s;h_1) - \mathcal{W}(t,s;h_2)]\, g(s)\, ds \right\|_{X^1} \leq \frac{2}{\varepsilon_*^{1/2}} \Lambda(t;h_1,h_2) \exp(\Lambda(t;h_1,h_2))\, (1 + e^{2\varepsilon_* t}) \left(\int_0^t \|g(s)\|^2_{X^0}\, ds \right)^{1/2}.$$

(b) If $t \in J$ and $\psi \in X^1_{\text{odd}}$, then

$$\|(\mathcal{W}(t,0;h_1) - \mathcal{W}(t,0;h_2))\psi\|_{X^1} \leq 2\Lambda(t;h_1,h_2) \exp(\Lambda(t;h_1,h_2))\, (1 + e^{2\varepsilon_* t})\, \|\psi\|_{X^1}.$$

Here,

$$\Lambda(t;h_1,h_2) = 2C_\Phi(\|h_1\|_{\infty,1} + \|h_2\|_{\infty,1}) \int_0^t \|h_1(\tau) - h_2(\tau)\|_{X^1}\, d\tau$$

with $\|g\|_{\infty,1} = \max\{\|g(t)\|_{X^1} : t \in J\}$.

Proof (a) By definition and by (C.90), (C.91) from Lemma C.17:

$$\begin{aligned}
&\left\| \int_0^t [\mathcal{W}(t,s;h_1) - \mathcal{W}(t,s;h_2)]\, g(s)\, ds \right\|^2_{X^1} \\
&= 16\pi^3 \sum_{k\in\mathbb{Z}} (1+k^2) \left\| \int_0^t [\mathcal{W}_k(t,s;h_1) - \mathcal{W}_k(t,s;h_2)]\, g_k(s)\, ds \right\|^2_{L^2_{\frac{1}{|Q'|}}(D)} \\
&= 16\pi^3 \sum_{k\in\mathbb{Z}} (1+k^2) \iint_D dI\, d\ell\, \ell \frac{1}{|Q'(e)|} \\
&\qquad \times \left| \int_0^t [\mathcal{W}_k(t,s;h_1)(I,\ell) - \mathcal{W}_k(t,s;h_2)(I,\ell)]\, g_k(s)(I,\ell)\, ds \right|^2 \\
&\leq 32\pi^3 \Lambda(t;h_1,h_2)^2 \exp(2\Lambda(t;h_1,h_2)) \\
&\qquad \times \sum_{k=2}^{\infty} (1+k^2) \iint_D dI\, d\ell\, \ell \frac{1}{|Q'(e)|} \left[\int_0^t \exp(-\varepsilon_* k^2(t-s))\, |g_k(s)(I,\ell)|\, ds \right]^2 \\
&\quad + 64\pi^3 \Lambda(t;h_1,h_2)^2 \exp(2\Lambda(t;h_1,h_2)) \\
&\qquad \times \iint_D dI\, d\ell\, \ell \frac{1}{|Q'(e)|} \left[\int_0^t \exp(2\varepsilon_*(t-s))\, |g_1(s)(I,\ell)|\, ds \right]^2.
\end{aligned}$$

If at this point we apply Lemma C.13, it follows that

$$\begin{aligned}
&\left\| \int_0^t [\mathcal{W}(t,s;h_1) - \mathcal{W}(t,s;h_2)]\, g(s)\, ds \right\|_{X^1}^2 \\
&\leq \frac{32\pi^3}{\varepsilon_*} \Lambda(t;h_1,h_2)^2 \exp(2\Lambda(t;h_1,h_2)) \sum_{k=2}^{\infty} \frac{1+k^2}{k^2} \int_0^t e^{-\varepsilon_* k^2 (t-s)} \, \|g_k(s)\|^2_{L^2_{\frac{1}{|Q'|}}(D)}\, ds \\
&+ \frac{16\pi^3}{\varepsilon_*} \Lambda(t;h_1,h_2)^2 \exp(2\Lambda(t;h_1,h_2))\, e^{4\varepsilon_* t} \left(1 - e^{-4\varepsilon_* t}\right) \int_0^t \|g_1(s)\|^2_{L^2_{\frac{1}{|Q'|}}(D)}\, ds \\
&\leq \frac{2}{\varepsilon_*} \Lambda(t;h_1,h_2)^2 \exp(2\Lambda(t;h_1,h_2)) \left(1 + e^{4\varepsilon_* t}\right) \int_0^t \|g(s)\|^2_{X^0}\, ds,
\end{aligned}$$

and this yields what is asserted.

(b) Here, we have similarly

$$\begin{aligned}
&\|(\mathcal{W}(t,0;h_1) - \mathcal{W}(t,0;h_2))\psi\|^2_{X^1} \\
&= 16\pi^3 \sum_{k\in\mathbb{Z}} (1+k^2) \iint_D dI\, d\ell\, \ell\, \frac{1}{|Q'(e)|} \,|\mathcal{W}_k(t,0;h_1)(I,\ell) - \mathcal{W}_k(t,0;h_2)(I,\ell)|^2\, |\psi_k(I,\ell)|^2 \\
&\leq 32\pi^3 \Lambda(t;h_1,h_2)^2 \exp(2\Lambda(t;h_1,h_2)) \sum_{k=2}^{\infty} (1+k^2) \exp(-2\varepsilon_* k^2 t) \iint_D dI\, d\ell\, \ell\, \frac{1}{|Q'(e)|}\, |\psi_k(I,\ell)|^2 \\
&+ 64\pi^3 \Lambda(t;h_1,h_2)^2 \exp(2\Lambda(t;h_1,h_2)) \exp(4\varepsilon_* t) \iint_D dI\, d\ell\, \ell\, \frac{1}{|Q'(e)|}\, |\psi_1(I,\ell)|^2 \\
&\leq 64\pi^3 \Lambda(t;h_1,h_2)^2 \exp(2\Lambda(t;h_1,h_2)) \left(1 + e^{4\varepsilon_* t}\right) \sum_{k=1}^{\infty} (1+k^2) \iint_D dI\, d\ell\, \ell\, \frac{1}{|Q'(e)|}\, |\psi_k(I,\ell)|^2 \\
&\leq 4\Lambda(t;h_1,h_2)^2 \exp(2\Lambda(t;h_1,h_2)) \left(1 + e^{4\varepsilon_* t}\right) \|\psi\|^2_{X^1},
\end{aligned}$$

and this suffices to completes the proof. □

Lemma C.19 *Let $(\Omega, \mathcal{F}, \mu)$ be a probability space and X a Banach space. Suppose that $f : \Omega \to X$ is (Bochner) integrable. Then*

$$\int_\Omega f(\omega)\, d\mu(\omega) \in \overline{\mathrm{co}}\{f(\omega) : \omega \in \Omega\},$$

where $\overline{\mathrm{co}}$ denotes the closure of the convex hull.

Proof See [13, Cor. 8, p. 48]. □

Appendix D
On Kato-Rellich Perturbation Theory

In this section, we derive a (likely non-optimal) result that might be known, although we have not been able to find a suitable reference. Usually, Kato-Rellich perturbation theory concerns analytic families of symmetric or self-adjoint operators A_z that are defined in an open neighborhood of $z = 0$, for instance, and depending on the multiplicity of an eigenvalue μ_0 of A_0, the existence of branches of eigenvalues and/or eigenfunctions close to $z = 0$ is discussed; see [44, 76], [71, Sect. XII.2] and [65, Sect. 5].

However, in the application that we are aiming for (see Chapter 4), the family of operators is analytic in $\Omega = \mathbb{C} \setminus [\delta_1^2, \infty[$ for some $\delta_1 > 0$, but it only has a continuous extension to $z = \delta_1^2$, in the sense that $A_{\delta_1^2} := \lim_{\lambda \in \mathbb{R},\, \lambda \to \delta_1^2 -} A_\lambda$ does exist in the operator norm. In this situation, one is not able to expand A_z in a (real or complex) neighborhood of $z = \delta_1^2$.

Let H be a Hilbert space. By $L(H)$ we denote the bounded linear operators in H, whereas $K(H)$ will stand for the compact linear operators in H. The spectrum of an operator A is $\sigma(A)$, whereas $\rho(A) = \mathbb{C} \setminus \sigma(A)$ is its resolvent.

Lemma D.1 *Let $\Omega = \mathbb{C} \setminus [\lambda_0, \infty[$ for some $\lambda_0 > 0$ and suppose that A_z is an analytic family of operators $A_z \in K(H)$ for $z \in \Omega$ such that $A_{\lambda_0} := \lim_{\lambda \in \mathbb{R},\, \lambda \to \lambda_0 -} A_\lambda$ does exist in the operator norm. In addition, we suppose that $A_z^* = A_{\bar{z}}$ for $z \in \Omega$ and that*

$$\langle A_{\tilde{\lambda}} \Psi, \Psi \rangle \geq \langle A_\lambda \Psi, \Psi \rangle \geq 0 \tag{D.1}$$

is satisfied for $\tilde{\lambda} \geq \lambda$, both in $]-\infty, \lambda_0[$, and $\Psi \in H$. Define $\mu_1(\lambda)$ to be the largest eigenvalue of A_λ. Then there are sequences $\lambda_k \nearrow \lambda_0$, $\varepsilon_k > 0$ and $\Psi_{k,\lambda} \in H$ for $\lambda \in]\lambda_k - \varepsilon_k, \lambda_k + \varepsilon_k[$ such that $\|\Psi_{k,\lambda}\| = 1$,

$$]\lambda_k - \varepsilon_k, \lambda_k + \varepsilon_k[\ni \lambda \mapsto \Psi_{k,\lambda} \in H$$

is real analytic for $k \in \mathbb{N}$, and $A_\lambda \Psi_{k,\lambda} = \mu_1(\lambda) \Psi_{k,\lambda}$ for $k \in \mathbb{N}$ and $\lambda \in]\lambda_k - \varepsilon_k, \lambda_k + \varepsilon_k[$. In addition, μ_1 is real analytic in $]\lambda_k - \varepsilon_k, \lambda_k + \varepsilon_k[$ and satisfies

M. Kunze, *A Birman-Schwinger Principle in Galactic Dynamics*,
Progress in Mathematical Physics 77, https://doi.org/10.1007/978-3-030-75186-9

$$\mu_1'(\lambda) = \langle A_\lambda' \Psi_{k,\lambda}, \Psi_{k,\lambda} \rangle \tag{D.2}$$

for $\lambda \in]\lambda_k - \varepsilon_k, \lambda_k + \varepsilon_k[$.

Proof From the spectral theory of compact positive self-adjoint operators, see [35, Sect. 6], it follows that the spectrum of each A_λ for $\lambda \in]-\infty, \lambda_0]$ consists of $\mu_1(\lambda) \geq \mu_2(\lambda) \geq \ldots \to 0$ (the eigenvalues are listed according to their finite multiplicities). Furthermore, due to (D.1) and using the Courant max-min principle, we have $\mu_k(\tilde{\lambda}) \geq \mu_k(\lambda)$ for $k \in \mathbb{N}$ and $\tilde{\lambda} \geq \lambda$, both in $]-\infty, \lambda_0]$; also see the proof of Lemma 4.3(e). Since $]-\infty, \lambda_0[\ni \lambda \mapsto A_\lambda \in L(H)$ is real analytic, this map is in particular locally Lipschitz continuous. From this fact, together with the Courant max-min principle, we deduce that each $\mu_k :]-\infty, \lambda_0[\to]0, \infty[$ is a continuous function, which is monotone increasing; once again cf. the proof of Lemma 4.3(e). Noting that $\mu_k(\lambda) \leq \mu_k(\lambda_0) \leq \mu_1(\lambda_0)$, the limits $\lim_{\lambda \to \lambda_0 -} \mu_k(\lambda)$ do exist and are finite. Hence if we define $\mu_k(\lambda_0) = \lim_{\lambda \to \lambda_0 -} \mu_k(\lambda)$, then the eigenvalues $\mu_k :]-\infty, \lambda_0] \to]0, \infty[$ are continuous.

For the 'Kato-Rellich-part' of the argument we follow [65, Thm. 5.8]. Let $m \in \mathbb{N}$ denote the multiplicity of the eigenvalue $\mu_1(\lambda_0)$ of A_{λ_0}. The case $m = 1$ is easier (and in fact a better result can be obtained, see Lemma D.2 below), so we consider $m \geq 2$. For illustration, we restrict ourselves to $m = 2$, and the general case is not much more difficult. Hence, we know that $\mu_1(\lambda_0) = \mu_2(\lambda_0) > \mu_3(\lambda_0) \geq \ldots$. Let $\eta > 0$ be such that $\mu_1(\lambda_0) - \eta > \mu_3(\lambda_0) + \eta$. From the continuity of μ_1 and μ_2, we infer that there is $\varepsilon > 0$ with the property that $\mu_1(\lambda), \mu_2(\lambda) \in]\mu_1(\lambda_0) - \eta, \mu_1(\lambda_0)]$ for $\lambda \in [\lambda_0 - \varepsilon, \lambda_0]$. For $k \geq 3$ and $\lambda \in]-\infty, \lambda_0]$ we also have $\mu_k(\lambda) \leq \mu_k(\lambda_0) \leq \mu_3(\lambda_0)$, i.e., the eigenvalue groups $\{\mu_1(\lambda), \mu_2(\lambda)\}$ and $\{\mu_3(\lambda), \ldots\}$ are strictly separated for $\lambda \in [\lambda_0 - \varepsilon, \lambda_0]$. Let $r = \mu_1(\lambda_0) - (\mu_3(\lambda_0) + \eta) > \eta > 0$ and $\Gamma = \{z \in \mathbb{C} : |z - \mu_1(\lambda_0)| = r\}$. Then $\mu_1(\lambda), \mu_2(\lambda)$ are inside of Γ, whereas $\mu_3(\lambda), \ldots$ are outside of Γ for $\lambda \in [\lambda_0 - \varepsilon, \lambda_0]$; we have $\mathrm{dist}(\mu_k(\lambda), \Gamma) \geq \mu_1(\lambda_0) - \mu_3(\lambda_0) - 2\eta > 0$ for $k = 1, 2$, whereas $\mathrm{dist}(\mu_k(\lambda), \Gamma) \geq \eta$ for $k \geq 3$. For $\lambda \in [\lambda_0 - \varepsilon, \lambda_0]$ consider the Riesz projection operator

$$P_\lambda = \frac{1}{2\pi i} \int_\Gamma (A_\lambda - \zeta)^{-1} \, d\zeta$$

on H. Since $\Gamma \cap \sigma(A_\lambda) = \emptyset$, the resolvent $R_{A_\lambda}(\zeta) = (A_\lambda - \zeta)^{-1}$ is well-defined for $\zeta \in \Gamma$; in fact R_{A_λ} is analytic in a neighborhood of Γ and satisfies

$$\|R_{A_\lambda}(\zeta)\| = \mathrm{dist}(\zeta, \sigma(A_\lambda))^{-1} \leq \min\{\eta, \mu_1(\lambda_0) - \mu_3(\lambda_0) - 2\eta\}^{-1} =: C_0$$

for $\lambda \in [\lambda_0 - \varepsilon, \lambda_0]$ and $\zeta \in \Gamma$. Hence for $\lambda, \tilde{\lambda} \in [\lambda_0 - \varepsilon, \lambda_0]$ the second resolvent identity yields

$$\sup_{\zeta \in \Gamma} \|R_{A_\lambda}(\zeta) - R_{A_{\tilde{\lambda}}}(\zeta)\| = \sup_{\zeta \in \Gamma} \|R_{A_\lambda}(\zeta)(A_\lambda - A_{\tilde{\lambda}}) R_{A_{\tilde{\lambda}}}(\zeta)\| \leq C_0^2 \|A_\lambda - A_{\tilde{\lambda}}\|,$$

which implies that $[\lambda_0 - \varepsilon, \lambda_0] \ni \lambda \mapsto P_\lambda \in L(H)$ is continuous. Defining more generally

$$P_z = \frac{1}{2\pi i} \int_\Gamma (A_z - \zeta)^{-1} \, d\zeta,$$

it may moreover be shown as in analytic Kato-Rellich theory (see [44, Chapter 7, Sect. 1], [71, Sect. XII.2], [65, Thm. 5.8]) that every $\lambda \in]\lambda_0 - \varepsilon, \lambda_0[$ has a complex neighborhood $U_\lambda \subset \mathbb{C}$ such that $U_\lambda \ni z \mapsto P_z \in L(H)$ is well-defined and analytic. The map $P_{\lambda_0} : H \to H$ is the orthogonal projection onto the eigenspace $\ker(A_{\lambda_0} - \mu_1(\lambda_0))$, which is $(m = 2)$-dimensional; see [65, Cor. 5.6]. If $\lambda \in [\lambda_0 - \varepsilon, \lambda_0[$, then $\sigma(A_\lambda) \cap]\mu_1(\lambda_0) - \eta, \mu_1(\lambda_0) + \eta[= \{\mu_1(\lambda), \mu_2(\lambda)\}$. Denoting the spectral resolution of the symmetric operator A_λ by $E^{(\lambda)}$, we thus have $P_\lambda = E^{(\lambda)}_{\mu_1(\lambda_0)+\eta} - E^{(\lambda)}_{\mu_1(\lambda_0)-\eta}$, and in particular P_λ is an orthogonal projection; cf. [65, Prop. 5.5 & Prop. 3.4]. Hence, it follows from the continuity of $[\lambda_0 - \varepsilon, \lambda_0] \ni \lambda \mapsto P_\lambda \in L(H)$ and [71, Lemma, p. 14] that $\dim \operatorname{ran} P_\lambda = 2$ for $\lambda < \lambda_0$ sufficiently close to λ_0, which we can assume to hold for $\lambda \in [\lambda_0 - \varepsilon, \lambda_0[$.

What is next is the reduction to a two-dimensional problem. According to [37, Prop. 6.9] with $\sigma_1 = \{\mu_1(\lambda), \mu_2(\lambda)\}$, we know that

$$P_\lambda A_\lambda = A_\lambda P_\lambda, \quad \sigma(A_\lambda P_\lambda) = \{\mu_1(\lambda), \mu_2(\lambda)\}, \quad \sigma(A_\lambda(I - P_\lambda)) = \{\mu_3(\lambda), \ldots\},$$

for $\lambda \in [\lambda_0 - \varepsilon, \lambda_0]$. Denoting $L_\lambda = \operatorname{ran} P_\lambda = P_\lambda H$, thus $A_\lambda : L_\lambda \to L_\lambda$ is well-defined and $\sigma(A_\lambda|_{L_\lambda}) = \{\mu_1(\lambda), \mu_2(\lambda)\}$. Let

$$C_\lambda = P_\lambda P_{\lambda_0} + (I - P_\lambda)(I - P_{\lambda_0}).$$

Then $C_\lambda : H \to H$ is an orthogonal projection and $C_{\lambda_0} = I$. Also $C_\lambda : L_{\lambda_0} \to L_\lambda$ is well-defined. In fact, if $\Psi \in L_{\lambda_0}$, then $\Psi = P_{\lambda_0}\Phi$ for some $\Phi \in H$. Therefore, $(I - P_{\lambda_0})\Psi = 0$ and $C_\lambda \Psi = P_\lambda P_{\lambda_0}^2 \Phi = P_\lambda \Psi \in L_\lambda$. In addition, $[\lambda_0 - \varepsilon, \lambda_0] \ni \lambda \mapsto C_\lambda \in L(H)$ is continuous. Hence, by decreasing $\varepsilon > 0$ further if necessary, we may assume that C_λ is invertible for $\lambda \in [\lambda_0 - \varepsilon, \lambda_0]$. In other words, C_λ is a linear bijection between the two-dimensional spaces L_{λ_0} and L_λ. Let $\{\Psi_1, \Psi_2\}$ be an orthonormal basis of $L_{\lambda_0} = \ker(A_{\lambda_0} - \mu_1(\lambda_0))$. Then $\{C_\lambda \Psi_1, C_\lambda \Psi_2\}$ is a basis of L_λ, and we obtain an orthonormal basis $\{\varphi_{1\lambda}, \varphi_{2\lambda}\}$ of L_λ by the Gram-Schmidt orthonormalization; explicitly,

$$\varphi_{1\lambda} = \frac{1}{\|C_\lambda \Psi_1\|} C_\lambda \Psi_1,$$

$$\varphi_{2\lambda} = \frac{1}{(\|C_\lambda \Psi_2\|^2 - \langle C_\lambda \Psi_2, \varphi_{1\lambda}\rangle^2)^{1/2}} (C_\lambda \Psi_2 - \langle C_\lambda \Psi_2, \varphi_{1\lambda}\rangle \, \varphi_{1\lambda}).$$

Since $C_{\lambda_0}\Psi_1 = \Psi_1$ and $C_{\lambda_0}\Psi_2 = \Psi_2$, we have $\varphi_{1\lambda_0} = \Psi_1$ and $\varphi_{2\lambda_0} = \Psi_2$. If we define a matrix $A(\lambda) = (a_{jk}(\lambda))_{j,k=1,2}$ by means of

$$a_{jk}(\lambda) = \langle A_\lambda \varphi_{j\lambda}, \varphi_{k\lambda}\rangle, \quad j, k = 1, 2,$$

then $A(\lambda)$ is symmetric and represents $A_\lambda : L_\lambda \to L_\lambda$ w.r.t. to the orthonormal basis $\{\varphi_{1\lambda}, \varphi_{2\lambda}\}$. In particular, $\sigma(A(\lambda)) = \{\mu_1(\lambda), \mu_2(\lambda)\}$, and $A_{\lambda_0}\Psi_j = \mu_1(\lambda_0)\Psi_j$ for $j = 1, 2$ implies that $A(\lambda_0) = \mu_1(\lambda_0)I_2$, with I_2 denoting the identity matrix in two dimensions. For a fixed $\lambda \in]\lambda_0 - \varepsilon, \lambda_0[$ we can extend all of the above definitions and relations analytically to $z \in U_\lambda$, the complex neighborhood of λ, where $z \mapsto P_z$ is analytic; in particular, $z \mapsto A(z)$ is analytic on U_λ.

Let $\Lambda = \{\lambda \in [\lambda_0 - \varepsilon, \lambda_0[: \mu_1(\lambda) \neq \mu_2(\lambda)\}$. Case 1: λ_0 is an accumulation point of Λ. Then there is a sequence $(\lambda_k) \subset]\lambda_0 - \varepsilon, \lambda_0[$ such that $\lambda_k \nearrow \lambda_0$ and $\mu_1(\lambda_k) > \mu_2(\lambda_k)$ for $k \in \mathbb{N}$. Thus, $\mu_1(\lambda_k)$ is a simple eigenvalue of $A(\lambda_k)$ for each $k \in \mathbb{N}$. Owing to a theorem by Rellich [76, p. 42], in particular there is $\varepsilon_k > 0$ such that $]\lambda_k - \varepsilon_k, \lambda_k + \varepsilon_k[\ni \lambda \mapsto \mu_1(\lambda)$ is real analytic, and moreover there is a real analytic function $]\lambda_k - \varepsilon_k, \lambda_k + \varepsilon_k[\ni \lambda \mapsto \xi_k(\lambda) \in \mathbb{R}^2$ such that $A(\lambda)\xi_k(\lambda) = \mu_1(\lambda)\xi_k(\lambda)$ and $|\xi_k(\lambda)| = 1$ for $\lambda \in]\lambda_k - \varepsilon_k, \lambda_k + \varepsilon_k[$. Let

$$\tilde{\Psi}_{k,\lambda} = \xi_k^{(1)}(\lambda)\varphi_{1\lambda} + \xi_k^{(2)}(\lambda)\varphi_{2\lambda} \in L_\lambda \subset H, \quad \xi_k = (\xi_k^{(1)}, \xi_k^{(2)}). \tag{D.3}$$

Then

$$\begin{aligned}
A_\lambda \tilde{\Psi}_{k,\lambda} &= \xi_k^{(1)}(\lambda)A_\lambda\varphi_{1\lambda} + \xi_k^{(2)}(\lambda)A_\lambda\varphi_{2\lambda} \\
&= \xi_k^{(1)}(\lambda)[a_{11}(\lambda)\varphi_{1\lambda} + a_{12}(\lambda)\varphi_{2\lambda}] + \xi_k^{(2)}(\lambda)[a_{21}(\lambda)\varphi_{1\lambda} + a_{22}(\lambda)\varphi_{2\lambda}] \\
&= [\xi_k^{(1)}(\lambda)a_{11}(\lambda) + \xi_k^{(2)}(\lambda)a_{21}(\lambda)]\varphi_{1\lambda} + [\xi_k^{(1)}(\lambda)a_{12}(\lambda) + \xi_k^{(2)}(\lambda)a_{22}(\lambda)]\varphi_{2\lambda} \\
&= \mu_1(\lambda)\xi_k^{(1)}\varphi_{1\lambda} + \mu_1(\lambda)\xi_k^{(2)}\varphi_{2\lambda} \\
&= \mu_1(\lambda)\tilde{\Psi}_{k,\lambda}
\end{aligned}$$

is verified. Since $|\xi_k(\lambda)| = 1$ we must have $\tilde{\Psi}_{k,\lambda} \neq 0$, as $\{\varphi_{1\lambda}, \varphi_{2\lambda}\}$ is linearly independent. Thus is we define $\Psi_{k,\lambda} = \|\tilde{\Psi}_{k,\lambda}\|^{-1}\tilde{\Psi}_{k,\lambda}$, then we obtain the claim. Case 2: λ_0 is not an accumulation point of Λ. Then there is a left-sided neighborhood of λ_0, which we assume to be $[\lambda_0 - \varepsilon, \lambda_0]$, so that $\mu_1(\lambda) = \mu_2(\lambda)$ for all $\lambda \in [\lambda_0 - \varepsilon, \lambda_0]$, i.e., the multiplicity of the first eigenvalue equals two over the whole interval. Let $(\lambda_k) \subset]\lambda_0 - \varepsilon, \lambda_0[$ be a sequence such that $\lambda_k \nearrow \lambda_0$. By a theorem of Rellich [76, p. 42], in particular there are $\varepsilon_k > 0$ and real analytic functions $]\lambda_k - \varepsilon_k, \lambda_k + \varepsilon_k[\ni \lambda \mapsto s_k(\lambda), t_k(\lambda)$ as well as real analytic functions $]\lambda_k - \varepsilon_k, \lambda_k + \varepsilon_k[\ni \lambda \mapsto \xi_k(\lambda), \zeta_k(\lambda) \in \mathbb{R}^2$ such that

$$A(\lambda)\xi_k(\lambda) = s_k(\lambda)\xi_k(\lambda), \quad A(\lambda)\zeta_k(\lambda) = t_k(\lambda)\zeta_k(\lambda),$$

and $\xi_k(\lambda), \zeta_k(\lambda)$ are orthonormal for $\lambda \in]\lambda_k - \varepsilon_k, \lambda_k + \varepsilon_k[$. Using part (2) of Rellich's result, by decreasing ε_k further we can additionally make sure that $\{s_k(\lambda), t_k(\lambda)\} = \sigma(A(\lambda)) = \{\mu_1(\lambda), \mu_2(\lambda)\}$, which means that actually $s_k(\lambda) = t_k(\lambda) = \mu_1(\lambda)$ holds for $\lambda \in]\lambda_k - \varepsilon, \lambda_k + \varepsilon[$. Hence, we can proceed as in Case 1 with (D.3) to complete the argument.

The proof of (D.2), which was already used by Rellich [75, p. 471, footnote], is well-known but included for completeness. In fact for μ_1 to be differentiable at some

λ and the formula to hold at λ, it suffices that the operator family is differentiable at λ, but the eigenvalue family needs only be continuous at λ; we found this observation in [3]. Let $\lambda \in]\lambda_k - \varepsilon_k, \lambda_k + \varepsilon_k[$ and $h > 0$ be small. Then

$$\begin{aligned}\langle (A_{\lambda+h} - A_\lambda)\Psi_{k,\lambda+h}, \Psi_{k,\lambda}\rangle &= \mu_1(\lambda+h)\langle \Psi_{k,\lambda+h}, \Psi_{k,\lambda}\rangle - \langle \Psi_{k,\lambda+h}, A_\lambda \Psi_{k,\lambda}\rangle \\ &= (\mu_1(\lambda+h) - \mu_1(\lambda))\langle \Psi_{k,\lambda+h}, \Psi_{k,\lambda}\rangle,\end{aligned}$$

so dividing by h and taking the limit $h \to 0+$ yields (D.2); recall that $\|\Psi_{k,\lambda}\| = 1$. □

Lemma D.2 *In the setting of Lemma D.1, suppose that additionally $\mu_1(\lambda_0)$ is a simple eigenvalue of A_{λ_0}. Then there is $\varepsilon > 0$ such that $]\lambda_0 - \varepsilon, \lambda_0[\ni \lambda \mapsto \mu_1(\lambda)$ is real analytic. In addition, there are $\Psi_\lambda \in H$ satisfying $\|\Psi_\lambda\| = 1$, $A_\lambda \Psi_\lambda = \mu_1(\lambda)\Psi_\lambda$ and $]\lambda_0 - \varepsilon, \lambda_0[\ni \lambda \mapsto \Psi_\lambda$ is real analytic. Furthermore,*

$$\mu_1'(\lambda) = \langle A_\lambda' \Psi_\lambda, \Psi_\lambda\rangle \tag{D.4}$$

for $\lambda \in]\lambda_0 - \varepsilon, \lambda_0[$.

Proof Since $\mu_1(\lambda_0) > \mu_2(\lambda_0)$, there is $\eta > 0$ and $\varepsilon > 0$ such that $\mu_1(\lambda_0) - \eta > \mu_2(\lambda_0) + \eta$, $\mu_1(\lambda) \in]\mu_1(\lambda_0) - \eta, \mu_1(\lambda_0)]$ as well as $\mu_k(\lambda) \le \mu_2(\lambda_0)$ for $\lambda \in [\lambda_0 - \varepsilon, \lambda_0]$ and $k \ge 2$; cf. the proof of Lemma D.1. Let $r = \mu_1(\lambda_0) - (\mu_2(\lambda_0) + \eta) > \eta > 0$ and $\Gamma = \{z \in \mathbb{C} : |z - \mu_1(\lambda_0)| = r\}$. Then $\mu_1(\lambda)$ is inside of Γ, whereas $\mu_2(\lambda), \ldots$ are outside of Γ for $\lambda \in [\lambda_0 - \varepsilon, \lambda_0]$. Once again we consider the Riesz projection operator

$$P_\lambda = \frac{1}{2\pi i}\int_\Gamma (A_\lambda - \zeta)^{-1}\, d\zeta$$

on H for $\lambda \in [\lambda_0 - \varepsilon, \lambda_0]$. Then $\lambda \in [\lambda_0 - \varepsilon, \lambda_0] \ni \lambda \mapsto P_\lambda \in L(H)$ is continuous. Defining more generally

$$P_z = \frac{1}{2\pi i}\int_\Gamma (A_z - \zeta)^{-1}\, d\zeta,$$

it may moreover be shown that every $\lambda \in]\lambda_0 - \varepsilon, \lambda_0[$ has a complex neighborhood $U_\lambda \subset \mathbb{C}$ such that $U_\lambda \ni z \mapsto P_z \in L(H)$ is well-defined and analytic. The map $P_{\lambda_0} : H \to H$ is the orthogonal projection onto the one-dimensional eigenspace $\ker(A_{\lambda_0} - \mu_1(\lambda_0))$. As before it follows that $\dim \operatorname{ran} P_\lambda = 1$ for $\lambda < \lambda_0$ sufficiently close to λ_0, which we can assume to hold for $\lambda \in [\lambda_0 - \varepsilon, \lambda_0[$. For $\Psi \in H$ such that $\|\Psi\| = 1$ and $A_{\lambda_0}\Psi = \mu_1(\lambda_0)\Psi$ we define $\tilde{\Psi}_\lambda = P_\lambda \Psi$ for $\lambda \in [\lambda_0 - \varepsilon, \lambda_0]$. Then $\tilde{\Psi}_\lambda \in \operatorname{ran} P_\lambda$ and

$$P_\lambda = \frac{1}{2\pi i}\int_{|\zeta - \mu_1(\lambda)| = \delta} (A_\lambda - \zeta)^{-1}\, d\zeta$$

for $\delta > 0$ small, by the Cauchy theorem. From [71, Theorem XII.5(d)], it follows that $(A_\lambda - \mu_1(\lambda))^n \tilde{\Psi}_\lambda = 0$ for some $n \in \mathbb{N}$. Since A_λ is symmetric, all eigenvalues are semisimple, which means that in fact $(A_\lambda - \mu_1(\lambda))\tilde{\Psi}_\lambda = 0$. Owing to $\Psi \in$ ran P_{λ_0} we have $\|\tilde{\Psi}_{\lambda_0}\| = \|P_{\lambda_0}\Psi\| = \|\Psi\| = 1$. Therefore, due to the continuity of $[\lambda_0 - \varepsilon, \lambda_0] \ni \lambda \mapsto P_\lambda \in L(H)$, we may suppose that $\tilde{\Psi}_\lambda \neq 0$ for $\lambda \in [\lambda_0 - \varepsilon, \lambda_0]$. Thus if we define $\Psi_\lambda = \|\tilde{\Psi}_\lambda\|^{-1}\tilde{\Psi}_\lambda$, then we are done for what concerns the eigenfunctions, as $]\lambda_0 - \varepsilon, \lambda_0[\ni \lambda \mapsto P_\lambda \in L(H)$ is real analytic. To establish that μ_1 is real analytic as well, let $\hat{\lambda} \in]\lambda_0 - \varepsilon, \lambda_0[$. Then [71, Theprem XII.8] implies that for z near $\hat{\lambda}$, say $z \in V_{\hat{\lambda}} \subset \mathbb{C}$, there is exactly one point $E(z) \in \sigma(A_z)$ near $\mu_1(\hat{\lambda})$, and this point is isolated and nondegenerate. In addition, the map $V_{\hat{\lambda}} \ni z \mapsto E(z)$ is analytic. Restricting to real $z = \lambda \in V_{\hat{\lambda}} \cap]\lambda_0 - \varepsilon, \lambda_0[$, the choice of ε above implies that $E(\lambda) = \mu_1(\lambda)$, which shows that μ_1 is real analytic on $V_{\hat{\lambda}} \cap]\lambda_0 - \varepsilon, \lambda_0[$. The relation (D.4) can be derived as before. □

References

1. Abramov, A.A., Aslanyan, A., Davies, E.B.: Bounds on complex eigenvalues and resonances. J. Phys. A **34**, 57–72 (2001)
2. Andréasson, H.: The Einstein-Vlasov system/kinetic theory. Living Rev. Relativ. **5**(33), 2002–7 (2002)
3. Andrew, A.L.: Finite and continuous perturbations of matrix eigenvalues. Appl. Math. Lett. **11**, 47–51 (1998)
4. Antonov, V.A.: Remarks on the problem of stability in stellar dynamics (in Russian). Astronom. Ž. **37**, 918-926 (1960); translated in Soviet Astronom. AJ **4**, 859–867 (1960)
5. Antonov, V.A.: Solution of the problem of stability of a stellar system with the Emden density law and spherical velocity distribution, (in Russian). J. Leningr. Univ. Ser. Mekh. Astron. **7**, 135–146 (1962)
6. Arnold, V.I.: Mathematical Methods of Classical Mechanics, 2nd edn. Springer, Berlin (1989)
7. Batt, J., Faltenbacher, W., Horst, E.: Stationary spherically symmetric models in stellar dynamics. Arch. Rational Mech. Anal. **93**, 159–183 (1986)
8. Behrndt, J.; ter Elst, A.F.M.; Gesztesy, F.: The generalized Birman-Schwinger principle. arXiv:2005.01195v4
9. Binney, J., Tremaine, S.: Galactic Dynamics, 2nd edn. Princeton University Press, Princeton (2008)
10. Birman M.Š.: On the spectrum of singular boundary-value problems (in Russian). Mat. Sb. (N.S.) **55**(97), 125–174 (1961); translated in Am. Math. Soc. Transl. **53**, 23–80 (1966)
11. Chicone, C.: The monotonicity of the period function for planar Hamiltonian vector fields. J. Differ. Equ. **69**, 310–321 (1987)
12. Cuenin, J.-C.: Eigenvalue bounds for Dirac and fractional Schrödinger operators with complex potentials. J. Funct. Anal. **272**, 2987–3018 (2017)
13. Diestel, J., Uhl Jr., J.J.: Vector Measures. American Mathematical Society, Providence (1977)
14. Donoghue Jr.,W.F.: Distributions and Fourier Transforms, Pure and Applied Mathematics, vol. 32. Academic Press, New York (1969)
15. Doremus, J.P., Baumann, G., Feix, M.R.: Stability of a self gravitating system with phase space density function of energy and angular momentum. Astron. Astrophys. **29**, 401–407 (1973)
16. Dyson, F.J.; Lenard, A.: Stability of matter. I and II. J. Math. Phys. **8**, 423-434 (1967); ibid. **9**, 698–711 (1968)

M. Kunze, *A Birman-Schwinger Principle in Galactic Dynamics*,
Progress in Mathematical Physics 77, https://doi.org/10.1007/978-3-030-75186-9

17. Eddington, A.S.: On the pulsations of a gaseous star and the problem of the Cepheid variables. Part I. Monthly Notices Roy. Astron. Soc. **79**, 2–22 (1918)
18. Fackerell, E.D.: Relativistic, spherically symmetric star clusters. IV. A sufficient condition for instability of isotropic clusters against radial perturbations. Astrophys. J. **160**, 859–874 (1970)
19. Fackerell, E.D.: Relativistic, spherically symmetric star clusters. V. A relativistic version of Plummer's model. Astrophys. J. **165**, 489–493 (1971)
20. Fanelli, L., Krejčiřík, D., Vega, L.: Spectral stability of Schrödinger operators with subordinated complex potentials. J. Spectr. Theory **8**, 575–604 (2018)
21. Fasano, A., Marmi, S.: Analytical Mechanics. Oxford University Press, Cambridge (2006)
22. Frank, R.L.: Eigenvalue bounds for Schrödinger operators with complex potentials. Bull. Lond. Math. Soc. **43**, 745–750 (2011)
23. Frank, R.L., Laptev, A., Safronov, O.: On the number of eigenvalues of Schrödinger operators with complex potentials. J. Lond. Math. Soc. **94**, 377–390 (2016)
24. Gallavotti, G.: The Elements of Mechanics. Springer, Berlin (1983)
25. Gidas, B., Ni, W.M., Nirenberg, L.: Symmetry and related properties via the maximum principle. Comm. Math. Phys. **68**, 209–243 (1979)
26. Gillon, D., Cantus, M., Doremus, J.P., Baumann, G.: Stability of self-gravitating spherical systems in which phase space density is a function of energy and angular momentum, for spherical perturbations. Astron. Astrophys. **50**, 467–470 (1976)
27. Glassey, R.T.: The Cauchy Problem in Kinetic Theory. SIAM, Philadelphia (1996)
28. Guo, Y.: On the generalized Antonov stability criterion. In: Nonlinear Wave Equations. American Mathematical Society, Providence, RI (1998). Contemp. Math. **263**, 85–107 (2000)
29. Guo, Y., Liu, Z.: Unstable and stable galaxy models. Comm. Math. Phys. **279**, 789–813 (2008)
30. Guo, Y., Rein, G.: A non-variational approach to nonlinear stability in stellar dynamics applied to the King model. Comm. Math. Phys. **271**, 489–509 (2007)
31. Hadžić, M., Rein, G.: Stability for the spherically symmetric Einstein-Vlasov system: a coercivity estimate. Math. Proc. Cambridge Philos. Soc. **155**, 529–556 (2013)
32. Hadžić, M., Rein, G.: On the small redshift limit of steady states of the spherically symmetric Einstein-Vlasov system and their stability. Math. Proc. Cambridge Philos. Soc. **159**, 529–546 (2015)
33. Hainzl, C., Hamza, E., Seiringer, R., Solovej, J.P.: The BCS functional for general pair interactions. Comm. Math. Phys. **281**, 349–367 (2008)
34. Hansmann, M.; Krejcirik ,D.: The abstract Birman-Schwinger principle and spectral stability. arXiv:2010.15102v1
35. Helffer, B.: Spectral Theory and Its Applications. Cambridge University Press, Cambridge (2013)
36. Hénon, M.: Vlasov equation? Astron. Astrophys. **114**, 211–212 (1982)
37. Hislop, P.D., Sigal, I.M.: Introduction to Spectral Theory. With Applications to Schrödinger Operators. Springer, Berlin (1996)
38. Ipser, J.R.; Thorne, K.S.: Relativistic, spherically symmetric star clusters. I. Stability theory for radial perturbations. Astrophys. J. **154**, 251–270 (1968)
39. Ipser, J.R.: Relativistic, spherically symmetric star clusters. II. Sufficient conditions for stability against radial perturbations. Astrophys. J. **156**, 509–527 (1969)
40. Ipser, J.R.: Relativistic, spherically symmetric star clusters. III. Stability of compact isotropic models. Astrophys. J. **158**, 17–43 (1969)
41. Jeans, J.H.: On the theory of star-streaming and the structure of the universe. Monthly Notices Roy. Astron. Soc. **76**, 70–84 (1915)
42. Kandrup, H.E.: A stability criterion for any collisionless stellar equilibrium and some concrete applications thereof. Astrophys. J. **370**, 312–317 (1991)
43. Kandrup, H.E., Sygnet, J.F.: A simple proof of dynamical stability for a class of spherical clusters. Astrophys. J. **298**, 27–33 (1985)
44. Kato, T.: Perturbation Theory for Linear Operators. Springer, Berlin (1966)
45. Kenic, C.E., Ruiz, A., Sogge, C.D.: Uniform Sobolev inequalities and unique continuation for second order constant coefficient differential operators. Duke Math. J. **55**, 329–347 (1987)

46. Klaus, M.: Some applications of the Birman-Schwinger principle. Helv. Phys. Acta **55**, 49–68 (1982/83)
47. Latushkin, Y., Vasudevan, S.: Eigenvalues of the linearized 2D Euler equations via Birman-Schwinger and Lin's operators. J. Math. Fluid Mech. **20**, 1667–1680 (2018)
48. Lemou, M., Méhats, F., Raphaël, P.: Structure of the linearized gravitational Vlasov-Poisson system close to a polytropic ground state. SIAM J. Math. Anal. **39**, 1711–1739 (2008)
49. Lemou, M., Méhats, F., Raphaël, P.: The orbital stability of the ground states and the singularity formation for the gravitational Vlasov Poisson system. Arch. Rational Mech. Anal. **189**, 425–468 (2008)
50. Lemou, M., Méhats, F., Raphaël, P.: A new variational approach to the stability of gravitational systems. Comm. Math. Phys. **302**, 161–224 (2011)
51. Lemou, M., Méhats, F., Raphaël, P.: Orbital stability of spherical galactic models. Invent. Math. **187**, 145–194 (2012)
52. Leonard, I.E.: Banach sequence spaces. J. Math. Anal. Appl. **54**, 245–265 (1976)
53. Lieb, E.H., Seiringer, R.: The Stability of Matter in Quantum Mechanics. Cambridge University Press, Cambridge (2010)
54. Lieb, E.H.; Thirring, W.: Bound for the kinetic energy of fermions which proves the stability of matter. Phys. Rev. Lett. **35**, 687–689 (1975); Errata ibid., 1116 (1975)
55. Lions, P.-L.; Perthame, B: Propagation of moments and regularity for the 3-dimensional Vlasov-Poisson system. Invent. Math. **105** 415–430 (1991)
56. Louis, P., Gerhard, O.: Can galaxies oscillate? A self-consistent model of a non-stationary stellar system. Monthly Notices Roy. Astron. Soc. **233**, 337–365 (1988)
57. Lynden-Bell, D.: The Hartree-Fock exchange operator and the stability of galaxies. Monthly Notices Roy. Astron. Soc. **144**, 189–217 (1969)
58. Lynden-Bell, D., Sanitt, N.: The Schrödinger operator criterion for the stability of galaxies and gas spheres. Monthly Notices Roy. Astron. Soc. **143**, 167–187 (1969)
59. Makino, T.: On spherically symmetric motions of a gaseous star governed by the Euler-Poisson equations. Osaka J. Math. **52**, 545–580 (2015)
60. Maréchal, L., Perez, J.: Radial orbit instability as a dissipation-induced phenomenon. Monthly Notices Roy. Astron. Soc. **405**, 2785–2790 (2010)
61. Mathur, S.D.: Existence of oscillation modes in collisionless gravitating systems. Monthly Notices Roy. Astron. Soc. **243**, 529–536 (1990)
62. Morrison, P.J.: Hamiltonian description of the ideal fluid. Rev. Modern Phys. **70**, 467–521 (1998)
63. Mouhot C.: Stabilité orbitale pour le système de Vlasov-Poisson gravitationnel (d'après Lemou-Méhats-Raphaël, Guo, Lin, Rein et al.), Astérisque **352**, 35–82 (2013)
64. Opial, Z.: Sur les périodes des solutions de l'équation différentielle $x'' + g(x) = 0$. Ann. Polon. Math. **10**, 49–72 (1961)
65. Pankov, A.: Introduction to Spectral Theory of Schrödinger Operators. ftp://ftp.math.ethz.ch/EMIS/journals/AMEN/posters/pankov.pdf
66. Perez, J.; Aly, J.-J.: Stability of spherical stellar systems. I. Analytical results. Monthly Notices Roy. Astron. Soc. **280**, 689–699 (1996)
67. Pfaffelmoser, K.: Global classical solutions of the Vlasov-Poisson system in three dimensions for general initial data. J. Differ. Equ. **95**, 281–303 (1992)
68. Pushnitski, A.: The Birman-Schwinger principle on the essential spectrum. J. Funct. Anal. **261**, 2053–2081 (2011)
69. Ramming, T., Rein, G.: Spherically symmetric equilibria for self-gravitating kinetic or fluid models in the nonrelativistic and relativistic case-a simple proof for finite extension. SIAM J. Math. Anal. **45**, 900–914 (2013)
70. Ramming, T., Rein, G.: Oscillating solutions of the Vlasov-Poisson system-a numerical investigation. Phys. D **365**, 72–79 (2018)
71. Reed, M., Simon, B.: Methods of Modern Mathematical Physics IV: Analysis of Operators. Academic Press, New York (1978)

72. Rein, G.: Stability of spherically symmetric steady states in galactic dynamics against general perturbations. Arch. Ration. Mech. Anal. **161**, 27–42 (2002)
73. Rein, G.: Collisionless kinetic equations from astrophysics-the Vlasov-Poisson system, in Handbook of Differential Equations: Evolutionary Equations, vol. III, pp. 383–476. Elsevier/North-Holland, Amsterdam (2007)
74. Rein, G., Rendall, A.D.: Compact support of spherically symmetric equilibria in non-relativistic and relativistic galactic dynamics. Math. Proc. Cambridge Philos. Soc. **128**, 363–380 (2000)
75. Rellich, F.: Störungstheorie der Spektralzerlegung. V. Math. Ann. **118**, 462–484 (1942)
76. Rellich, F.: Perturbation Theory of Eigenvalue Problems. New York University, Institute of Mathematical Sciences, New York (1954)
77. Rojas, D.: The monotonicity of the apsidal angle using the theory of potential oscillators. Qual. Theory Dyn. Syst. **17**, 631–635 (2018)
78. Rothe, F.: Remarks on periods of planar Hamiltonian systems. SIAM J. Math. Anal. **24**, 129–154 (1993)
79. Schaaf, R.: A class of Hamiltonian systems with increasing periods. J. Reine Angew. Math. **363**, 96–109 (1985)
80. Schaeffer, J.: Global existence of smooth solutions to the Vlasov-Poisson system in three dimensions. Comm. Partial Differ. Equ. **16**, 1313–1335 (1991)
81. Schwinger, J.: On the bound states of a given potential. Proc. Nat. Acad. Sci. U.S.A. **47**, 122–129 (1961)
82. Simon, B.: Quantum Mechanics for Hamiltonians Defined as Quadratic Forms. Princeton University Press, Princeton (1971)
83. Simon, B.: Functional Integration and Quantum Physics. Academic Press, London (1979)
84. Simon, B.: On the absorption of eigenvalues by continuous spectrum in regular perturbation problems. J. Funct. Anal. **25**, 338–344 (1977)
85. Simon, B.: A Comprehensive Course in Analysis, Part 2A: Basic Complex Analysis. American Mathematical Society, Providence (2015)
86. Simon, B.: A Comprehensive Course in Analysis. Part 4: Operator Theory. American Mathematical Society, Providence (2015)
87. Stein, E.M.: Singular Integrals and Differentiability Properties of Functions. Princeton University Press, Princeton (1970)
88. Straub, C.: Stability of the King Model: A Coercivity-Based Approach. Master's thesis, Universität Bayreuth (2019)
89. Sygnet, J.F.; des Forets, G.; Lachieze-Rey, M.; Pellat, R.: Stability of gravitational systems and gravothermal catastrophe in astrophysics. Astrophys. J. **276**, 737–745 (1984)
90. Thirring, W.: Lehrbuch der Mathematischen Physik, Band 1: Klassische Dynamische Systeme, 2nd edn. Springer, Berlin (1988)
91. Vlasov, A.A.: The vibrational properties of the electron gas, Zh. Eksp. Teor. Fiz. **8**, 291 (1938); Usp. Fiz. Nauk **93**, 444 (1967); see Sov. Phys. Usp. **10**, 721 (1968)

The manufacturer's authorised representative in the EU is Springer Nature Customer Service Centre GmbH, Europaplatz 3, 69115 Heidelberg, Germany. If you have any concerns regarding our products, please contact ProductSafety@springernature.com

Printed and bound by CPI Group (UK) Ltd, Croydon, CR0 4YY
28/07/2026
02179231-0001